Environmental Analysis
and Technology for the
Refining Industry

CHEMICAL ANALYSIS

A SERIES OF MONOGRAPHS ON ANALYTICAL CHEMISTRY
AND ITS APPLICATIONS

Series Editor
J. D. WINEFORDNER

VOLUME 168

Environmental Analysis and Technology for the Refining Industry

JAMES G. SPEIGHT

WILEY-INTERSCIENCE

A JOHN WILEY & SONS, INC., PUBLICATION

Copyright © 2005 by John Wiley & Sons, Inc. All rights reserved.

Published by John Wiley & Sons, Inc., Hoboken, New Jersey.
Published simultaneously in Canada.

No part of this publication may be reproduced, stored in a retrieval system, or transmitted in any form or by any means, electronic, mechanical, photocopying, recording, scanning, or otherwise, except as permitted under Section 107 or 108 of the 1976 United States Copyright Act, without either the prior written permission of the Publisher, or authorization through payment of the appropriate per-copy fee to the Copyright Clearance Center, Inc., 222 Rosewood Drive, Danvers, MA 01923, (978) 750-8400, fax (978) 750-4470, or on the web at www.copyright.com. Requests to the Publisher for permission should be addressed to the Permissions Department, John Wiley & Sons, Inc., 111 River Street, Hoboken, NJ 07030, (201) 748-6011, fax (201) 748-6008, or online at http://www.wiley.com/go/permission.

Limit of Liability/Disclaimer of Warranty: While the publisher and author have used their best efforts in preparing this book, they make no representations or warranties with respect to the accuracy or completeness of the contents of this book and specifically disclaim any implied warranties of merchantability or fitness for a particular purpose. No warranty may be created or extended by sales representatives or written sales materials. The advice and strategies contained herein may not be suitable for your situation. You should consult with a professional where appropriate. Neither the publisher nor author shall be liable for any loss of profit or any other commercial damages, including but not limited to special, incidental, consequential, or other damages.

For general information on our other products and services or for technical support, please contact our Customer Care Department within the United States at (800) 762-2974, outside the United States at (317) 572-3993 or fax (317) 572-4002.

Wiley also publishes its books in a variety of electronic formats. Some content that appears in print may not be available in electronic formats. For more information about Wiley products, visit our web site at www.wiley.com.

Library of Congress Cataloging-in-Publication Data:

Speight, J. G.
    Environmental analysis and technology for the refining industry / James G. Speight.
        p. cm.
    Includes bibliographical references and index.
    ISBN-13 978-0-471-67942-4 (acid-free paper)
    ISBN-10 0-471-67942-9 (acid-free paper)
  1. Petroleum refineries—Waste disposal—Environmental aspects. 2. Petroleum waste—Management. I. Title.
    TD899.P4S68 2005
    665.5'38—dc22
                                                                            2005001262

Printed in the United States of America.

10 9 8 7 6 5 4 3 2 1

CONTENTS

Preface		xi
PART I	**PETROLEUM TECHNOLOGY**	1
Chapter 1	**Definitions and Terminology**	3
	1.1 The Environment	4
	1.2 Petroleum	9
	1.3 Classification	11
	1.3.1 Chemical Composition	12
	1.3.2 Correlation Index	13
	1.3.3 Density	14
	1.3.4 Carbon Distribution	15
	1.3.5 Viscosity–Gravity Constant	15
	1.3.6 UOP Characterization Factor	16
	1.4 Petroleum Products	16
	1.4.1 Boiling Range	19
	1.4.2 Environmental Behavior	19
	1.5 Refinery Waste	20
	1.5.1 Chemical Characteristics	21
	1.5.2 Environmental Behavior	25
	References	27
Chapter 2	**Composition and Properties**	31
	2.1 Composition	32
	2.1.1 Elemental Composition	32
	2.1.2 Chemical Composition	33
	2.1.3 Composition by Volatility	35
	2.1.4 Composition by Fractionation	37
	2.2 Properties	40
	2.2.1 Density and Specific Gravity	40
	2.2.2 Elemental (Ultimate) Analysis	41
	2.2.3 Fractionation by Chromatography	42

	2.2.4	Liquefaction and Solidification	44
	2.2.5	Metals Content	44
	2.2.6	Spectroscopic Properties	45
	2.2.7	Surface and Interfacial Tension	47
	2.2.8	Viscosity	49
	2.2.9	Volatility	51
	References		53

Chapter 3 Refinery Products and By-Products — 57

- 3.1 Refinery Products — 60
  - 3.1.1 Liquefied Petroleum Gas — 64
  - 3.1.2 Naphtha, Gasoline, and Solvents — 68
  - 3.1.3 Kerosene and Diesel Fuel — 71
  - 3.1.4 Fuel Oil — 71
  - 3.1.5 Lubricating Oil — 74
  - 3.1.6 White Oil, Insulating Oil, and Insecticides — 75
  - 3.1.7 Grease — 75
  - 3.1.8 Wax — 76
  - 3.1.9 Asphalt — 77
  - 3.1.10 Coke — 77
- 3.2 Petrochemicals — 78
- 3.3 Refinery Chemicals — 80
  - 3.3.1 Alkalis — 80
  - 3.3.2 Acids — 81
  - 3.3.3 Catalysts — 83
- References — 86

Chapter 4 Refinery Wastes — 87

- 4.1 Process Wastes — 90
  - 4.1.1 Desalting — 92
  - 4.1.2 Distillation — 93
  - 4.1.3 Visbreaking and Coking — 96
  - 4.1.4 Fluid Catalytic Cracking — 98
  - 4.1.5 Hydrocracking and Hydrotreating — 99
  - 4.1.6 Alkylation and Polymerization — 101
  - 4.1.7 Catalytic Reforming — 105
  - 4.1.8 Isomerization — 105
  - 4.1.9 Deasphalting and Dewaxing — 106
- 4.2 Entry into the Environment — 108

		4.2.1	Storage and Handling of Petroleum Products	108
		4.2.2	Release into the Environment	110
	4.3	Toxicity		113
		4.3.1	Lower-Boiling Constituents	114
		4.3.2	Higher-Boiling Constituents	117
		4.3.3	Total Petroleum Hydrocarbons	119
		4.3.4	Wastewater	120
	References			121

PART II ENVIRONMENTAL TECHNOLOGY AND ANALYSIS   123

Chapter 5 Environmental Regulations   125

- 5.1 Environmental Impact of Refining   131
  - 5.1.1 Air Pollution   131
  - 5.1.2 Water Pollution   132
  - 5.1.3 Soil Pollution   132
- 5.2 Environmental Regulations in the United States   132
  - 5.2.1 Clean Air Act   133
  - 5.2.2 Resource Conservation and Recovery Act   137
  - 5.2.3 Clean Water Act   140
  - 5.2.4 Safe Drinking Water Act   141
  - 5.2.5 Comprehensive Environmental Response, Compensation, and Liability Act   142
  - 5.2.6 Oil Pollution Act   143
  - 5.2.7 Occupational Safety and Health Act   144
  - 5.2.8 Toxic Substances Control Act   144
  - 5.2.9 Hazardous Materials Transportation Act   146
- 5.3 Refinery Outlook   146
  - 5.3.1 Hazardous Waste Regulations   146
  - 5.3.2 Regulatory Background   147
  - 5.3.3 Requirements   147
- 5.4 Management of Refinery Waste   148
- References   149

Chapter 6 Sample Collection and Preparation   151

- 6.1 Petroleum Chemicals   151
- 6.2 Sample Collection and Preparation   153
  - 6.2.1 Sample Collection   154
  - 6.2.2 Extract Concentration   168

	6.2.3	Sample Cleanup	170
6.3	Measurement		170
6.4	Accuracy		172
6.5	Precision		173
6.6	Method Validation		174
6.7	Quality Control and Quality Assurance		179
	6.7.1	Quality Control	179
	6.7.2	Quality Assurance	181
6.8	Method Detection Limit		182
References			183

Chapter 7 Analytical Methods   185

7.1	Leachability and Toxicity		186
7.2	Total Petroleum Hydrocarbons		186
	7.2.1	Gas Chromatographic Methods	191
	7.2.2	Infrared Spectroscopy Methods	195
	7.2.3	Gravimetric Methods	196
	7.2.4	Immunoassay Methods	198
7.3	Petroleum Group Analysis		198
	7.3.1	Thin-Layer Chromatography	200
	7.3.2	Immunoassay	201
	7.3.3	Gas Chromatography	201
	7.3.4	High-Performance Liquid Chromatography	203
	7.3.5	Gas Chromatography–Mass Spectrometry	204
7.4	Petroleum Fractions		205
References			206

Chapter 8 Total Petroleum Hydrocarbons   207

8.1	Petroleum Constituents		209
8.2	Analytical Methods		210
	8.2.1	Environmental Samples	210
	8.2.2	Biological Samples	217
	8.2.3	Semivolatile and Nonvolatile Hydrocarbons	228
8.3	Assessment of the Methods		230
References			234

Chapter 9	**Analysis of Gaseous Effluents**		**237**
	9.1 Gaseous Products		239
		9.1.1 Liquefied Petroleum Gas	239
		9.1.2 Natural Gas	240
		9.1.3 Refinery Gas	241
		9.1.4 Sulfur Oxides, Nitrogen Oxides, Hydrogen Sulfide, and Carbon Dioxide	244
		9.1.5 Particulate Matter	244
	9.2 Environmental Effects		245
	9.3 Sampling		247
	9.4 Analysis		247
		9.4.1 Calorific Value (Heat of Combustion)	248
		9.4.2 Composition	249
		9.4.3 Density	252
		9.4.4 Sulfur	253
		9.4.5 Volatility and Vapor Pressure	253
	References		254
Chapter 10	**Analysis of Liquid Effluents**		**257**
	10.1 Naphtha		258
		10.1.1 Composition	261
		10.1.2 Density (Specific Gravity)	266
		10.1.3 Evaporation Rate	266
		10.1.4 Flash Point	266
		10.1.5 Odor and Color	267
		10.1.6 Volatility	267
	10.2 Fuel Oil		268
		10.2.1 Asphaltene Content	269
		10.2.2 Composition	270
		10.2.3 Density (Specific Gravity)	271
		10.2.4 Elemental Analysis	272
		10.2.5 Flash Point	276
		10.2.6 Metals Content	276
		10.2.7 Pour Point and Viscosity	277
		10.2.8 Stability	278
	10.3 Wastewaters		279
	References		280

Chapter 11   Analysis of Solid Effluents — 283

11.1 Residua and Asphalt — 283
    11.1.1 Acid Number — 286
    11.1.2 Asphaltene Content — 287
    11.1.3 Carbon Disulfide Insoluble Constituents — 288
    11.1.4 Composition — 289
    11.1.5 Density (Specific Gravity) — 290
    11.1.6 Elemental Analysis — 292
    11.1.7 Float Test — 293
    11.1.8 Softening Point — 293
    11.1.9 Viscosity — 294
    11.1.10 Weathering — 294

11.2 Coke — 294
    11.2.1 Ash — 296
    11.2.2 Composition — 297
    11.2.3 Density — 299
    11.2.4 Dust Control Material — 300
    11.2.5 Hardness — 300
    11.2.6 Metals — 300
    11.2.7 Sulfur — 301

References — 302

Chapter 12   Pollution Prevention — 305

12.1 Refinery Wastes and Treatment — 306
    12.1.1 Air Emissions — 306
    12.1.2 Wastewater and Treatment — 309
    12.1.3 Other Waste and Treatment — 311

12.2 Pollution Prevention — 311
    12.2.1 Pollution Prevention Options — 312
    12.2.2 Recycling — 315
    12.2.3 Treatment Options — 316

12.3 Adoption of Pollution Reduction Options — 317

References — 319

Glossary — 321

Index — 343

PREFACE

There are many areas of the chemical industry that are responsible for the release of pollutants into the environment. Petroleum refining is one such industry that has seen inadvertent spillage of unrefined petroleum and petroleum products. Since the beginning of the environmental movement in the 1960s, the continuing question relates to the relative condition of the environment.

The capacity of the environment to absorb the air emissions and waste products as well as the other impacts of process technologies is limited. The petroleum refining industry is keeping pace with environmental legislation to ensure that air emissions, effluents, and waste products are handled without maximum expediency and without environmental disruption. In fact, expenditures by the refining industry have risen remarkable over a very short period that speaks for industry's efforts to protect the environment. Although dramatic improvements have been made in pollution control by the industry, there is work to be done. Perhaps the place most in need of further work is an understanding the nature of the waste materials; and to understand these materials, strong analytical programs are necessary.

Time is showing that environmental analysts and the various test methods are no longer a mere ancillary adjunct associate to any environmental program. The analyst is a necessary part of any effort to meet the environmental challenges. Thus, the intent of this book is to focus on the analytical issues that become the focus of any environmental monitoring or cleanup program for petroleum refineries. Even though the prime focus is on refining operations, many of the principles and test methods can also be applied to the release of petroleum during recovery operations and transport as well as to the release of petroleum products during storage, transport, and utilization.

The book will serve as a reference for analysts in detailing the steps required to identify petroleum and petroleum products (although more detailed texts are available for this purpose) and those test methods that should be applied for monitoring and cleanup of petroleum and petroleum products. As such, the book offers a ready guide to the many issues that are related to ecosystems as well as to pollutant mitigation and cleanup.

To accomplish this goal, the book focuses on the various aspects of environmental science and engineering as applied to the petroleum refining industry. Part I presents an introduction to, and a description of, the nomenclature used by refiners and by environmental scientists and engineers. This part includes a description of petroleum, petroleum refining, and petroleum products. Part II includes a discussion of the relevant environmental regulations in the United

States and descriptions of the various refining processes and the emissions from these processes. Examples are given of the application of environmental regulations to petroleum refining and petroleum products as well as current and proposed methods of the mitigation of environmental effects and waste management. Additional chapters cover the methods of analysis that might be used for samples recovered from the environment and for typical petroleum products, and how these analyses might be used to predict potential effects on the environment.

JAMES G. SPEIGHT

Laramie, Wyoming
January 2005

PART I
PETROLEUM TECHNOLOGY

CHAPTER

1

DEFINITIONS AND TERMINOLOGY

There is probably no one who can testify with any degree of accuracy (although there is always someone who can testify with a high degree of uncertainty) as to the last time Earth was pristine and unpolluted. Yet, to attempt to return the environment to such a mythical time might have a severe effect on current indigenous life, perhaps a form of pollution in reverse!

However, there is the possibility that through the judicious use of resources and application of the principles of environmental analysis, environmental science, and environmental (disciplines involved in the study of the environment) (Speight, 1996; Manahan, 1999; Woodside, 1999), we can come to a state where pollution is minimal and not a threat to the future. Such a program will involve not only well-appointed suites of analytical tests but also subsequent studies, that range from the effects of changes in the environmental conditions on the flora and fauna of a region to the more esoteric studies of animals in laboratories. These studies can include aspects of chemistry, chemical engineering, microbiology, and hydrology as they can be applied to solve environmental problems (Pickering and Owen, 1994; Speight, 1996; Schwarzenbach et al., 2003; Tinsley, 2004). As an historical aside, environmental engineering (formerly known as sanitary engineering) originally developed as a subdiscipline of civil engineering.

Despite numerous safety protocols that are in place and the care taken to avoid environmental incidents (EPA, 2004), virtually every industry suffers accidents that lead to environmental problems, complexities, and chemical contamination. The petroleum industry is no exception to such accidents. It is therefore helpful to be aware of the nature of the raw material and the products arising therefrom, in order to understand the nature of any contamination and thus the best cleanup methods to choose.

Frequently, the existence and source of such information is unknown thus the data are not examined. Even when the existence and sources of information are known, decisions must be made in order to make an informed, and often quick decision on the next steps, even if later, one decides not to use it for a particular application. Knowing about the relevant data gives investigators and analysts the ability to assess the data based on quality assurance criteria. This is especially true for users near the end of long decision processes, such as site cleanup, ecological risk assessments, and natural resource damage assessments.

Environmental Analysis and Technology for the Refining Industry, by James G. Speight
Copyright © 2005 John Wiley & Sons, Inc.

Considering the composition of petroleum and petroleum products (Speight, 1994, 1999), it is not surprising that petroleum and petroleum-derived chemicals are environmental pollutants (Loeher, 1992; Olschewsky and Megna, 1992). The world's economy is highly dependent on petroleum for energy production, and widespread use has led to enormous releases to the environment of petroleum, petroleum products, exhaust from internal combustion engines, emissions from oil-fired power plants, and industrial emissions where fuel oil is employed.

The toxicity of polynuclear aromatic hydrocarbons is perhaps one of the most serious long-term problems associated with the use of petroleum. They comprise a large class of petroleum compounds containing two or more benzene rings. Polynuclear aromatic hydrocarbons are formed in nature by long-term, low-temperature chemical reactions in sedimentary deposits of organic materials and in high-temperature events such as volcanoes and forest fires. The major source of this pollution is, however, human activity. Polynuclear aromatic hydrocarbons accumulate in soil, sediment, and biota. At high concentrations, they can be acutely toxic by disrupting membrane function. Many cause sunlight-induced toxicity in humans and fish and other aquatic organisms. In addition, long-term chronic toxicity has been demonstrated in a wide variety of organisms. Through metabolic activation, some polynuclear aromatic hydrocarbons *form* reactive intermediates that bind to deoxyribonucleic acid (DNA). For this reason, many of these hydrocarbons are *mutagenic* (tending to cause mutations), *teratogenic* (tending to cause developmental malformations), or *carcinogenic* (tending to cause cancer).

The terminology found in the various areas of petroleum and environmental technology can be extremely confusing to the uninitiated; excellent examples of the confusion that abounds are to be found (Speight, 1990, 1994, 1999). As a beginning of this process of data examination and ingestion, in this chapter we introduce the terminology of environmental technology and petroleum. Chemical waste, as it pertains to petroleum, is also defined and classified into various subgroups.

1.1. THE ENVIRONMENT

To start with an extremely relevant definition, *environmental technology* is the application of scientific and engineering principles to the study of the environment, with the goal of improvement of the environment. Furthermore, issues related to the pollution of the environment are relative.

Any organism is exposed to an *environment*, even if the environment is predominantly many members of the same organism. An example is a bacterium in a culture that is exposed to many members of the same species. Thus, the environment is all external influences, abiotic (physical factors) and biotic (actions of other organisms), to which an organism is exposed. The environment affects basic life functions, growth, and the reproductive success of organisms, and determines their local and geographic distribution patterns. A fundamental idea in *ecology* is that the environment changes in time and space and that living organisms respond to these changes.

Since ecology is that branch of science related to the study of the relationship of organisms to their environment, an *ecosystem* is an ecological community (or living unit) considered together with the nonliving factors of its environment as a unit. By way of brief definition, *abiotic factors* include such influences as light radiation (from the sun), ionizing radiation (cosmic rays from outer space), temperature (local and regional variations), water (seasonal and regional distributions), atmospheric gases, wind, soil (texture and composition), and catastrophic disturbances. The latter phenomena are usually unpredictable and infrequent, such as fire, hurricanes, volcanic activity, landslides, major floods, and any disturbance that drastically alters the environment and thus changes the species composition and activity patterns of the inhabitants.

On the other hand, *biotic factors* include natural interactions (e.g., predation and parasitism) and anthropogenic stress (e.g., the effect of human activity on other organisms). Because of the abiotic and biotic factors, the environment to which an organism is subjected can affect the life functions, growth, and reproductive success of the organism and can determine the local and geographic distribution patterns of the organism.

Living organisms respond to changes in the environment either by adapting or by becoming extinct. The basic principles of the concept that living organisms respond to changes in the environment were put forth by Darwin and Lamarck. The former noted the slower adaptation (evolutionary trends) of living organisms, while the latter noted the more immediate adaptation of living organisms to the environment. Both essentially espoused the concept of the survival of the fittest, alluding to the ability of an organism to live in harmony with its environment. This was assumed to indicate that organisms that competed successfully with environmental forces would survive. However, there is the alternative thought: that organisms that can live in a harmonious symbiotic relationship with their environment have an equally favorable chance of survival. The influence of the environment on organisms can be viewed on a large scale (i.e., the relationship between regional climate and geographic distribution of organisms) or on a smaller scale (i.e., some highly localized conditions determine the precise location and activity of individual organisms).

Organisms may respond differently to the frequency and duration of a given environmental change. For example, if some individual organisms in a population have adaptations that allow them to survive and to reproduce under new environmental conditions, the population will continue but the genetic composition will have changed (Darwinism). On the other hand, some organisms have the ability to adapt to the environment (i.e., to adjust their physiology or morphology in response to the immediate environment) so that the new environmental conditions are less (certainly no more) stressful than the previous conditions. Such changes may not be genetic (Lamarckism).

In terms of *anthropogenic stress* (the effect of human activity on other organisms), there is a need for the identification and evaluation of the potential impacts of proposed projects, plans, programs, policies, or legislative actions on the physical–chemical, biological, cultural, and socioeconomic components of the environment. This activity, also known as *environmental impact assessment* (EIA), refers to the interpretation of the significance of anticipated changes related to a proposed project. The activity encourages consideration of the environment and arriving at actions that are environmentally compatible.

Identifying and evaluating the potential impact of human activities on the environment requires the identification of mitigation measures. *Mitigation* is the sequential consideration of the following measures: (1) avoiding the impact by not taking a certain action or partial action; (2) minimizing the impact by limiting the degree or magnitude of the action and its implementation; (3) rectifying the impact by repairing, rehabilitating, or restoring the affected environment; (4) reducing or eliminating the impact over time by preservation and maintenance operations during the life of the action; and (5) compensating for the impact by replacing or providing substitute resources or environments.

Nowhere is the effect of anthropogenic stress felt more than in the development of natural resources of the Earth. Natural resources are varied in nature and often require definition. For example, in relation to mineral resources, for which there is also descriptive nomenclature (ASTM C294), the terms related to the available quantities of the resource must be defined. In this instance, the term *resource* refers to the total amount of the mineral that has been estimated to be available ultimately. The term *reserves* refers to well-identified resources that can profitably be extracted and utilized by means of existing technology. In many countries, fossil fuel resources are often classified as a subgroup of the total mineral resources.

In some cases, environmental pollution is a clear-cut phenomenon, whereas in others it remains a question of degree. The ejection of various materials into the environment is often cited as pollution, but there is the ejection of beneficial chemicals that can assist the air, water, and land to perform their functions. However, it must be emphasized that even though certain chemicals are indigenous to an environment, their ejection into the environment, in quantities above the naturally occurring limits can be extremely harmful. In fact, the timing and the place of a chemical release are influential in determining whether a chemical is beneficial, benign, or harmful! Thus, what may be regarded as a pollutant in one instance can be a beneficial chemical in another instance. The phosphates in fertilizers are examples of useful (beneficial) chemicals, whereas phosphates generated as by-products in the metallurgical and mining industries may, depending on the specific industry, be considered pollutants (Chenier, 1992). In this case, the means by which such pollution can be prevented must be recognized (Breen and Dellarco, 1992). Thus, increased use of Earth's resources as well as the use of a variety of chemicals that are nonindigenous to the Earth have put a burden on the ability of the environment to tolerate such materials.

Finally, some recognition must be made of the term *carcinogen* since many of the environmental effects noted in this book can lead to cancer. *Carcinogens* are cancer-causing substances, and there is a growing awareness of the presence of carcinogenic materials in the environment. A classification scheme is provided for such materials (Table 1.1). The number of substances with which a person

Table 1.1. Weight-of-Evidence Carcinogenicity Classification Scheme as Determined by the U.S. Environmental Protection Agency

Group	Description
A	Human carcinogen
B1	Probable human carcinogen; limited human data are available
B2	Probable human carcinogen; carcinogen in animals but inadequate evidence in humans
C	Possible human carcinogen
D	Not classifiable as a human carcinogen
E	No carcinogenic activity in humans

Source: Zakrzewski, 1991; Milman and Weisburger, 1994.

comes in contact are in the tens of thousands and there is not a full understanding of the long-term effects of these substances in their possible propensity to cause genetic errors that ultimately lead to carcinogenesis. *Teratogens* are substances that tend to cause developmental malformations.

Pollution is the introduction of indigenous (beyond the natural abundance) and nonindigenous (artificial) gaseous, liquid, and solid contaminants into an ecosystem. The atmosphere and water and land systems have the ability to cleanse themselves of many pollutants within hours or days, especially when the effects of the pollutant are minimized by the natural constituents of the ecosystem. For example, the atmosphere might be considered to be self-cleaning as a result of rain. However, removal of some pollutants from the atmosphere (e.g., sulfates and nitrates) by rainfall results in the formation of *acid rain*, which can cause serious environmental damage to ecosystems within the water and land systems (Johnson and Gordon, 1987; Pickering and Owen, 1994).

Briefly, lakes in some areas of the world are now registering a low pH (acidic) reading because of excess acidity in rain. This was first noticed in Scandinavia and is now prevalent in eastern Canada and the northeastern United States. Normal rainfall has a pH of 5.6, and thus slight acidity (neutral water has a pH equal to 7.0), because the carbon dioxide (CO_2) in the air combines with water to form carbonic acid (H_2CO_3):

$$CO_2 + H_2O \rightarrow H_2CO_3$$

The increased use of hydrocarbon fuels in the last five decades is slowly increasing the concentration of carbon dioxide in the atmosphere, which produces more carbonic acid, leading to an imbalance in the natural carbon dioxide content of the atmosphere, which, in turn, leads to more acidity in the rain. In addition, there is a *greenhouse effect*, and the average temperature of the Earth may be increasing.

In addition, excessive use of fuels with a high sulfur and nitrogen content causes sulfuric and nitric acids in the atmosphere from sulfur dioxide and nitrogen oxide products of combustion, which can be represented simply as

$$SO_2 + H_2O \rightarrow H_2SO_3 \quad \text{(sulfurous acid)}$$

$$2SO_2 + O_2 \rightarrow 2SO_3$$

$$SO_3 + H_2O \rightarrow H_2SO_4 \quad \text{(sulfuric acid)}$$

$$NO + H_2O \rightarrow HNO_2 \quad \text{(nitrous acid)}$$

$$2NO + O_2 \rightarrow NO_2$$

$$NO_2 + H_2O \rightarrow HNO_3 \quad \text{(nitric acid)}$$

A *pollutant* is a substance (for simplicity most are referred to as *chemicals*) present in a particular location when it is not indigenous to the location or is in a greater-than-natural concentration. The substance is often the product of human activity. By virtue of its name, the pollutant has a detrimental effect on the environment, in part or *in toto*. Pollutants can also be subdivided into two classes, primary and secondary:

$$\text{source} \rightarrow \text{primary pollutant} \rightarrow \text{secondary pollutant}$$

Primary pollutants are those pollutants emitted directly from the source. In terms of atmospheric pollutants by petroleum constituents, examples are hydrogen sulfide, carbon oxides, sulfur dioxide, and nitrogen oxides from refining operations (see above). The question of classifying nitrogen dioxide and sulfur trioxide as primary pollutants often arises, as does the origin of the nitrogen. In the former case, these higher oxides can be formed in the upper levels of the combustors. The nitrogen from which the nitrogen oxides are formed does not originate solely from the fuel but may also originate from the air used for the combustion.

Secondary pollutants are produced by interaction of primary pollutants with another chemical or by dissociation of a primary pollutant, or by other effects within a particular ecosystem. Again, using the atmosphere as an example, formation of the constituents of acid rain is an example of the formation of secondary pollutants (see above).

A *contaminant*, which is not usually classified as a pollutant unless it has some detrimental effect, can cause deviation from the normal composition of an environment.

A *receptor* is an object (animal, vegetable, or mineral) or a locale that is affected by a pollutant.

A *chemical waste* is any solid, liquid, or gaseous waste material that if improperly managed or disposed of, may pose substantial hazards to human health and the environment (Table 1.2). At any stage of the management process, a chemical waste may be designated by law as a *hazardous waste* (Chapter 12). Improper disposal of such waste streams in the past has created a need for very expensive cleanup operations (Tedder and Pohland, 1993). Correct handling of such

Table 1.2. Types of Chemical Waste

Source	Waste Type
Chemical manufacturers	Strong acids and bases
	Spent solvents
	Reactive materials
Vehicle maintenance shops	Heavy-metal paints
	Ignitable materials
	Used lead–acid batteries
	Spent solvents
Printing industry	Heavy-metal solutions
	Waste ink
	Spent solvents
	Spent electroplating wastes
	Ink sludge containing heavy metals
Leather products	Waste toluene and benzene
Paper industry	Paint wastes containing heavy metals
Construction industry	Ignitable paint wastes
	Spent solvents
	Strong acids and bases
Cleaning agents and cosmetics manufacturing	Heavy-metal dusts
	Ignitable materials
	Flammable solvents
	Strong acids and bases
Furniture and wood manufacturing and refinishing	Ignitable materials
	Spent solvents
Metal manufacturing	Paint wastes containing heavy metals
	Strong acids and bases
	Cyanide wastes
	Sludge containing heavy metals

chemicals (NRC, 1981), can, in addition to dispensing with many of the myths related to chemical processing (Kletz, 1990), mitigate some of the problems that occur when incorrect handling is the norm!

1.2. PETROLEUM

Petroleum, and the equivalent term *crude oil*, cover a wide assortment of materials consisting of mixtures of hydrocarbons and other compounds that contain variable amounts of sulfur, nitrogen, and oxygen and which may vary widely in volatility, specific gravity, and viscosity. Metal-containing constituents, notably those compounds that contain vanadium and nickel, usually occur in the more viscous crude oils in amounts up to several thousand parts per million and can have serious consequences during processing of these feedstocks (Speight, 1999, and references cited therein). Because petroleum is a mixture of widely varying constituents and proportions, its physical properties also vary widely, as does its color, from colorless to black.

Indeed, petroleum reservoirs have been found in vastly different parts of the world and their chemical composition varies greatly. Consequently, no single petroleum composition can be defined. Thus, petroleum-derived inputs to the environment vary considerably in composition, and the complexity of petroleum composition is matched by the range of properties of the components and the physical, chemical, and biochemical processes that contribute to the distributive pathways and determine the fate of the inputs.

Put simply, petroleum is a naturally occurring mixture of hydrocarbons, generally in a liquid state, which may include compounds of sulfur, nitrogen, oxygen, metals, and other elements (ASTM, 2004). In more specific terms, petroleum has also been defined (ITAA, 1936) as:

1. Any naturally occurring hydrocarbon, whether in a liquid, gaseous, or solid state;
2. Any naturally occurring mixture of hydrocarbons, whether in a liquid, gaseous, or solid state; or
3. Any naturally occurring mixture of one or more hydrocarbons, whether in a liquid, gaseous, or solid state, and one or more of the following: hydrogen sulfide, helium, and carbon dioxide.

The definition includes any petroleum as defined above that has been returned to a natural reservoir.

In terms of the elemental composition of petroleum, the carbon content is relatively constant; it is the hydrogen and heteroatom contents that are responsible for the major differences. Nitrogen, oxygen, and sulfur are present in only trace amounts in some petroleum, which thus consists primarily of hydrocarbons. On the other hand, a crude oil containing 9.5% heteroatoms may contain essentially no true hydrocarbon constituents insofar as the constituents contain *at least one or more* nitrogen, oxygen, and/or sulfur atoms within the molecular structures.

There are also other *types* of petroleum that differ from conventional petroleum insofar as they are much more difficult to recover from subsurface reservoirs. These materials have a much higher viscosity (and lower API gravity) than those of conventional petroleum, and primary recovery of these petroleum types usually requires thermal stimulation of the reservoir (Speight, 1999, and references cited therein).

When petroleum occurs in a reservoir that allows the crude material to be recovered by pumping operations as a free-flowing dark- to light-colored liquid, it is often referred to as *conventional petroleum*. Heavy oils comprise the other *types* of petroleum; they differ from conventional petroleum in being much more difficult to recover from subsurface reservoirs. The definition of heavy oils is usually based on the API gravity or viscosity value, and the definition is quite arbitrary, although there have been attempts to rationalize the definition based on viscosity, API gravity, and density.

In addition to attempts to define petroleum, heavy oil, bitumen, and residua, there have been several attempts to classify these materials by the use of properties such as API gravity, sulfur content, or viscosity (Speight, 1999). However, any attempt to classify petroleum, heavy oil, and bitumen on the basis of a single property is no longer sufficient to define the nature and properties of petroleum and petroleum-related materials, perhaps even being an exercise in futility.

For many years, petroleum and heavy oil were very generally defined in terms of physical properties. For example, heavy oil was considered to be a crude oil that had gravity between 10 and 20° API. For example, Cold Lake heavy crude oil (Alberta, Canada) has an API gravity equal to 12°, but extra-heavy oil (such as tar sand bitumen), which requires recovery by nonconventional and nonenhanced methods, has an API gravity in the range 5 to 10°. Residua would vary depending on the temperature at which distillation was terminated, but vacuum residua were usually in the range 2 to 8° API.

However, to define conventional *petroleum, heavy oil*, and *bitumen*, the use of a single physical parameter such as API gravity or viscosity is not sufficient and is only a general indicator of the nature of the material. Other properties, such as the method of recovery, composition, and most of all, the properties of the bulk deposit, must also be included in any definition of these materials. Only then will it be possible to classify petroleum and its derivatives (Speight, 1999).

A *residuum* (pl. *residua*, also shortened to *resid*, pl. *resids*) is the residue obtained from petroleum after nondestructive distillation has removed all the volatile materials. The temperature of the distillation is usually maintained below 350°C (660°F), since the rate of thermal decomposition of petroleum constituents is minimal below this temperature but the rate of thermal decomposition of petroleum constituents is substantial above 350°C (660°F) (Speight, 1999, and references cited therein).

Residua are black, viscous materials obtained by distillation of a crude oil under atmospheric pressure (atmospheric residuum) or under reduced pressure (vacuum residuum). They may be liquid at room temperature (generally, atmospheric residua) or almost solid (generally, vacuum residua) depending on the cut point of the distillation or depending on the nature of the crude oil (Speight, 1999; Speight and Ozum, 2002).

1.3. CLASSIFICATION

By definition, *petroleum* (also called *crude oil*) is a mixture of gaseous, liquid, and solid hydrocarbon compounds. Petroleum occurs in sedimentary rock deposits throughout the world and contains small quantities of nitrogen-, oxygen-, and sulfur-containing compounds as well as trace amounts of metallic constituents (Long and Speight, 1998; Reynolds, 1998; Speight, 1999, and references cited therein). Thus, the classification of petroleum as a hydrocarbon mixture should follow from this definition, but some clarification is required.

The original methods of classification arose because of commercial interest in petroleum type and were a means of providing refinery operators with a rough guide to processing conditions. It is therefore not surprising that systems based on a superficial inspection of a physical property, such as specific gravity or API (Baumé) gravity, are easily applied and are actually used to a large extent in expressing the quality of crude oils. Such a system is approximately indicative of the general character of a crude oil as long as materials of one general type are under consideration. For example, among crude oils from a particular area, an oil of 40° API (specific gravity = 0.825) is usually more valuable than one of 20° API (specific gravity = 0.934) because it contains more light fractions (e.g., gasoline) and fewer heavy, undesirable asphaltic constituents.

1.3.1. Chemical Composition

Composition refers to the specific mixture of chemical compounds that constitute petroleum. The composition of these materials is related to the nature and mix of the organic material that generated the hydrocarbons. Composition is also subject to the influence on that composition of natural processes such as migration (movement of oil from source rock to reservoir rock), biodegradation (alteration by the action of microbes), and water washing (effect of contact with water flowing in the subsurface) (Speight, 1993, 1999). Thus, petroleum is the result of the metamorphosis of natural products as a result of chemical and physical changes imparted by the prevailing conditions at a particular locale.

Petroleum varies in appearance from a thin (mobile), nearly colorless liquid to a thick (viscous), almost black oil. The specific gravity at 15.6°C (60°F) varies correspondingly from about 0.75 to 1.00 (57 to 10° API), with the specific gravity of most crude oils falling in the range 0.80 to 0.95 (45 to 17° API). Thus, it is not surprising that petroleum varies in composition from one oil field to another, from one well to another in the same field, and even from one level to another in the same well. This variation can be in both molecular weight and the types of molecules present in petroleum. Petroleum may well be described as a mixture of organic molecules drawn from a wide distribution of molecular types that lie within a wide distribution of molecular weights.

By definition, *a hydrocarbon contains carbon and hydrogen only*. On the other hand, if an organic compound contains nitrogen, and/or sulfur, and/or oxygen, and/or metals, it is a heteroatomic compound and not a hydrocarbon. Organic compounds containing heteroelements (elements such as nitrogen, oxygen, and sulfur), in addition to carbon and hydrogen, are defined in terms of the locations of these heteroelements within the molecule. In fact, it is, to a large extent, the heteroatomic function that determines the chemical and physical reactivity of the heteroatomic compounds; and the chemical and physical reactivity of the heteroatomic compounds is quite different from the chemical and physical reactivity of the hydrocarbons.

Petroleum is a *naturally occurring hydrocarbon insofar* as it contains compounds that are composed of carbon and hydrogen only which do not contain any heteroatoms (nitrogen, oxygen, and sulfur as well as compounds containing metallic constituents, particularly vanadium, nickel, iron, and copper). The hydrocarbons found in petroleum are classified into the following types (Chapter 2):

1. *Paraffins*, saturated hydrocarbons with straight or branched chains but without a ring structure.
2. *Cycloparaffins* (*naphthenes*), saturated hydrocarbons containing one or more rings, each of which may have one or more paraffinic side chains (more correctly known as *alicyclic hydrocarbons*).
3. *Aromatics*, hydrocarbons containing one or more aromatic nuclei, such as benzene, naphthalene, or phenanthrene ring systems, that may be linked up with (substituted) naphthalene rings and/or paraffinic side chains.

On this basis, petroleum may have some value in the crude state but, when refined, provides fuel gas, petrochemical gas (methane, ethane, propane, and butane), transportation fuel (gasoline, diesel fuel, aviation fuel), solvents, lubricants, asphalt, and many other products. In addition to the hydrocarbon constituents, petroleum does contain heteroatomic (nonhydrocarbon) species, but they are in the minority compared to the number of carbon and hydrogen atoms. They do, nevertheless, impose a major influence on the behavior of petroleum and petroleum products as well as on the refining processes (Speight and Ozum, 2002).

A widely used classification of petroleum distinguishes between crude oils either on a *paraffin base* or on an *asphalt base* and arose because paraffin wax separates from some crude oils on cooling, whereas other oils show no separation of paraffin wax on cooling. The terms *paraffin base* and *asphalt base* were introduced and have remained in common use (van Nes and van Westen, 1951).

The presence of paraffin wax is usually reflected in the paraffinic nature of the constituent fractions, and a high asphaltic content corresponds with the naphthenic properties of the fractions. As a result, the misconception has arisen that paraffin-base crude oils consist mainly of paraffins and asphalt-base crude oils mainly of cyclic (or *naphthenic*) hydrocarbons. In addition to paraffin- and asphalt-base oils, a mixed base had to be introduced for those oils that leave a mixture of bitumen and paraffin wax as a residue by nondestructive distillation.

1.3.2. Correlation Index

An early attempt to give the classification system a quantitative basis suggested that a crude should be called asphaltic if the distillation residue contained less than 2% wax and paraffinic if it contained more than 5%. A division according to the chemical composition of the 250 to 300°C (480 to 570°F) fraction has also been suggested (Speight, 1999, and references cited therein). Difficulties arise in using such a classification because in fractions boiling above 200°C (390°F), the molecules can no longer be placed in a single group because most of them are of a typically mixed nature. Purely naphthenic or aromatic molecules occur very seldom; cyclic compounds generally contain paraffinic side chains, and often even aromatic and naphthenic rings side by side. More direct chemical information is often desirable and can be supplied by means of the correlation index (CI).

The correlation index, developed by the U.S. Bureau of Mines, is based on a plot of specific gravity versus the reciprocal of the boiling point in kelvin ($K = {}^\circ C + 273$). For pure hydrocarbons, the line described by the constants of the individual members of the normal paraffin series is given the value CI = 0, and a parallel line passing through the point for the values of benzene is given as CI = 100; thus,

$$CI = 473.7d - 456.8 + \frac{48{,}640}{K}$$

where K for a petroleum fraction is the average boiling point determined by the standard Bureau of Mines distillation method and d is the specific gravity.

Values for the index between 0 and 15 indicate a predominance of paraffinic hydrocarbons in the fraction. A value from 15 to 50 indicates a predominance of either naphthenes or of mixtures of paraffins, naphthenes, and aromatics. An index value above 50 indicates a predominance of aromatic species.

1.3.3. Density

Since the early years of the industry, density (specific gravity) has been, the principal and often the only specification of crude oil products and was taken as an index of the proportion of gasoline and, particularly, kerosene present. As long as only one type of petroleum was in use, the relations were approximately true, but as crude oils having other properties were discovered and came into use, the significance of density measurements disappeared. Nevertheless, crude oils of particular types are still rated by gravity, as are gasoline and naphtha within certain limits of other properties. The use of density values has been advocated for quantitative application using a scheme based on the American Petroleum Institute (API) gravity of the 250 to 275°C (480 to 525°F, 1760 mm) and 275 to 300°C (525 to 570°F, 40 mm) distillation fractions (Speight, 1999, and references cited therein). Indeed, investigation of crude oils from worldwide sources showed that 85% fell into one of the three classes: paraffin, intermediate, or naphthene base.

It has also been proposed to classify heavy oils according to *characterization gravity*. This is defined as the arithmetic average of the instantaneous specific gravity of the distillates boiling at 177°C (350°F), 232°C (450°F), and 288°C (550°F) vapor line temperature at 25 mm pressure in a true boiling-point distillation.

In addition, a method of petroleum classification based on other properties as well as the density of selective fractions has been developed. The method consists of a preliminary examination of the aromatic content of the fraction boiling up to 145°C (295°F), as well as that of the asphaltene content, followed by a more detailed examination of the chemical composition of the naphtha (bp < 200°C < 390°F). For this examination a graph is used that is a composite of curves expressing the relation among the percentage distillate from the naphtha, aniline point, refractive index, specific gravity, and boiling point. The aniline point after acid extraction is included to estimate the paraffin-to-naphthene ratio.

1.3.4. Carbon Distribution

A method for the classification of crude oils can only be efficient (1) if it indicates the distribution of components according to volatility, and (2) if it indicates one or more characteristic properties of the various distillate fractions. The distribution according to volatility has been considered the main property of petroleum, and any fractionating column with a sufficient number of theoretical plates may be used for recording a curve in which the boiling point of each fraction is plotted against the percentage by weight.

However, for characterization of the various fractions of petroleum, use of the n–d–M method (n = refractive index, d = density, M = molecular weight) is suggested. This method enables determination of the carbon distribution and thus indicates the percentage of carbon in aromatic structure ($\%C_A$), the percentage of carbon in naphthenic structure ($\%C_N$), and the percentage of carbon in paraffinic structure ($\%C_P$). The yields over the various boiling ranges can also be estimated; for example, in the lubricating oil fractions the percentage of carbon in paraffinic structure can be divided into two parts, giving the percentage of carbon in normal paraffins ($\%C_{nP}$) and the percentage of carbon in paraffinic side chains. The percentage of normal paraffins present in lubricating oil fractions can be calculated from the percentage of normal paraffinic carbon ($\%C_{nP}$) by multiplication by a factor that depends on the hydrogen content of the fractions.

It is also possible to extrapolate the carbon distribution to the gasoline range on the one hand and to the residue on the other hand. A high value of $\%C_A$ at 500°C (930°F) boiling point usually indicates a high content of asphaltenes in the residue, whereas a high value of $\%C_{nP}$ at a 500°C (930°F) boiling point usually indicates a waxy residue.

1.3.5. Viscosity–Gravity Constant

The viscosity–gravity constant and the Universal Oil Products characterization factor have been used to some extent as a means of classifying crude oils. Both parameters are generally employed to give an indication of the paraffinic character of the crude oil, and both have been used, if a subtle differentiation can be made, as a means of petroleum characterization rather than for petroleum classification.

Nevertheless, the viscosity–gravity constant (VGC) was one of the early indexes proposed to characterize (or classify) oil types:

$$\text{VGC} = 10d - \frac{1.0752 \log(v - 38)}{10 - \log(v - 38)}$$

where d is the specific gravity 60/60°F and v is the Saybolt viscosity at 39°C (100°F). For heavy oil, where the low-temperature viscosity is difficult to measure, an alternative formula,

$$\text{VGC} = d - 0.24 - \frac{0.022 \log(v - 35.5)}{0.755}$$

has been proposed in which the 99°C (210°F) Saybolt viscosity is used. The two do not agree well for low-viscosity oils. However, the viscosity–gravity constant is of particular value in indicating a predominantly paraffinic or cyclic composition. The lower the index number, the more paraffinic is the stock; for example, naphthenic lubricating oil distillates have VGC = 0.876, and the raffinate obtained by solvent extraction of lubricating oil distillate has VGC $\sim$ 0.840.

1.3.6. UOP Characterization Factor

The UOP characterization factor is perhaps one of the more widely used derived characterization or classification factors and is defined by the formula

$$K = \sqrt[3]{\frac{T_B}{d}}$$

where T_B is the average boiling point in degrees Rankine (°F + 460) and d is the specific gravity 60°/60°F. This factor has been shown to be additive on a weight basis. It was devised originally to show the thermal cracking characteristics of heavy oils; thus, highly paraffinic oils have $K \sim$ 12.5 to 13.0, and cyclic (naphthenic) oils have $K \sim$ 10.5 to 12.5.

1.4. PETROLEUM PRODUCTS

Petroleum is rarely used in the form produced at the well, but is converted in refineries into a wide range of products, such as gasoline, kerosene, diesel fuel, jet fuel, and domestic and industrial fuel oils, together with petrochemical feedstocks such as ethylene, propylene, butene, butadiene, and isoprene. Petroleum is refined, that is, it is separated into useful products (Figure 1.1; Chapter 3).

Unless properties dictate otherwise (Speight, 1999; Speight and Ozum, 2002), refining consists of initially dividing the petroleum into fractions of different boiling ranges by distillation. Other forms of treatment are utilized during the refining process to remove undesirable components of the crude oil. The fractions themselves are often distilled further to produce the desired commercial product. A variety of additives may be incorporated into some of the refined products to adjust the octane ratings or improve engine performance characteristics.

The lowest-boiling (lightest) constituents of petroleum are gases at room temperature, which are collected and used as heating gas mixtures and in the petrochemical industry or as a refinery fuel. The next-lightest hydrocarbons, which occur in molecules that contain four to nine carbon atoms and have a boiling range (also known as the light and heavy naphtha fraction), are used in

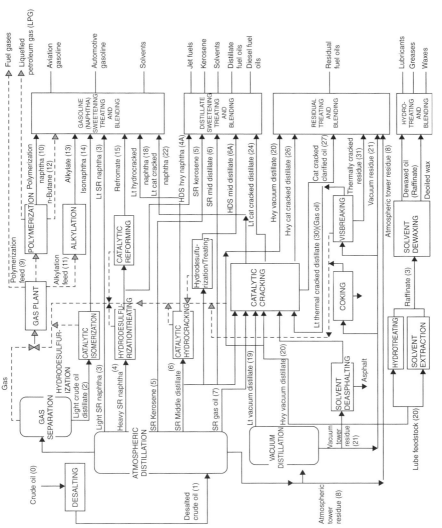

Figure 1.1. General layout of a petroleum refinery.

gasoline formulation. Constituents boiling in the middle ranges, *middle distillates*, are used for the production of kerosene diesel fuel, jet fuel, and fuel oil. These fuels contain paraffins (alkanes), cycloparaffins (cycloalkanes), aromatics, and olefins from approximately the nine- to 20-carbon molecular range.

The highest-boiling molecular-weight compounds that do not distill under refinery conditions or vaporize at all are asphalts or paraffins, depending on the source of the crude oil. The highest-boiling fractions are high-molecular-weight hydrocarbons suitable for lubricants and heating oil. Lubricants may contain hydrocarbons ranging from 18 to 25 carbon atoms per molecule. Paraffin wax and petroleum jelly typically contain 28 to 38 carbon atoms per molecule. Other petroleum products include a wide variety of solvents that can have a considerable influence on living organisms, particularly the human organism (Table 1.3). Refined oils may also have a number of additives, such as gelling inhibitors, which are added to diesel fuels during cold weather. Certain additives may be of special concern in an injury assessment, either because they are toxic themselves or because they significantly change the behavior of the oil products.

Petroleum products have a vast array of uses. In approximate order of importance the uses are: as fuels for vehicles and industry, as heating oils, as lubricants, as raw materials in manufacturing petrochemicals and pharmaceuticals, and as solvents. By a wide margin, most products derived from petroleum find use as fossil fuels to run vehicles, to produce electricity, and to heat homes and business. About 65% of the petroleum used as fuel is consumed as gasoline in automobiles. Thus, petroleum products are ubiquitous in the modern environment, which leads to contamination problems for both the environment and in sampling activities.

Table 1.3. Effects of Organic Solvents

Solvent	Affected Parts of Human Body
Aliphatic hydrocarbons	
Pentanes, hexanes, heptanes, octanes	Central nervous system and liver
Halogenated aliphatic hydrocarbons	
Methylene chloride	Central nervous system, respiratory system
Chloroform	Liver
Carbon tetrachloride	Liver and kidneys
Aromatic hydrocarbons	
Benzene	Blood, immune system
Toluene	Central nervous system
Xylene	Central nervous system
Alcohols	
Methyl alcohol (methanol and toxic metabolites)	Optic nerve
Isopropyl alcohol	Central nervous system
Glycols	
Ethylene glycol (and toxic metabolites)	Central nervous system

1.4.1. Boiling Range

There are several ways to classify or group various petroleum products. Refined oils are sometimes characterized by the approximate boiling-point range, which corresponds with the size (such as the number of carbon atoms) of the petroleum hydrocarbons in the refined oil:

1. 1 to 205°C (32 to 400°F): *naphtha* or *straight-run gasoline* (meaning, not produced through catalytic decomposition).
2. 205-345°C (400 to 655°F): *middle distillates*, including kerosene, jet fuel, heating oil, and diesel fuel.
3. 345 to 565°C (655 to 1050°F): *gas oil*, including lubricating (lube) oil and wax.
4. 565°C+ (1050°F+): residuum, which may be cut with lighter oil to produce bunker oil and other fuel oil.

1.4.2. Environmental Behavior

Another way to describe or characterize petroleum products is by generalized spill cleanup categories, and the following categories are in use by the National Oceanic and Atmospheric Administration (NOAA) to identify cleanup options:

1. *Gasoline products* are highly volatile products that evaporate quickly (often completely) within one or two days. They are narrow cut fractions with no residue and low viscosity, which spread rapidly to a thin sheen on water or onto the land. They are highly toxic to biota, will penetrate the substrate, and are nonadhesive.
2. *Diesel-like products* (jet fuel, diesel, No. 2 fuel oil, kerosene) are moderately volatile products that can evaporate with no residue. They have a low-to-moderate viscosity, spread rapidly into thin slicks, and form stable emulsions. They have a moderate-to-high (usually, high) toxicity to biota, and the specific toxicity is often related to type and concentration of aromatic compounds. They have the ability to penetrate substrate, but fresh (unoxidized) spills are nonadhesive.
3. *Intermediate products* (No. 4 fuel oil, lube oil) are products that are loss volatile than the two previous categories; up to one-third will evaporate within 24 hours. They have a moderate-to-high viscosity and a variable toxicity that depends on the amount of the lower-boiling components. These products may penetrate the substrate, and therefore cleanup is most effective if conducted quickly.
4. *Low-API fuel oil* (heavy industrial fuel oil) is a medium-viscosity product that are highly variable and often blended with lower-boiling products. The blends may be unstable and the oil may separate when spilled onto the ground or onto a waterway. The oil may be buoyant or sink in water,

depending on water density. The sunken oil has little potential for evaporation and may accumulate on the bottom (of the waterway) under calm conditions. However, the sunken oil may be resuspended during storm events, providing shoreline oiling (contamination). These products weather (oxidize) slowly.

5. *Residual products* (No. 6 fuel oil, bunker C oil): these products have little (usually, no) ability to evaporate. When spilled, persistent surface and intertidal area contamination is likely with long-term contamination of the sediment. The products are very viscous to semisolid and often become less viscous when warmed. They weather (oxidize) slowly and may form tar balls that can sink in waterways (depending on product density and water density). They are highly adhesive to soil. Heavy oil, a viscous petroleum, and bitumen from tar sand deposits also come into this category of contaminant.

1.5. REFINERY WASTE

The pollution of ecosystems, either inadvertently or deliberately, has been a fact of life for millennia (Pickering and Owen, 1994). In recent times, the evolution of industrial operations has led to issues related to the disposal of a wide variety of chemical contaminants (Easterbrook, 1995). Chemical wastes that were once exotic have become commonplace and hazardous (Tedder and Pohland, 1993). Recognition of this makes it all the more necessary that steps be taken to terminate the pollution, preferably at the source or before it is discharged into the environment. It is also essential that the necessary tests be designed to detect the pollution and its effect on living forms.

Any chemical substance, if improperly managed or disposed of, may pose a danger to living organisms, materials, structures, or the environment, by explosion or fire hazards, corrosion, toxicity to organisms, or other detrimental effects. In addition, when released to the environment, many chemical substances can be classified as hazardous or nonhazardous. Consideration must be given to the distribution of chemical wastes on land systems, in water systems, and in the atmosphere.

In general terms, the origin of chemical wastes refers to their points of entry into the environment. Point-source leaks and spills (i.e., sources that release emissions) through a confined vent (stack) or opening and non-point-source emissions have resulted in environmental contamination from petroleum and petroleum products. Spills of crude oil and fuels have caused wide-ranging damage in marine and freshwater environments. Oil slicks and tars in shore areas and beaches can ruin the aesthetic value of entire regions. Other sources of environmental leakage as it affects the petroleum industry may consist of (1) deliberate addition to soil, water, or air by humans: for example, the disposal of used engine oil; (2) evaporation or wind erosion from emissions into the atmosphere; (3) leaching from waste dumps into groundwater, streams, and bodies of water; (4) leakage, such as from underground storage tanks or pipelines;

(5) accidents, such as fire or explosion; and (6) emissions waste treatment or storage facilities.

In terms of waste definition, there are three basic approaches (as it pertains to petroleum, petroleum products, and nonpetroleum chemicals) to defining petroleum or a petroleum product as hazardous: (1) a qualitative description of the waste by origin, type, and constituents; (2) classification by characteristics based on testing procedures; and (3) classification as a result of the concentration of specific chemical substances.

However, various countries use different definitions of chemical waste and there are often several inconsistencies in the definitions. Usually, the definition involves qualification of whether or not the material is hazardous. For example, in some countries, a hazardous waste is any material that is especially hazardous to human health, air, or water, or which is explosive, flammable, or may cause disease. Poisonous waste is material that is poisonous, noxious, or polluting and whose presence on the land is liable to give rise to an environmental hazard. But in more general terms (in any country), hazardous waste is waste material that is unsuitable for treatment or disposal in municipal treatment systems, incinerators, or landfills and which therefore requires special treatment.

Moreover, and somewhat paradoxically, measures taken to reduce air and water pollution may actually increase the production of chemical wastes. As examples, disposal of petroleum wastes by water treatment processes can yield a chemical sludge or concentrated liquor that requires stabilization and disposal (Cheremisinoff, 1995). Scrubbing to remove hydrogen sulfide, sulfur oxides, and low-boiling organic sulfides as well as carbon dioxide (gas cleaning) are not immune to process waste, even though the chemistry of the cleaning processes is, in theory, reversible (Speight, 1993). Sludge is often produced and the disposal of this material became a major environmental issue that cannot be ignored. In addition, electrostatic precipitators, used to remove metals from flue gases (Speight, 1993), also yield significant quantities of solid by-products, some of which are hazardous.

1.5.1. Chemical Characteristics

A chemical waste is considered hazardous if it exhibits one or more of the following characteristics: *ignitability, corrosivity, reactivity,* and *toxicity*. Under the authority of the Resource Conservation and Recovery Act (RCRA) and the U.S. Environmental Protection Agency (EPA), a hazardous substance has one or more of the foregoing characteristics.

Briefly, *ignitability* is that characteristic of chemicals that are volatile liquids and the vapors are prone to ignition in the presence of an ignition sources. Nonliquids that may catch fire from friction or contact with water and which burn vigorously or are persistently ignitable compressed gases and oxidizers also fall under the mantle of ignitable chemicals. Examples include solvents, friction-sensitive substances, and pyrophoric solids that may include catalysts and metals isolated from various refining processes. Organic solvents are indigenous to the petroleum industry and release to the atmosphere as vapor and can

Table 1.4. Flammability of Selected Organic Liquids

Liquid	Flash Point (°C)[a]	Volume Percent in Air[b]	
		LFL	UFL
Diethyl ether	−43	1.9	36
Pentane	−40	1.5	7.8
Acetone	−20	2.6	13
Toluene	4	1.3	7.1
Methanol	12	6.0	37
Gasoline (2,2,4-trimethyl-pentane)	—	1.4	7.6
Naphthalene	157	0.9	5.9

[a] Closed-cup flash point test.
[b] LFL, lower flammability limit; UFL, upper flammability limit at 25°C (77°F).

pose a significant inhalation hazard. Improper storage, use, and disposal can result in the contamination of land systems as well as groundwater and drinking water (Barcelona et al., 1990).

Often, the term *ignitable chemical* (i.e., naphtha or gasoline) is used in the same sense as the term *flammable chemical* (Table 1.4) insofar as it is a chemical that will burn readily, but a *combustible chemical* (any higher-boiling hydrocarbon product of refining but which can include naphtha or gasoline) often requires relatively more persuasion to burn (i.e., the chemical is less flammable). Most petroleum products that are likely to burn accidentally are low-boiling liquids that form vapors that are usually denser than air and thus tend to settle in low spots. The tendency of a liquid to ignite is measured by a test in which the liquid is heated and exposed to a flame periodically until the mixture of vapor and air ignites at the liquid's surface. The temperature at which this occurs is called the *flash point*.

There are several standard tests for determining the flammability of materials (ASTM, 2004). For example, the upper and lower concentration limits for the *flammability* of chemicals and waste can be determined by standard test methods (ASTM D4982, E681), as can the *combustibility* and the *flash point* (ASTM D1310, E176, E502). With these definitions in mind it is possible to divide ignitable materials into two subclasses:

1. A *flammable solid* is a solid that can ignite from friction or from heat remaining from its manufacture, or which may cause a serious hazard if ignited. Explosive materials are not included in this classification.
2. A *flammable liquid* is a liquid having a flash point below 37.8°C (100°F) (ASTM D92, D1310). A *combustible liquid* has a flash point in excess of 37.8°C (100°F) but below 93.3°C (200°F). Gases are substances that exist

entirely in the gaseous phase at 0°C (32°F) and 1 atm pressure (14.7 psi) pressure. A *flammable compressed gas* [such as liquefied petroleum gas (LPG), or any liquefied hydrocarbon gas or petroleum product] meets specified criteria for lower flammability limit, flammability range, and flame projection.

In considering the ignition of vapors, two important concepts are *flammability limit* and *flammability range*. Values of the vapor/air ratio below which ignition cannot occur because of insufficient fuel define the lower flammability limit. Similarly, values of the vapor/air ratio above which ignition cannot occur because of insufficient air define the upper flammability limit. The difference between upper and lower flammability limits at a specified temperature is the flammability range.

Dust explosions (ASTM E789) that can occur during catalytic reactor shutdown and cleaning are due to the production of finely divided solids through attrition. Many catalyst dusts can burn explosively in air. Thus, control of dust generated by catalyst attrition is essential (Mody and Jakhete, 1988).

Substances that catch fire spontaneously in air without an ignition source are called *pyrophoric*. These include phosphorus, the alkali metals and powdered forms of magnesium, calcium, cobalt, manganese, iron, zirconium, and aluminum—all of which may occur at one time or another at a refinery site. Moisture in air is often a factor in *spontaneous ignition*.

Corrosivity is that characteristic of chemicals that exhibits extremes of acidity or basicity or a tendency to corrode steel. Such chemicals, used in various refining (treating) processes, are acidic and are capable of corroding metal such as tanks, containers, drums, and barrels. On the other hand, *reactivity* is a violent chemical change (an explosive substance is an obvious example) that can result in pollution and/or harm to indigenous flora and fauna. Such wastes are unstable under ambient conditions insofar as they can create explosions, toxic fumes, gases, or vapors when mixed with water.

Finally, *toxicity* (defined in terms of a standard extraction procedure followed by chemical analysis for specific substances) is a characteristic of all chemicals, whether petroleum or nonpetroleum in origin. Toxic wastes are harmful or fatal when ingested or absorbed, and when such wastes are disposed of on land, the chemicals may drain (leach) from the waste and pollute groundwater. Leaching of such chemicals from contaminated soil may be particularly evident when the area is exposed to acid rain. The acidic nature of the water may impart mobility to the waste by changing the chemical character of the waste or the character of the minerals to which the waste species are adsorbed.

As with flammability, there are many tests that can be used to determine corrosivity (ASTM D1838, D2251). Most corrosive substances belong to at least one of the following four chemical classes: strong acids, strong bases, oxidants, or dehydrating agents (Table 1.5). All are used in the refining industry. For example, sulfuric acid is a prime example of a corrosive substance (ASTM C694). As well as being a strong acid (ASTM E1011), concentrated sulfuric acid is also a

Table 1.5. Examples of Corrosive Substances

Name (Formula)	Properties and Effects
Nitric acid (HNO_3)	Strong acid, strong oxidizer, corrodes metals, reacts with protein in tissue
Hydrochloric acid (HCl)	Strong acid, corrodes metals, HCl gas damages respiratory tract
Hydrofluoric acid (HF)	Corrodes metals, dissolves glass, causes bad burns
Alkali metal hydroxides (e.g., NaOH)	Corrode zinc, lead, and NaOH and KOH; dissolve tissue; cause severe burns
Hydrogen peroxide (H_2O_2)	Causes severe burns
Interhalogen compounds	Corrosive irritants, dehydrate tissue
Halogen oxides (OF_2, Cl_2O, Cl_2O_7)	Corrosive irritants, dehydrate tissue
Halogens (F_2, Cl_2, Br_2)	Corrosive to mucous membranes, strong irritants

dehydrating agent and oxidant. The heat generated when water and concentrated sulfuric acid are mixed illustrates the high affinity of sulfuric acid for water. If this is done incorrectly by adding water to the acid, localized boiling and spattering can occur and result in personal injury. The major destructive effect of sulfuric acid on skin tissue is the removal of water with an accompanying release of heat. Contact of sulfuric acid with tissue results in tissue destruction at the point of contact. Inhalation of sulfuric acid fumes or mists damages tissues in the upper respiratory tract and eyes. Long-term exposure to sulfuric acid fumes or mists has caused erosion of teeth as well as destruction of other parts of the body!

Reactive chemicals are those that tend to undergo rapid or violent reactions under certain conditions. Such substances include those that react violently or form potentially explosive mixtures with water, such as some of the common oxidizing agents (Table 1.6). Explosives (Sudweeks et al., 1983; Austin, 1984) constitute another class of reactive chemicals. For regulatory purposes, those substances are also classified as reactive that react with water, acid, or base to produce toxic fumes, particularly hydrogen sulfide or hydrogen cyanide.

Heat and temperature are usually very important factors in reactivity since many reactions require energy of activation to get them started. The rates of most reactions tend to increase sharply with increasing temperature, and most chemical reactions give off heat. Therefore, once a reaction is started in a reactive mixture lacking an effective means of heat dissipation, the rate will increase exponentially with time (doubling with every 10° rise in temperature), leading to an uncontrollable event. Other factors that may affect the reaction rate include the physical form of reactants, the rate and degree of mixing of reactants, the degree of dilution with a nonreactive medium (e.g., an inert solvent), the presence of a catalyst, and pressure.

Toxicity is of the utmost concern in dealing with chemicals and their disposal (ASTM D4447). This includes both long-term chronic effects from continual or

Table 1.6. Common Oxidizing Agents

Name	Formula	Gas/Liquid/Solid
Ammonium nitrate	NH_4NO_3	Solid
Ammonium perchlorate	NH_4ClO_4	Solid
Bromine	Br_2	Liquid
Chlorine	Cl_2	Gas (stored as liquid)
Fluorine	F_2	Gas
Hydrogen peroxide	H_2O_2	Solution in water
Nitric acid	HNO_3	Concentrated solution
Nitrous oxide	N_2O	Gas (stored as liquid)
Ozone	O_3	Gas
Perchloric acid	$HClO_4$	Concentrated solution
Potassium permanganate	$KMnO_4$	Solid
Sodium dichromate	$Na_2Cr_2O_7$	Solid

periodic exposures to low levels of toxic chemicals and acute effects from a single large exposure (Zakrzewski, 1991). Not all toxins are immediately apparent. For example, living organisms require certain metals for physiological processes. When present at concentrations above the level of homeostatic regulation, these metals can be toxic (ASTM E1302). In addition, there are metals that are chemically similar to, but higher in molecular weight than, the essential metals (heavy metals). Metals can exert toxic effects by direct irritant activity, blocking functional groups in enzymes, altering the conformation of biomolecules, or displacing essential metals in a metalloenzyme.

1.5.2. Environmental Behavior

In addition to the classification of petroleum-related chemicals by the characteristics described above, the U.S. Environmental Protection Agency designates more than 450 chemicals or chemical wastes that are specific substances or classes of substances known to be hazardous. Each such chemical or waste is assigned a hazardous waste number in the format of a letter followed by three numerals, where a different letter is assigned to substances from each of the following list:

1. F-type: chemicals or chemical wastes from nonspecific sources (Table 1.7)
2. K-type: chemicals or chemical wastes from specific sources (Table 1.8)
3. P-type: chemicals or chemical wastes that are hazardous and that are mostly specific chemical species such as fluorine
4. U-type: generally hazardous chemicals or chemical wastes that are predominantly specific compounds

Table 1.7. Chemical Wastes Designated as F-Category Wastes

Number	Waste Material
F001	Spent halogenated solvents chlorinated fluorocarbons; sludge from solvent recovery processes
F004	Spent nonhalogenated solvents (cresols, nitrobenzene); still bottoms from solvent recovery operations
F007	Spent solution from electroplating operations
F010	Sludge from metal heat-treating operations

Table 1.8. Chemical Wastes Designated as K-Category Wastes

Number	Waste Material
K001	Sediment/sludge from wastewater treatment from wood-preserving processes (especially creosote and pentachlorophenol sediment/sludge)
K002	Wastewater treatment sludge from chrome yellow and orange pigments
K020	Residue from vinyl chloride distillation
K034	2,6-Dichlorophenol waste
K047	Pink/red water from trichloroethane manufacture
K049	Waste oil/emulsion/solids from petroleum refining
K060	Ammonia lime still sludge
K067	Electrolytic sludge from zinc production

The Comprehensive Environmental Response, Compensation, and Liability Act (CERCLA) gives a broader definition of hazardous substances, which includes the following:

1. Any element, compound, mixture, solution, or substance whose release may substantially endanger public health, public welfare, or the environment
2. Any element, compound, mixture, solution, or substance in reportable quantities designated by CERCLA Section 102
3. Certain substances or toxic pollutants designated by the Water Pollution Control Act.
4. Any hazardous air pollutant listed under Section 112 of the Clean Air Act

5. Any imminently hazardous chemical substance or mixture that has been the subject of government action under Section 7 of the Toxic Substances Control Act (TSCA)
6. Any hazardous chemical or chemical waste listed or having characteristics identified by the Resource Conservation Recovery Act, with the exception of those suspended by Congress under the Solid Waste Disposal Act

In terms of quantity by weight, more wastes than all others combined are those from categories designated by hazardous waste numbers preceded by F and K. The F categories are those wastes from nonspecific sources (Table 1.7). K-type hazardous wastes are those from specific sources produced by industries, such as the manufacture of inorganic pigments, organic chemicals, pesticides, explosives, iron and steel, and nonferrous metals, and from processes such as petroleum refining or wood preservation (Table 1.8).

Some refinery wastes that might exhibit a degree of hazard are exempt from the Resource Conservation Recovery Act regulation by legislation and include the following:

1. Ash and scrubber sludge from thermal generation or power generation by utilities
2. Oil and gas field drilling mud
3. By-product brine from petroleum production
4. Catalyst dust

Eventual reclassification of these types of low-hazard wastes could increase the quantities of regulated wastes severalfold.

REFERENCES

ASTM. 2004. *Annual Book of ASTM Standards*, Vol. 04.02. American Society for Testing and Materials, West Conshohocken, PA.

Austin, G. T. 1984. *Shreve's Chemical Process Industries*, 5th ed. McGraw-Hill, New York, Chap. 22.

Barcelona, M., Wehrmann, A., Keeley, J. F., and Pettyjohn, J. 1990. *Contamination of Ground Water*. Noyes Data Corp., Park Ridge, NJ.

Breen, J. J., and Dellarco, M. J. (Eds.). 1992. *Pollution Prevention in Industrial Processes*, Symposium Series No. 508. American Chemical Society, Washington, DC.

Chenier, P. J. 1992. *Survey of Industrial Chemistry*, 2nd ed. VCH Publishers, New York.

Cheremisinoff, P. 1995. *Handbook of Water and Wastewater Treatment Technology*. Marcel Dekker, New York.

Easterbrook, G. 1995. *A Moment on the Earth: The Coming Age of Environmental Optimism*. Viking Press, New York.

EPA. 2004. Environmental Protection Agency, Washington, DC. Web site: http://www.epa.gov.

ITAA. 1936. Income Tax Assessment Act. Government of the Commonwealth of Australia.

Johnson, R. W., and Gordon, G. E. (Eds.). 1987. *The Chemistry of Acid Rain: Sources and Atmospheric Processes*, Symposium Series No. 349. American Chemical Society, Washington, DC.

Kletz, T. A. 1990. *Improving Chemical Engineering Practices*, 2nd ed. Hemisphere Publishing, New York.

Loeher, R. C. 1992. In *Petroleum Processing Handbook*, J. J. McKetta (Ed.). Marcel Dekker, New York, p. 190.

Long, R. B., and Speight, J. G. 1998. In *Petroleum Chemistry and Refining*, J. G. Speight (Ed.). Taylor & Francis, Philadelphia, Chap. 1.

Manahan, S. E. 1999. *Environmental Chemistry*, 7th ed. Lewis Publishers, Chelsea, MI.

Milman, H. A., and Weisburger, E. K. 1994. *Handbook of Carcinogen Testing*, 2nd ed. Noyes Data Corp., Park Ridge, NJ.

Mody, V., and Jakhete, R. 1988. *Dust Control Handbook*. Noyes Data Corp., Park Ridge, NJ.

NRC. 1981. *Prudent Practices for Handling Hazardous Chemicals in Laboratories*. National Academy Press, National Research Council, Washington, DC.

Olschewsky, D., and Megna, A. 1992. In *Petroleum Processing Handbook*, J. J. McKetta (Ed.). Marcel Dekker, New York, p. 179.

Pickering, K. T., and Owen, L. A. 1994. *Global Environmental Issues*. Routledge, New York.

Reynolds, J. G. 1998. In *Petroleum Chemistry and Refining*, J. G. Speight (Ed.). Taylor & Francis, Philadelphia, Chap. 3.

Schwarzenbach, R. P., Gschwend, P. M., and Imboden, D. M. 2003. *Environmental Organic Chemistry*, 2nd ed. Wiley, Hoboken, NJ.

Speight, J. G. 1990. *Fuel Science and Technology Handbook*. Marcel Dekker, New York.

Speight, J. G. 1993. *Gas Processing: Environmental Aspects and Methods*. Butterworth-Heinemann, Oxford.

Speight, J. G. 1994. *The Chemistry and Technology of Coal*, 2nd ed. Marcel Dekker, New York.

Speight, J. G. 1996. *Handbook of Environmental Technology*. Taylor & Francis, Philadelphia.

Speight, J. G. 1999. *The Chemistry and Technology of Petroleum*, 3rd ed. Marcel Dekker, New York.

Speight, J. G., and Ozum, B. 2002. *Petroleum Refining Processes*. Marcel Dekker, New York.

Sudweeks, W. B., Larsen, R. D., and Balli, F. K. 1983. In *Riegel's Handbook of Industrial Chemistry*, J. A. Kent (Ed.). Van Nostrand Reinhold, New York, p. 700.

Tedder, D. W., and Pohland, F. G. (Eds.). 1993. *Emerging Technologies in Hazardous Waste Management III*, Symposium Series No. 518. American Chemical Society, Washington, DC.

Tinsley, I. J. 2004. *Chemical Concepts in Pollutant Behavior*, 2nd ed. Wiley Hoboken, NJ.

van Nes, K., and van Westen, H. A. 1951. *Aspects of the Constitution of Mineral Oils*. Elsevier, Amsterdam.

Woodside, G. 1999. *Hazardous Materials and Hazardous Waste Management*, 2nd ed. Wiley, New York.

Zakrzewski, S. F. 1991. *Principles of Environmental Toxicology, ACS Professional Reference Book*. American Chemical Society, Washington, DC.

CHAPTER

2

COMPOSITION AND PROPERTIES

Just as the chemical and physical properties of petroleum have offered challenges in selecting and designing optimal upgrading schemes, they also introduce challenges when determining the effects of petroleum and its product on the environment. In particular, predicting the fate of the polynuclear aromatic systems, the heteroatom systems (principally, compounds containing nitrogen and sulfur), and the metal-containing systems (principally, compounds of vanadium, nickel, and iron) in the feedstocks is the subject of many studies and migration models. These constituents generally cause processing problems, and knowledge of the behavior of these elements is essential for process improvements, process flexibility, and environmental compliance.

Because of the variation in the amounts of chemical types and bulk fractions, it should not be surprising that petroleum exhibits a wide range of physical properties, and several relationships can be made between various physical properties. Whereas properties such as viscosity, density, boiling point, and color of petroleum may vary widely, the ultimate or elemental analysis varies, as already noted, over a narrow range for a large number of petroleum samples. The carbon content is relatively constant, and the hydrogen and heteroatom contents are responsible for the major differences between petroleum. Nitrogen, oxygen, and sulfur can be present in only trace amounts in some petroleum, which as a result consists primarily of hydrocarbons. On the other hand, petroleum containing 9.5% heteroatoms may contain essentially no true hydrocarbon constituents insofar as the constituents contain *at least one or more* nitrogen, oxygen, and/or sulfur atoms within the molecular structures. Coupled with the changes to the feedstock constituents brought about by refinery operations, it is not surprising that petroleum characterization is a monumental task.

Thus, initial inspection of the nature of the petroleum will provide deductions about the most logical means of cleanup and any subsequent environmental effects. Indeed, careful evaluation of petroleum from physical property data is a major part of the initial study of any petroleum that has been released to the environment. Proper interpretation of the data resulting from the inspection of crude oil requires an understanding of their significance. Consequently, various standards organizations, such as the American Society for Testing and Materials (ASTM, 2004) in North America and the Institute of Petroleum in the United Kingdom (IP, 2004), have devoted considerable time and effort to the correlation

---

Environmental Analysis and Technology for the Refining Industry, by James G. Speight
Copyright © 2005 John Wiley & Sons, Inc.

and standardization of methods for the inspection and evaluation of petroleum and petroleum products.

The data derived from any one, or more, of the evaluation techniques present an indication of the nature of petroleum and its products. The data can be employed to give the environmental scientist or engineer an indication of the means by which the spilled material can be, or should be, recovered. Other properties (Speight, 1999) may also be required for further evaluation, or, more likely, for comparison of before and after scenarios even though they may not play any role in dictating which cleanup operations are necessary.

To take this one step further, it may then be possible to develop preferred cleanup methods from one (but preferably more) of the physical properties, as determined by the evaluation test methods.

2.1. COMPOSITION

An important step in assessing the effects of petroleum products that have been released into the environment is to evaluate the nature of the particular mixture and eventually, select an optimum remediation technology for that mixture. As a general rule, crude oil (unrefined petroleum) is a complex mixture composed of the same compounds, but the quantities of the individual compounds differ in crude oils from different locations. This rule of thumb implies that the quantities of some compounds can be zero in a given mixture of compounds that comprise crude oil from a specific location.

In very general terms, petroleum is a mixture of (1) hydrocarbon types, (2) nitrogen compounds, (3) oxygen compounds, (4) sulfur compounds, and (5) metallic constituents. Petroleum products are less well defined in terms of heteroatom compounds and are better defined in terms of the hydrocarbon types present. However, this general definition is not adequate to describe the true composition as it relates to the behavior of the petroleum, and its products, in the environment. For example, the occurrence of amphoteric species (i.e., compounds having a mixed acid–base nature) is not always addressed, nor is the phenomenon of molecular size or the occurrence of specific functional types that can play a major role in petroleum behavior.

In the present context, petroleum composition is defined in terms of (1) the elemental composition, (2) the chemical composition, and (3) the fractional composition. All three are interrelated, although the closeness of the elemental composition makes it difficult to relate precisely to the chemical and fractional composition. The chemical and fractional composition are somewhat easier to relate because of the quantities of heteroatoms (nitrogen, oxygen, sulfur, and metals) that occur in the higher-boiling fractions.

2.1.1. Elemental Composition

The elemental composition of petroleum varies greatly from crude oil to crude oil. Most compounds in petroleum (usually more than 75%) are types of hydrocarbons, and the majority of the chemical components in petroleum are made up of five main elements: carbon 82 to 87% w/w, hydrogen 11 to 15% w/w, sulfur 0 to 8% w/w, nitrogen 0 to 1% w/w, and oxygen 0.0 to 0.5% w/w. The elements are combined to form a complex mixture of organic compounds that range in molecular weight from 16 (methane; CH_4) to several hundred, perhaps even to several thousand. A wide range of metals is also found in trace amounts in crude oil. All metals through the atomic number 42 (molybdenum) have been found, with the exception of rubidium and niobium; a few heavier elements also have been detected. Nickel and vanadium are the most important, because they are present in all crude oil, usually at concentrations far higher than those of any other metal.

2.1.2. Chemical Composition

The chemical composition of petroleum also varies over a wide range. A broad functional definition of petroleum hydrocarbons is that hydrocarbons are composed primarily of many organic compounds of natural origin and low water solubility.

Remembering that hydrocarbons are (by definition) compounds containing carbon and hydrogen *only*, the hydrocarbon content may be as high as 97% by weight in a lighter paraffinic crude oil, about 50% by weight in heavy crude oil, and less than 30% by weight in tar sand bitumen. However, within the hydrocarbon constituents, there is a variation of chemical type. Petroleum hydrocarbons may be paraffinic, alicyclic, or aromatic and occur in varying concentrations within the various fractions of a single crude oil. Thus, the constituents of petroleum that occur in varying amounts depend on the source and character of the oil (Table 2.1):

1. *Alkanes* (also called *normal paraffins* or *n-paraffins*). These constituents are characterized by branched or unbranched chains of carbon atoms with attached hydrogen atoms, and contain no carbon–carbon double bonds (they are saturated). Examples of alkanes are pentane (C_5H_{12}) and heptane (C_7H_{16}).

2. *Cycloalkanes* or *cycloparaffins* (also called *naphthenes*). These constituents are characterized by the presence of simple closed rings of carbon atoms (such as the cyclopentane ring or the cyclohexane ring). Naphthenes are generally stable and relatively insoluble in water.

3. *Alkenes* (also called *olefins*). These constituents are characterized by the presence of branched or unbranched chains of carbon atoms. Alkenes are not generally found in crude oil but are common in refined products, such as naphtha (a precursor to gasoline). Common gaseous alkenes include ethylene ($CH_2=CH_2$) and propene (also called propylene, $CH_3CH=CH_2$).

4. *Single-ring aromatics*. Aromatic constituents are characterized by the presence of rings with six carbon atoms and are considered to be the most acutely toxic component of crude oil constituents because of their association with chronic and carcinogenic effects. Many low-molecular-weight aromatics are also soluble in water, increasing the potential for exposure to aquatic resources.

Table 2.1. Compound Types in Petroleum and Petroleum Products

Class	Compound Types
Saturated hydrocarbons	n-Paraffins
	Isoparaffins and other branched paraffins
	Cycloparaffins (naphthenes)
	Condensed cycloparaffins (including steranes, hopanes)
	Alkyl side chains on ring systems
Unsaturated hydrocarbons	Olefins not indigenous to petroleum; present in products of thermal reactions
Aromatic hydrocarbons	Benzene systems
	Condensed aromatic systems
	Condensed aromatic–cycloalkyl systems
	Alkyl side chains on ring systems
Saturated heteroatomic systems	Alkyl sulfides
	Cycloalkyl sulfides
	Alkyl side chains on ring systems
Aromatic heteroatomic systems	Furans (single- and multiring systems)
	Thiophenes (single- and multiring systems)
	Pyrroles (single- and multiring systems)
	Pyridines (single- and multiring systems)
	Mixed heteroatomic systems
	Amphoteric (acid–base) systems
	Alkyl side chains on ring systems

Source: Speight, 2001.

The number of rings, which may range from one to five, further distinguishes aromatics.

5. *Multiring aromatics.* Aromatics with two or more condensed rings are referred to as polynuclear aromatic hydrocarbons (PNAs) and naphthalene (two-ring) and phenanthrene (three-ring). The most abundant aromatic hydrocarbon families have two and three fused rings with one to four carbon atom alkyl group substitutions. Condensed aromatic constituents with more than two condensed rings are also present in the higher-boiling fractions of petroleum.

A typical crude oil contains, generally speaking, approximately 1% polynuclear aromatic hydrocarbons. Concentrations of total carcinogenic polynuclear aromatic hydrocarbons (like benzo[*a*]pyrene) range from 12 to <100 ppm. Fresh crude oil will contain a fraction of volatile hydrocarbons, such as benzene, toluene, xylenes, and other aromatics, some of which may pose a threat to human health. However, the relative mass fraction of volatile hydrocarbons in crude oil is significantly less than that found in crude oil distillate products such as gasoline. The volume percentage of benzene in gasoline may range up to 3% (30,000 ppm), while the benzene content of crude oil is approximately 0.2% (2000 ppm). As a result of the lower percentage of volatile aromatics, vapor emissions from crude oil–contaminated soils are expected to be much less than potential emissions from gasoline-contaminated soils.

If the only constituents of petroleum were the hydrocarbons, the complexity is further illustrated by the number of potential isomers (i.e., molecules having the same atomic formula) that can exist for a given number of paraffinic carbon atoms and that increases rapidly as molecular weight increases:

Carbon Atoms per Hydrocarbon	Number of Isomers
4	2
8	18
12	355
18	60,523

This same increase in the number of isomers with molecular weight also applies to the other molecular types present. Since the molecular weights of the molecules found in petroleum can vary from that of methane (CH_4; molecular weight = 16) to several thousand (Speight, 1999, and references cited therein), it is clear that the heavier nonvolatile fractions can contain virtually unlimited numbers of molecules. However, in reality the number of molecules in any specified fraction is limited by the nature of the precursors of petroleum, their chemical structures, and the physical conditions that are prevalent during the maturation (conversion of the precursors) processes.

In addition to hydrocarbons, petroleum also contains compounds that consist of nitrogen, oxygen, and sulfur (in the minority) as well as trace amounts of metals such as vanadium, nickel, iron, and copper. Porphyrins, the major organometallic compounds present in petroleum, are large, complex cyclic carbon structures derived from chlorophyll and characterized by the ability to contain a central metal atom (trace metals are commonly found within these compounds).

2.1.3. Composition by Volatility

Distillation is a common method for the fractionation of petroleum that is used in the laboratory as well as in refineries. The technique of distillation has been practiced for many centuries, and the stills that have been employed have taken many forms (Speight, 1999). Distillation is the first and the most fundamental step in the refining process (after the crude oil has been cleaned and any remnants of brine removed) (Bland and Davidson, 1967; Speight, 1999, and references cited therein; Speight and Ozum, 2002, and references cited therein), which is often referred to as the *primary refining process*. Distillation involves the separation of the various hydrocarbon compounds that occur naturally in a crude oil into a number of different fractions (a fraction is often referred to as a *cut*).

In the atmospheric distillation process (Figure 2.1), heated crude oil is separated in a distillation column (distillation tower, fractionating tower, atmospheric pipe still) into streams which are then purified, transformed, adapted, and treated

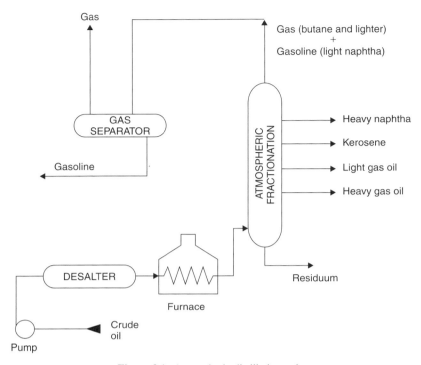

Figure 2.1. Atmospheric distillation unit.

in a number of subsequent refining processes, resulting in products for the refinery's market. The lower-boiling products separate out higher up the column, whereas the heavier, less volatile products settle out toward the bottom of the distillation column. Fractions produced in this manner, known as *straight-run fractions*, range from (atmospheric tower) gas, gasoline, and naphtha, to kerosene, gas oils, and light diesel, and to (vacuum tower) lubricating oil and residuum.

The atmospheric residuum is then fed to the vacuum distillation unit at a pressure of 10 mmHg, with light vacuum gas oil, heavy vacuum gas oil, and vacuum residue as the products (Figure 2.2). The fractions obtained by vacuum distillation of the reduced crude (atmospheric residuum) from an atmospheric distillation unit depend on whether or not the unit is designed to produce lubricating or vacuum gas oils. In the former case, the fractions include (1) heavy gas oil, which is an overhead product and is used as a catalytic cracking stock or, after suitable treatment, a light lubricating oil, (2) lubricating oil (usually, three fractions: light, intermediate, and heavy), which is obtained as a sidestream product, and (3) asphalt (or residuum), which is the bottom product and may be used directly as, or to produce, asphalt, and which may also be blended with gas oils to produce a heavy fuel oil. Operating conditions for vacuum distillation are usually 50 to 100 mmHg (atmospheric pressure = 760 mmHg).

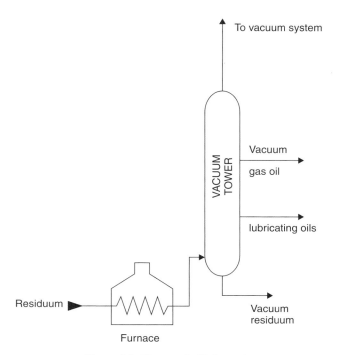

Figure 2.2. Vacuum distillation unit.

Fractions from the atmospheric and vacuum towers are often used as feedstocks to these second-stage refinery processes that break down the fractions or bring about a basic chemical change in the nature of a particular hydrocarbon compound to produce specific products.

2.1.4. Composition by Fractionation

An understanding of the chemical types (or composition) in petroleum can lead to an understanding of the chemical aspects of petroleum behavior. Indeed, this is not only a matter of knowing the elemental composition of a feedstock; it is also a matter of understanding the bulk properties as they relate to the chemical or physical composition of the material. For example, it is difficult to understand, a priori, the behavior of petroleum and petroleum products from the elemental composition alone, and more information is necessary to understand environmental behavior.

Fractionation of petroleum by volatility, informative as it might be, does not give any indication of the physical nature of petroleum. This is more often achieved by subdivision of the petroleum into bulk fractions that are separated by a variety of solvent and adsorption methods.

Thus, in the simplest sense, petroleum and petroleum products can be considered to be composites of four major fractions (saturates, aromatics, resins, and asphaltenes) (SARA analysis) (Figure 2.3) in varying amounts, depending on the type of petroleum and the type of petroleum product (Speight, 2001, 2002). In fact, the lower-boiling product may only be compositions of one or two fractions (saturates and aromatics). Whatever the case, it must never be forgotten that the nomenclature of these fractions lies within the historical development of petroleum science and that the fraction names are operational and are related more to the general characteristics of the fraction than to the identification of specific compound types.

Asphaltenes and resins can comprise a large fraction of crude oils and heavy fuel oils, making those oils very dense and viscous. Composition is dependent on source (these structures have the highest individual molecular weight of all crude oil components and are basically colloidal aggregates). Asphaltenes are substances in petroleum that are insoluble in solvents of low molecular weight such as pentane or heptane. These compounds are composed of very polynuclear aromatic and heterocyclic molecules and are solids at normal temperatures. Consequently, oils that have high asphaltene contents are very viscous, have a high pour point, and are generally nonvolatile in nature. The porphyrins, asphaltene, and resin compounds are considered the residual oil, or residuum. During the weathering process, this fraction is the last to degrade, and its persistence over years has been noted.

There are two other operational definitions that should be noted at this point: the terms *carbenes* and *carboids* (Figure 2.3). Both such fractions are, by definition, insoluble in benzene (or toluene), but the carbenes are soluble in carbon disulfide whereas the carboids are insoluble in carbon disulfide (Speight, 1999). Only traces of these materials occur in petroleum and none occur in the products, unless the product is a high-boiling product of thermal treatment (such as visbroken feedstocks). On the other hand, oxidized petroleum and oxidized high-boiling

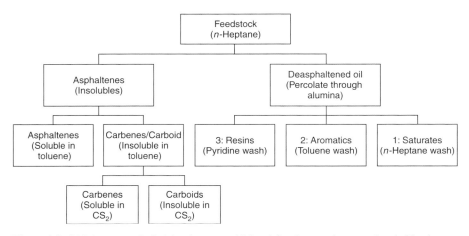

Figure 2.3. SARA-type analysis (showing two additional fractions, carbenes and carboids, that are generally recognized as the results of thermal processes).

products that have been susceptible to oxidation through a spill may contain such fractions. But again, it must be remembered that the fractions separated by the various techniques are based on solubility or adsorption properties and not on specific chemical types.

Thus, the separation of crude oil into two fractions, asphaltenes and maltenes, is conveniently brought about by means of low-molecular-weight paraffinic hydrocarbons (ASTM D2006, D2007, D3279, D4124; IP 143) that were recognized to have selective solvency for hydrocarbons, and simple relatively low-molecular-weight heteroatom derivatives. The more complex, higher-molecular-weight compounds are precipitated particularly well by the addition of 40 volumes of *n*-pentane or *n*-heptane in the methods generally preferred at present (Speight, 1999, and references cited therein). The insoluble fraction, the asphaltenes, should correctly be called *n-pentane asphaltenes* or *n-heptane asphaltenes*, and the method is qualitatively and quantitatively reproducible.

After removal of the asphaltene fraction, further fractionation of petroleum is also possible and there are three standard (ASTM) methods that provide for the separation of a feedstock into four or five constituent fractions (Speight, 1999, and references cited therein). It is interesting to note that as the methods have evolved, there has been a change from the use of pentane (ASTM D2006, D2007) to heptane (ASTM D4124) to separate asphaltenes. This is, in fact, in keeping with the production of a more consistent fraction that represents the higher-molecular-weight constituents of petroleum.

Two of the methods (ASTM D2007, D4124) use adsorbents to fractionate the deasphaltened oil, but the third method (ASTM D2006) advocates the use of various grades of sulfuric acid to separate the material into compound types. Caution is advised in the application of this method since the method does not work well with all feedstocks. For example, when the *sulfuric acid* method (ASTM D2006) is applied to the separation of heavy feedstocks, complex emulsions can be produced.

There are precautions that must be taken when attempting to separate heavy feedstocks or polar feedstocks into constituent fractions. The disadvantages in using ill-defined adsorbents are that adsorbent performance differs with the same feed and in certain instances may even cause chemical and physical modification of the feed constituents. The use of a chemical reactant such as sulfuric acid should only be advocated with caution since feeds react differently and may even cause irreversible chemical changes and/or emulsion formation. These advantages may be of little consequence when it is not, for various reasons, the intention to recover the various product fractions *in toto* or in the original state, but in terms of the compositional evaluation of different feedstocks, the disadvantages are very real.

In summary, the terminology used for the identification of the various methods might differ. However, in general terms, group-type analysis of petroleum is often identified by the acronyms for the names: PONA (paraffins, olefins, naphthenes, and aromatics), PIONA (paraffins, isoparaffins, olefins, naphthenes, and aromatics), PNA (paraffins, naphthenes, and aromatics), PINA (paraffins, isoparaffins, naphthenes, and aromatics), or SARA (saturates, aromatics, resins, and asphaltenes). However, it must be recognized that the fractions produced by the use of different adsorbents will differ in content and will also differ from fractions produced by solvent separation techniques.

The variety of fractions isolated by these methods and the potential for the differences in composition of the fractions makes it even more essential that the method be described accurately and that it be reproducible not only in a particular laboratory but also among laboratories (Speight, 2001).

2.2. PROPERTIES

Petroleum is typically described in terms of its physical properties (such as density and pour point) and chemical composition (such as percent composition of various petroleum hydrocarbons, asphaltenes, and sulfur). Although very complex in makeup, crude can be broken down into four basic classes of petroleum hydrocarbons. Each class is distinguished on the basis of molecular composition. In addition, properties important for characterizing the behavior of petroleum and petroleum products when spilled into waterways or onto land and/or released into the air include flash point, density (read: specific gravity and/or API gravity), viscosity, emulsion formation in waterways, and adhesion to soil.

Generally, the properties of petroleum constituents varying over the boiling range 0 to >565°C (32 to 1050°F). Variations in density also occur over the range 0.6 to 1.3, and pour points can vary from <0 to >100°C. Although these properties may seem to be of lower consequence in the grand scheme of environmental cleanup, they are important insofar as these properties influence (1) the evaporation rate, (2) the ability of the petroleum constituents or petroleum product to float on water, and (3) the fluidity or mobility of the petroleum or petroleum product at various temperatures.

2.2.1. Density and Specific Gravity

The *density* and *specific gravity* of crude oil (ASTM D287, D1298, D941, D1217, D1555) are two properties that have found wide use in the industry for preliminary assessment of the character of the crude oil (Table 2.2). *Density* is the mass of a unit volume of material at a specified temperature and has the dimensions of grams per cubic centimeter (a close approximation to grams per milliliter). *Specific gravity* is the ratio of the mass of a volume of the substance to the mass of the same volume of water and is dependent on two temperatures, those at which the masses of the sample and the water are measured. When the water temperature is 4°C (39°F), the specific gravity is equal to the density in the centimeter–gram–second (cgs) system, since the volume of 1 g of water at that temperature is, by definition, 1 mL. Thus, the density of water, for example, varies with temperature, and its specific gravity at equal temperatures is always unity. The standard temperatures for a specific gravity in the petroleum industry in North America are 60°/60°F (15.6°/15.6°C).

Table 2.2. Variation of Density and API Gravity, and Residuum Content for Selected Crude Oils

Crude Oil	Specific Gravity	API Gravity	Residuum, $>1000°F$
U.S. domestic			
California	0.858	33.4	23.0
Oklahoma	0.816	41.9	20.0
Pennsylvania	0.800	45.4	2.0
Texas	0.827	39.6	15.0
Texas	0.864	32.3	27.9
Foreign			
Bahrain	0.861	32.8	26.4
Iran	0.836	37.8	20.8
Iraq	0.844	36.2	23.8
Kuwait	0.860	33.0	31.9
Saudi Arabia	0.840	37.0	27.5
Venezuela	0.950	17.4	33.6

Although density and specific gravity are used extensively, the API (American Petroleum Institute) gravity is the preferred property:

$$°API = 141.5/\text{sp gr at } 60°/60°F - 131.5$$

The specific gravity of petroleum usually ranges from about 0.8 (45.3° API) for the light crude oil and heavy crude oil to over 1.0 (less than 10° API) for tar sand bitumen. Density or specific gravity or API gravity may be measured by means of a hydrometer (ASTM D287, D1298) or by means of a pycnometer (ASTM D941, D1217). The variation of density with temperature, effectively the coefficient of expansion, is a property of great technical importance, since most petroleum products are sold by volume, and specific gravity is usually determined at the prevailing temperature [21°C (70°F)] rather than at the standard temperature [60°F (15.6°C)]. The tables of gravity corrections (ASTM D1555) are based on an assumption that the coefficient of expansion of all petroleum products is a function (at fixed temperatures) of density only. Recent work has focused on the calculation and predictability of density using new mathematical relationships (Gomez, 1989, 1992).

2.2.2. Elemental (Ultimate) Analysis

The analysis of petroleum for the percentages of carbon, hydrogen, nitrogen, oxygen, and sulfur is perhaps the first method used to examine the general nature, and perform an evaluation, of a feedstock. The atomic ratios of the various elements to carbon (i.e., H/C, N/C, O/C, and S/C) are frequently used for indications of the overall character of the feedstock. It is also of value to determine the amounts of trace elements, such as vanadium and nickel, in a feedstock since these materials can have serious deleterious effects on catalyst performance during refining by catalytic processes.

For example, *carbon content* can be determined by the method designated for coal and coke (ASTM D3178) or by the method designated for municipal solid waste (ASTM E777). There are also methods designated for:

1. *Hydrogen content* (ASTM D1018, D3178, D3343, D3701, E777)
2. *Nitrogen content* (ASTM D3179, D3228, D3431, E148, E258, E778)
3. *Oxygen content* (ASTM E385)
4. *Sulfur content* (ASTM D124, D1266, D1552, D1757, D2662, D3177, D4045, D4294)

Of the data that are available, the proportions of the elements in petroleum vary only slightly over narrow limits: carbon, 83.0 to 87.0%; hydrogen, 10.0 to 14.0%; nitrogen, 0.1 to 2.0%; oxygen, 0.1 to 1.5%; and sulfur, 0.1 to 6.0%. Of the ultimate analytical data, more has been made of the sulfur content than of any other property. For example, the sulfur content (ASTM D124, D1552, D4294) and the API gravity represent the two properties that have, in the past, had the greatest influence on determining the value of petroleum as a feedstock.

The sulfur content varies from about 0.1 wt% to about 3 wt% for the more conventional crude oils to as much as 5 to 6% for heavy oil and bitumen. Depending on the sulfur content of the crude oil feedstock, residua, may be of the same order or even have a substantially higher sulfur content.

2.2.3. Fractionation by Chromatography

The evaluation of petroleum must of necessity involve a study of composition because of the interrelationship of the physical properties and composition as part of the overall evaluation of different feedstocks. There are several ASTM procedures for feedstock/product evaluation:

1. Determination of aromatic content of olefin-free gasoline by silica gel adsorption (ASTM D936)
2. Separation of aromatic and nonaromatic fractions from high-boiling oils (ASTM D2549)
3. Determination of hydrocarbon groups in rubber extender oils by clay-gel adsorption (ASTM D2007)
4. Determination of hydrocarbon types in liquid petroleum products by a fluorescent indicator adsorption test (ASTM D1319)

Gel permeation chromatography is an attractive technique for determination of the number-average molecular weight (M_n) distribution of petroleum fractions, especially the heavier constituents, and petroleum products (Altgelt, 1968 1970;

Oelert, 1969; Baltus and Anderson, 1984; Hausler and Carlson, 1985; Reynolds and Biggs, 1988).

Ion-exchange chromatography is also widely used in the characterization of petroleum constituents and products. For example, cation-exchange chromatography can be used primarily to isolate the nitrogen constituents in petroleum (Snyder and Buell, 1965; Drushel and Sommers, 1966; McKay et al., 1974), thereby giving an indication of how the feedstock might behave during refining and an indication of any potential deleterious effects on catalysts.

Liquid chromatography (also called *adsorption chromatography*) has helped to characterize the group composition of crude oils and hydrocarbon products since the beginning of the twentieth century. The type and relative amount of certain hydrocarbon classes in the matrix can have a profound effect on the quality and performance of a hydrocarbon product. The *fluorescent indicator adsorption* (FIA) method (ASTM D1319) has been used to measure the paraffinic, olefinic, and aromatic content of gasoline, jet fuel, and liquid products in general (Suatoni and Garber, 1975; Miller et al., 1983; Norris and Rawdon, 1984).

High-performance liquid chromatography (HPLC) has found great utility in separating different hydrocarbon group types and identifying specific constituent types (Colin and Vion, 1983; Drushel, 1983; Miller et al., 1983; Chartier et al., 1986). Of particular interest is application of the HPLC technique to identification of the molecular types in the heavier feedstocks, especially molecular types in the asphaltene fraction. This technique is especially useful for studying such materials on a *before-* and *after-processing* basis (Chmielowiec et al., 1980; Alfredson, 1981; Bollet et al., 1981; Colin and Vion, 1983; George and Beshai, 1983; Felix et al., 1985; Coulombe and Sawatzky, 1986; Speight, 1986).

Several recent high-performance liquid chromatographic separation schemes are applicable since they also incorporate detectors not usually associated with conventional hydrocarbon group analyses (Matsushita et al., 1981; Miller et al., 1983; Rawdon, 1984; Lundanes and Greibokk, 1985; Schwartz and Brownlee, 1986; Hayes and Anderson, 1987).

The general advantages of the high-performance liquid chromatography method are: (1) Each sample may be analyzed *as received* even though the boiling range may vary over a considerable range; (2) the total time per analysis is usually on the order of minutes; and perhaps most important, (3) the method can be adapted for any recoverable petroleum sample or product.

In recent years, *supercritical fluid chromatography* has found use in the characterization and identification of petroleum constituents and products. A supercritical fluid is defined as a substance above its critical temperature. A primary advantage of chromatography using supercritical mobile phases results from the mass transfer characteristics of the solute. The increased diffusion coefficients of supercritical fluids compared with liquids can lead to greater speed in separations or greater resolution in complex mixture analyses. Another advantage of supercritical fluids compared with gases is that they can solubilize thermally labile and nonvolatile solutes and, upon expansion (decompression) of this solution, introduce the solute into the vapor phase for detection (Lundanes et al., 1986).

Currently, supercritical fluid chromatography is leaving the stages of infancy. The indications are that it will find wide applicability to the problems of characterization and identification of the higher-molecular-weight species in petroleum (Schwartz et al., 1988), thereby adding an extra dimension to our understanding of refining chemistry. It will still retain the option as a means of product characterization, although the use may be somewhat limited because of the ready availability of other characterization techniques.

2.2.4. Liquefaction and Solidification

Petroleum and the majority of petroleum products are liquids at ambient temperature, and problems that may arise from solidification during normal use are not common. Nevertheless, the *melting point* is a test (ASTM D87, D127) that often serves to determine the state of the petroleum or the product under various weather conditions or under applied conditions, such as steam stripping of the material from the soil. The reverse process, *solidification*, has received attention, again to determine the behavior of the material in nature. However, solidification of petroleum and petroleum products has been differentiated into four categories: freezing point, congealing point, cloud point, and pour point.

Petroleum becomes more or less a plastic solid when cooled to sufficiently low temperatures. This is due to the congealing of the various hydrocarbons that constitute the oil. The cloud point of petroleum (or a product) is the temperature at which paraffin wax or other solidifiable compounds present in the oil appear as a haze when the oil is chilled under definitely prescribed conditions (ASTM D2500, D3117). As cooling is continued, petroleum becomes more solid, and the pour point is the lowest temperature at which the oil pours or flows under definitely prescribed conditions when it is chilled without disturbance at a standard rate (ASTM D97).

The solidification characteristics of petroleum and its products depend on its grade or type. For pure or essentially pure hydrocarbons, the solidification temperature is the freezing point, the temperature at which a hydrocarbon passes from a liquid to a solid state (ASTM D1015, D1016). For grease and residua, the temperature of interest is that at which fluidity occurs, commonly known as the *dropping point*. The dropping point of grease is the temperature at which the grease passes from a plastic solid to a liquid state and begins to flow under the conditions of the test (ASTM D566, D2265). For another type of plastic solid, including petrolatum and microcrystalline wax, both melting point and congealing point are of interest.

2.2.5. Metals Content

Heteroatoms (*nitrogen, oxygen, sulfur*, and *metals*) are found in every crude oil, and the concentrations have to be reduced to convert the oil to transportation fuel. This is caused by the fact that if nitrogen and sulfur are present in the final fuel during combustion, nitrogen oxides (NO_x) and sulfur oxides (SO_x) form, respectively. In addition, metals affect many upgrading processes adversely, poisoning catalysts in refining and causing deposits in combustion. In addition, heteroatoms play a major role in environmental issues and can cause the petroleum to adhere to the soil, ensuring long-term contamination.

A variety of tests (ASTM D1026, D1262, D1318, D1368, D1548, D1549, D2547, D2599, D2788, D3340, D3341, D3605) have been designated for the determination of metals in petroleum and petroleum products. At the time of writing, the specific test for the determination of metals in whole feeds has not been designated. However, this task can be accomplished by combustion of the sample so that only inorganic ash remains. The ash can then be digested with an acid and the solution examined for metal species by atomic absorption (AA) spectroscopy or by inductively coupled argon plasma (ICP) spectrometry.

2.2.6. Spectroscopic Properties

Spectroscopic studies have played an important role in the evaluation of petroleum and of petroleum products for the last three decades, and many spectroscopic methods are now used as standard methods of analysis of petroleum and its products before and after a spill. Application of these methods to petroleum and its end products is a natural consequence for the environmental scientist and engineer.

Infrared Spectroscopy

Conventional infrared spectroscopy yields information about the functional features of various petroleum constituents. For example, infrared spectroscopy will aid in the identification of N–H and O–H functions, the nature of polymethylene chains, the C–H out-of-place bending frequencies, and the nature of any polynuclear aromatic systems (Speight, 1999, and references cited therein).

Infrared spectroscopy is used for the determination of benzene in motor and/or aviation gasoline (ASTM D4053), while ultraviolet spectroscopy is employed for the evaluation of mineral oils (ASTM D2269) and for determining the naphthalene content of aviation turbine fuels (ASTM D1840).

With the recent progress in *Fourier transform infrared* (FTIR) *spectroscopy*, quantitative estimates of the various functional groups can also be made. This is particularly important for application to the higher-molecular-weight solid constituents of petroleum (i.e., the asphaltene fraction).

Nuclear Magnetic Resonance

Nuclear magnetic resonance has frequently been employed for general studies and for the structural studies of petroleum constituents (Bouquet and Bailleul, 1982; Hasan et al., 1989). In fact, *proton magnetic resonance* (PMR) studies (along with infrared spectroscopic studies) were perhaps the first studies of the modern era that allowed structural inferences to be made about the polynuclear aromatic systems that occur in the high-molecular-weight constituents of petroleum.

Nuclear magnetic resonance spectroscopy has been developed as a standard method for the determination of hydrogen types in aviation turbine fuels (ASTM D3701). *X-ray fluorescence spectrometry* has been applied to the determination of lead in gasoline (ASTM D2599) as well as to the determination of sulfur in various petroleum products (ASTM D2622, D4294).

Carbon-13 magnetic resonance (CMR) can play a useful role. Since carbon magnetic resonance deals with analyzing the carbon distribution types, the obvious structural parameter to be determined is the aromaticity, f_a. Direct determination from the various types of carbon environments is one of the better methods for the determination of aromaticity. Thus, through a combination of proton and carbon magnetic resonance techniques, refinements can be made on the structural parameters, and for the solid-state high-resolution carbon magnetic resonance technique, additional structural parameters can be obtained.

Mass Spectrometry

Mass spectrometry can play a key role in the identification of the constituents of feedstocks and products (Aczel, 1989). The principal advantages of mass spectrometric methods are (1) high reproducibility of quantitative analyses, (2) the potential for obtaining detailed data on the individual components and/or carbon number homologs in complex mixtures, and (3) the minimal sample size required for analysis. The ability of mass spectrometry to identify individual components in complex mixtures is unmatched by any modern analytical technique; perhaps the exception is gas chromatography.

The methods include the use of *mass spectrometry* to determine (1) the hydrocarbon types in middle distillates (ASTM D2425); (2) the hydrocarbon types of gas oil saturate fractions (ASTM D2786); (3) the hydrocarbon types in low-olefin gasoline (ASTM D2789); and (d) the aromatic types of gas oil aromatic fractions (ASTM D3239).

However, there are disadvantages arising from the use of mass spectrometry, and these are (1) limitation of the method to organic materials that are volatile and stable at temperatures up to 300°C (570°F), and (2) the difficulty of separating isomers for absolute identification. The sample is usually destroyed, but this is seldom a disadvantage.

Other Techniques

Other techniques include the use of *flame emission spectroscopy* for determination of trace metals in gas turbine fuels (ASTM D3605) and the use of *absorption spectrophotometry* for determination of the alkyl nitrate content of diesel fuel (ASTM D4046). *Atomic absorption* has been employed as a means of measuring the lead content of gasoline (ASTM D3237) and for the manganese content of gasoline (ASTM D3831) as well as for determining the barium, calcium, magnesium, and zinc contents of lubricating oils (ASTM D4628). *Flame photometry* has been employed as a means of measuring the lithium and sodium content of lubricating greases (ASTM D3340) and the sodium content of residual fuel oil (ASTM D1318).

2.2.7. Surface and Interfacial Tension

Surface tension is a measure of the force acting at a boundary between two phases. If the boundary is between a liquid and a solid or between a liquid and a gas (air), the attractive forces are referred to as *surface tension*, but the attractive forces between two immiscible liquids are referred to as *interfacial tension*.

Temperature and molecular weight have a significant effect on surface tension (Table 2.3). For example, in the normal hydrocarbon series, a rise in temperature leads to a decrease in the surface tension, but an increase in molecular weight increases the surface tension. A similar trend, that is, an increase in molecular weight causing an increase in surface tension, also occurs in the acrylic series and, to a lesser extent, in the alkylbenzene series.

Table 2.3. Surface Tension of Selected Hydrocarbons

Hydrocarbon	Unit	Surface Tension		
		20°C (68°F)	38°C (100°F)	93°C (200°F)
n-Pentane	dyn/cm	16.0	14.0	8.0
	mN/m	16.0	14.0	8.0
n-Hexane	dyn/cm	18.4	16.5	10.9
	mN/m	18.4	16.5	10.9
n-Heptane	dyn/cm	20.3	18.6	13.1
	mN/m	20.3	18.6	13.1
n-Octane	dyn/cm	21.8	20.2	14.9
	mN/m	21.8	20.2	14.9
Cyclopentane	dyn/cm	22.4		
	mN/m	22.4		
Cyclohexane	dyn/cm	25.0		
	mN/m	25.0		
Tetralin	dyn/cm	35.2		
	mN/m	35.2		
Decalin	dyn/cm	29.9		
	mN/m	29.9		
Benzene	dyn/cm	28.8		
	mN/m	28.8		
Toluene	dyn/cm	28.5		
	mN/m	28.5		
Ethylbenzene	dyn/cm	29.0		
	mN/m	29.0		
n-Butylbenzene	dyn/cm	29.2		
	mN/m	29.2		

Source: Speight, 2001.

The surface tension of petroleum and petroleum products has been studied for many years. The narrow range of values (24 to 38 dyn/cm) for such widely diverse materials as gasoline (26 dyn/cm), kerosene (30 dyn/cm), and the lubricating fractions (34 dyn/cm) has rendered the surface tension of little value for any attempted characterization. However, it is generally acknowledged that nonhydrocarbon materials dissolved in an oil reduce the surface tension: polar compounds, such as soaps and fatty acids, are particularly active. The effect is marked at low concentrations, up to a critical value beyond which further additions cause little change; the critical value corresponds closely with that required for a monomolecular layer on the exposed surface, where it is adsorbed and accounts for the lowering. Recent work has focused on the predictability of surface tension using mathematical relationships (Gomez, 1987):

$$\text{dynamic surface tension} = \frac{681.3}{K(1 - T/13.488^{1.7654} \times sg^{2.1250})^{1.2056}}$$

where K is the Watson characterization factor, sg is the specific gravity, and T is the temperature in kelvin.

A high proportion of the complex phenomena shown by emulsions and foams, which are common when petroleum enters the environment, can be traced to these induced surface-tension effects. Dissolved gases, even hydrocarbon gases, lower the surface tension of oils, but the effects are less dramatic and the changes probably result from dilution. The matter is of some importance in environmental issues because the viscosity and surface tension of the petroleum govern the amount of oil that migrates or can be recovered under certain conditions.

On the other hand, although petroleum products show little variation in surface tension, within a narrow range the *interfacial tension* of petroleum, especially of petroleum products, against aqueous solutions provides valuable information (ASTM D971). Thus, the interfacial tension of petroleum is subject to the same constraints as those for surface tension: that is, differences in composition, molecular weight, and so on. When oil–water systems are involved, the pH of the aqueous phase influences tension at the interface; the change is small for highly refined oils, but increasing pH causes a rapid decrease for poorly refined, contaminated, or slightly oxidized oils.

A change in interfacial tension between oil and alkaline water has been proposed as an index for following the refining or deterioration of certain products, such as turbine and insulating oils. When surface or interfacial tensions are lowered by the presence of solutes, which tend to concentrate on the surface, some time is required to obtain the final concentration and hence the final value of the tension. In such systems dynamic and static tension must be distinguished; the first concerns the freshly exposed surface having nearly the same composition as the body of the liquid; it usually has a value only slightly less than that of the pure solvent. The static tension is that existing after equilibrium concentration has been reached at the surface.

The interfacial tension between oil and distilled water provides an indication of compounds in the oil that have an affinity for water. The measurement of interfacial tension has received special attention because of its possible use in predicting when an oil in constant use will reach the limit of its serviceability. This interest is based on the fact that oxidation decreases the interfacial tension of the oil. Furthermore, the interfacial tension of turbine oil against water is lowered by the presence of oxidation products, impurities from the air or rust particles, and certain antirust compounds intentionally blended in the oil. Thus, a depletion of the antirust additive may cause an increase in interfacial tension, whereas the formation of oxidation products or contamination with dust and rust lowers the interfacial tension.

2.2.8. Viscosity

Viscosity is the force in dynes required to move a plane 1 cm^2 in area at a distance of 1 cm from another plane 1 cm^2 in area through a distance of 1 cm in 1 s. In the centimeter–gram–second (cgs) system, the unit of viscosity is the poise (P) or centipoise (1 cP = 0.01 P). Two other terms in common use are kinematic viscosity and fluidity. The *kinematic viscosity* is the viscosity in centipoise divided by the specific gravity, and the unit is the stoke (cm^2/s), although the centistoke (0.01 st = 1 cSt) is in more common use; *fluidity* is simply the reciprocal of viscosity.

The viscosity (ASTM D445, D88, D2161, D341, D2270) of petroleum oils varies markedly over a very wide range (less than 10 cP at room temperature, to many thousands of centipoise at the same temperature). Many types of instruments have been proposed for the determination of viscosity. The simplest and most widely used are capillary types (ASTM D445), and the viscosity is derived from the equation

$$v = \frac{Br^4 P}{8nl}$$

where B is the quantity discharged in unit time, r the tube radius, P the pressure difference between the ends of a capillary, n the *coefficient of viscosity*, and l the tube length. Not only are such capillary instruments the simplest, but when designed in accordance with known principles and used with known necessary correction factors, they are probably the most accurate viscometers available. It is usually more convenient, however, to use relative measurements, and for this purpose the instrument is calibrated with an appropriate standard liquid of known viscosity.

Batch flow times are generally used; in other words, the time required for a fixed amount of sample to flow from a reservoir through a capillary is the datum actually observed. Any features of technique that contribute to longer flow times are usually desirable. Some of the principal capillary viscometers in use are those of Cannon–Fenske, Ubbelohde, Fitzsimmons, and Zeitfuchs.

The Saybolt universal viscosity (SUS) (ASTM D88) is the time in seconds required for the flow of 60 mL of petroleum from a container, at constant temperature, through a calibrated orifice. The Saybolt furol viscosity (SFS) (ASTM D88) is determined in a similar manner except that a larger orifice is employed.

As a result of the various methods for viscosity determination, it is not surprising that much effort has been spent on interconversion of the several scales, especially converting Saybolt to kinematic viscosity (ASTM D2161),

$$\text{kinematic viscosity} = a \times \text{Saybolt seconds} + b/\text{Saybolt seconds}$$

where a and b are constants.

The Saybolt universal viscosity equivalent to a given kinematic viscosity varies slightly with the temperature at which the determination is made because the temperature of the calibrated receiving flask used in the Saybolt method is not the same as that of the oil. Conversion factors are used to convert kinematic viscosity from 2 to 70 cSt at 38°C (100°F) and 99°C (210°F) to equivalent Saybolt universal viscosity in seconds. Appropriate multipliers are listed to convert kinematic viscosity over 70 cSt. For a kinematic viscosity determined at any other temperature, the equivalent Saybolt universal value is calculated by use of the Saybolt equivalent at 38°C (100°F) and a multiplier that varies with the temperature:

$$\text{Saybolt seconds at } 100°F(38°C) = \text{cSt} \times 4.635$$

$$\text{Saybolt seconds at } 210°F(99°C) = \text{cSt} \times 4.667$$

Viscosity decreases as the temperature increases, and the rate of change appears to depend primarily on the nature or composition of the petroleum, but other factors, such as volatility, may also have an effect. The effect of temperature on viscosity is generally represented by the equation

$$\log \log(n + c) = A + B \log T$$

where n is absolute viscosity, A and B are constants, and T is temperature. This equation has been sufficient for most purposes and has come into very general use. The constants A and B vary widely with different oils, but c remains fixed at 0.6 for all oils with a viscosity over 1.5 cSt; it increases only slightly at lower viscosity (0.75 at 0.5 cSt). When plotted, the viscosity–temperature characteristics of any oil thus create a straight line, and parameters A and B are equivalent to the intercept and slope of the line. To express the viscosity and viscosity–temperature characteristics of an oil, the slope and viscosity at one temperature must be known; the usual practice is to select 38°C (100°F) and 99°C (210°F) as the observation temperatures.

Suitable conversion tables are available (ASTM D341), and each table or chart is constructed such that for any given petroleum or petroleum product, the viscosity–temperature points result in a straight line over the applicable temperature range. Thus, only two viscosity measurements need be made at temperatures far enough apart to determine a line on the appropriate chart from which the approximate viscosity at any other temperature can be read.

Since the viscosity–temperature coefficient of high-boiling fractions, such as lubricating oil, is an important expression of its suitability, a convenient number to express this property is very useful, and hence a viscosity index (ASTM D2270) was derived:

$$\text{viscosity index} = L - \frac{U}{L} - H \times 100$$

where L and H are the viscosities of the zero and 100 index reference oils, both having the same viscosity at 99°C (210°F), and U is that of the unknown, all at 38°C (100°F). Originally, the viscosity index was calculated from Saybolt viscosity data, but subsequently, figures were provided for kinematic viscosity.

The viscosity of petroleum fractions increases on the application of pressure, and this increase may be very large. The pressure coefficient of viscosity correlates with the temperature coefficient even when oils of widely different types are compared. At higher pressures the viscosity decreases with increasing temperature, as at atmospheric pressure; in fact, viscosity changes of small magnitude are usually proportional to density changes, whether these are caused by pressure or by temperature.

Because of the importance of viscosity in determining the transport properties of petroleum, and this is particularly important in the migration of petroleum and petroleum products through soil, recent work has focused on the development of an empirical equation for predicting the dynamic viscosity of low- and high-molecular-weight hydrocarbon vapors at atmospheric pressure (Gomez, 1995). The equation uses molar mass and specific temperature as the input parameters and offers a means of estimation of the viscosity of a wide range of petroleum fractions. Other work has focused on the prediction of the viscosity of blends of lubricating oils as a means of accurately predicting the viscosity of the blend from the viscosity of the base oil components (Al-Besharah et al., 1989).

2.2.9. Volatility

The volatility of a liquid or liquefied gas may be defined as its tendency to vaporize, that is, to change from a liquid to a vapor or gaseous state. Because one of the three essentials for combustion in a flame is that the fuel be in the gaseous state, volatility is a primary characteristic of liquid fuels.

The vaporizing tendencies of petroleum and petroleum products are the basis for the general characterization of liquid petroleum fuels, such as liquefied petroleum gas, natural gasoline, motor and aviation gasoline, naphtha, kerosene, gas oil, diesel fuel, and fuel oil (ASTM D2715). A test (ASTM D6) also exists for determining the loss of material when crude oil and asphaltic compounds are heated. Another test (ASTM D20) is a method for the distillation of road tars that might also be applied to estimating the volatility of high-molecular-weight unknown residues.

For many environmental purposes it is necessary to have information on the initial stage of vaporization. To supply this need, flash and fire, vapor pressure, and evaporation methods are available. The data from the early stages of the several distillation methods are also useful. For other uses it is important to know the tendency of a product to vaporize partially or completely, and in some cases to know if small quantities of high-boiling components are present. For such purposes, chief reliance is placed on the distillation methods.

The *flash point* of petroleum or a petroleum product is the temperature to which the product must be heated under specified conditions to give off sufficient vapor to form a mixture with air that can be ignited momentarily by a specified flame (ASTM D56, D92, D93). The *fire point* is the temperature to which the product must be heated under the prescribed conditions of the method to burn continuously when the mixture of vapor and air is ignited by a specified flame (ASTM D92).

From the viewpoint of safety, information about the flash point is of most significance at or slightly above the maximum temperatures [30 to 60°C (86 to 140°F)] that may be encountered in storage, transportation, and use of liquid petroleum products, in either closed or open containers. In this temperature range the relative fire and explosion hazard can be estimated from the flash point. For products with flash points below 40°C (104°F), special precautions are necessary for safe handling. Flash points above 60°C (140°F) gradually lose their safety significance until they become indirect measures of some other quality.

The flash point of a petroleum product is also used to detect contamination. A substantially lower flash point than expected for a product is a reliable indicator that a product has become contaminated with a more volatile product, such as gasoline. The flash point is also an aid in establishing the identity of a particular petroleum product.

A further aspect of volatility that receives considerable attention is the vapor pressure of petroleum and its constituent fractions. The *vapor pressure* is the force exerted on the walls of a closed container by the vaporized portion of a liquid. Conversely, it is the force that must be exerted on the liquid to prevent it from vaporizing further (ASTM D323). The vapor pressure increases with temperature for any given gasoline, liquefied petroleum gas, or other product. The temperature at which the vapor pressure of a liquid, either a pure compound or a mixture of many compounds, equals 1 atm pressure (14.7 psi, absolute) is designated as the boiling point of the liquid.

In each homologous series of hydrocarbons, the boiling points increase with molecular weight; structure also has a marked influence: It is a general rule that branched paraffin isomers have lower boiling points than that of the corresponding n-alkane. Furthermore, in any specific aromatic series, steric effects notwithstanding, there is an increase in boiling point with an increase in the carbon number of the alkyl side chain. This applies particularly to alkyl aromatic compounds, where alkyl-substituted aromatic compounds can have higher boiling points than polycondensed aromatic systems. This fact is very meaningful when attempts are made to develop hypothetical structures for asphaltene constituents.

One of the main properties of petroleum that serves to indicate the comparative ease with which a material evaporates after a spill is its volatility. Investigation of the volatility of petroleum is usually carried out under standard conditions in which petroleum or the product is subdivided by distillation into a variety of

Table 2.4. Boiling Fractions from Petroleum

Product	Lower Carbon Limit	Upper Carbon Limit	Lower Boiling Point		Upper Boiling Point	
			°C	°F	°C	°F
Refinery gas	C_1	C_4	−161	−259	−1	31
Liquefied petroleum gas	C_3	C_4	−42	−44	−1	31
Naphtha	C_5	C_{17}	36	97	302	575
Gasoline	C_4	C_{12}	−1	31	216	421
Kerosene/diesel fuel	C_8	C_{18}	126	302	258	575
Aviation turbine fuel	C_8	C_{16}	126	302	287	548
Fuel oil	C_{12}	$>C_{20}$	216	>343	421	>649
Lubricating oil	$>C_{20}$		>343	>649		
Wax	C_{17}	$>C_{20}$	302	575	>343	>649
Asphalt	$>C_{20}$		>343	>649		
Coke	$>C_{50}$[a]		>1000[a]	>1832[a]		

Source: Speight, 2002.

[a] Carbon number and boiling point are difficult to assess; inserted for illustrative purposes only.

fractions of different *fractions*, sometimes referred to as *cut points* (Table 2.4). In fact, distillation involves the general procedure of vaporizing the petroleum liquid in a suitable flask either at *atmospheric pressure* (ASTM D86, D216, D285, D447, D2892) or at *reduced pressure* (ASTM D1160).

Simulated distillation (*simdis*) by gas chromatography is often applied to petroleum to obtain true boiling-point data for distillates and crude oils and may also be useful in predicting the amount of material that can be (or needs to be) recovered after a spill. Two standardized methods (ASTM D2887, D3710) are available for the boiling-point determination of petroleum fractions and gasoline, respectively. The ASTM D2887 method utilizes nonpolar, packed gas chromatographic columns in conjunction with flame ionization detection. The upper limit of the boiling range covered by this method is to approximately 540°C (1000°F), the atmospheric equivalent boiling point. Recent efforts in which high-temperature gas chromatography was used have focused on extending the scope of the ASTM D2887 method for higher-boiling petroleum materials to 800°C (1470°F), the atmospheric equivalent boiling point (Schwartz et al., 1988).

REFERENCES

Aczel, T. 1989. *Prepr. Div. Pet. Chem. Am. Chem. Soc.*, 34(2): 318.
Al-Besharah, J. M., Mumford, C. J., Akashah, S. A., and Salman, O. 1989. *Fuel*, 68: 809.
Alfredson, T. V. 1981. *J. Chromatogr.*, 218: 715.
Altgelt, K. H. 1968. *Prepr. Div. Petl. Chem. Am. Chem. Soc.*, 13(3): 37.
Altgelt, K. H. 1970. *Bitumen Teere Asphalte Peche*, 21: 475.

ASTM. 2004. *Annual Book of Standards.* American Society for Testing and Materials, West Conshohocken, PA.

Baltus, R. E., and Anderson, J. L. 1984. *Fuel*, 63: 530.

Bland, W. F., and Davidson, R. L. 1967. *Petroleum Processing Handbook.* McGraw-Hill, New York.

Bollet, C., Escalier, J. C., Souteyrand, C., Caude, M., and Rosset, R. 1981. *J. Chromatogr.*, 206: 289.

Bouquet, M., and Bailleul, A. 1982. In *Petroanalysis '81*, Advances in Analytical Chemistry in the Petroleum Industry, 1975–1982, G. B. Crump (Ed.). Wiley, Chichester, West Sussex, England.

Chartier, P., Gareil, P. Caude, M., Rosset, R., Neff, B., Bourgognon, H. F., and Husson, J. F. 1986. *J. Chromatogr.*, 357: 381.

Chmielowiec, J., Beshai, J. E., and George, A. E. 1980. *Fuel*, 59: 838.

Colin, J. M., and Vion, G. 1983. *J. Chromatogr.*, 280: 152.

Coulombe, S., and Sawatzky, H. 1986. *Fuel*, 65: 552.

Drushel, H. V. 1983. *J. Chromatogr. Sci.*, 21: 375.

Drushel, H. V., and Sommers, A. L. 1966. *Anal. Chem.*, 38: 19.

Felix, G., Bertrand, C., and Van Gastel, F. 1985. *Chromatographia*, 20(3): 155.

George, A. E., and Beshai, J. E. 1983. *Fuel*, 62: 345.

Gomez, J. V. 1987. *Oil Gas J.*, December 7, p. 68.

Gomez, J. V. 1989. *Oil Gas J.*, March 27, p. 66.

Gomez, J. V. 1992. *Oil Gas J.*, July 13, p. 49.

Gomez, J. V. 1995. *Oil Gas J.*, February 6, p. 60.

Hasan, M., Ali, M. F., and Arab, M. 1989. *Fuel*, 68: 801.

Hausler, D. W., and Carlson, R. S. 1985. *Prepr. Div. Pet. Chem. Am. Chem. Soc.*, 30(1): 28.

Hayes, P. C., and Anderson, S. D. 1987. *J. Chromatogr.*, 387: 333.

IP. 2004. *Standard Methods for Analysis and Testing of Petroleum and Related Products*, IP 143. Institute of Petroleum, London.

Lundanes, E., and Greibokk, T. 1985. *J. Chromatogr.*, 349: 439.

Lundanes, E., Iversen, B., and Greibokk, T. 1986. *J. Chromatogr.*, 366: 391.

Matsushita, S., Tada, Y., and Ikushige. 1981. *J. Chromatogr.*, 208: 429.

McKay, J. F., Cogswell, T. E., and Latham, D. R. 1974. *Prepr. Div. Pet. Chem. Am. Chem. Soc.*, 19(1): 25.

Miller, R. L., Ettre, L. S., and Johansen, N. G. 1983. *J. Chromatogr.*, 259: 393.

Norris, T. A., and Rawdon, M. G. 1984. *Anal. Chem.*, 56: 1767.

Oelert, H. H. 1969. *Erdoel Kohle*, 22: 536.

Rawdon, M. 1984. *Anal. Chem.*, 56: 831.

Reynolds, J. G., and Biggs, W. R. 1988. *Fuel Sci. Technol. Int.*, 6: 329.

Schwartz, H. E., and Brownlee, R. G. 1986. *J. Chromatogr.*, 353: 77.

Schwartz, H. E., Brownlee, R. G., Boduszynski, M. M., and Su, F. 1988. *Anal. Chem.*, 59: 1393.

Speight, J. G. 1986. *Prepr. Div. Pet. Chem. Am. Chem. Soc.*, 31(4): 818.

Speight, J. G. 1999. *The Chemistry and Technology of Petroleum*, 3rd ed. Marcel Dekker, New York.

Speight, J. G. 2001. *Handbook of Petroleum Analysis*. Wiley, New York.

Speight, J. G. 2002. *Handbook of Petroleum Product Analysis*. Wiley, Hoboken, NJ.

Suatoni, J. C., and Garber, H. R. 1975. *J. Chromatogr. Sci.*, 13: 367.

CHAPTER 3

REFINERY PRODUCTS AND BY-PRODUCTS

Like many industrial feedstocks, petroleum in unrefined or crude form has little or no direct use. Its value as an industrial commodity is realized only after the production of salable products by a series of refining steps (Figure 3.1). Each refining step is, in fact, a separate process, and thus a refinery is a series of integrated processes that generate the products desired according to the market demand (Meyers, 1997; Speight, 1999, and references cited therein). Therefore, the value of petroleum is related directly to the yield of products and is subject to the call of the market. In general, crude oil, once refined, yields three basic groupings of products that are produced when it is broken down into cuts or fractions (Table 3.1).

The gas and gasoline fractions form the lower-boiling products and are usually more valuable than the higher-boiling fractions and provide gas (liquefied petroleum gas), naphtha, aviation fuel, motor fuel, and feedstocks for the petrochemical industry. Naphtha, a precursor to gasoline and solvents, is extracted from both the light and middle range of distillate cuts and is also used as a feedstock for the petrochemical industry. *Middle distillates*, products from the middle-boiling range of petroleum, include kerosene, diesel fuel, distillate fuel oil, and light gas oil; waxy distillate and lower-boiling lubricating oils are sometimes include in the middle distillates. The remainder of the crude oil includes the higher-boiling lubricating oils, gas oil, and residuum (the nonvolatile fraction of the crude oil). The residuum can also produce heavy lubricating oils and waxes but is more often used for asphalt production. The complexity of petroleum is emphasized insofar as the actual proportions of light, medium, and heavy fractions vary significantly from one crude oil to another.

The current trend throughout the refining industry is to produce more fuel products from each barrel of petroleum and to process those products in different ways to meet product specifications for use in various (automobile, diesel, aircraft, and marine) engines. Overall, the demand for liquid fuels has expanded rapidly and demand has developed for gas oils and fuels for domestic central heating and fuel oil for power generation, as well as for light distillates and other inputs, derived from crude oil, for the petrochemical industries.

As the need for the lower-boiling products developed, petroleum yielding the desired quantities of the lower-boiling products became less available and refineries had to introduce conversion processes to produce greater quantities

Environmental Analysis and Technology for the Refining Industry, by James G. Speight
Copyright © 2005 John Wiley & Sons, Inc.

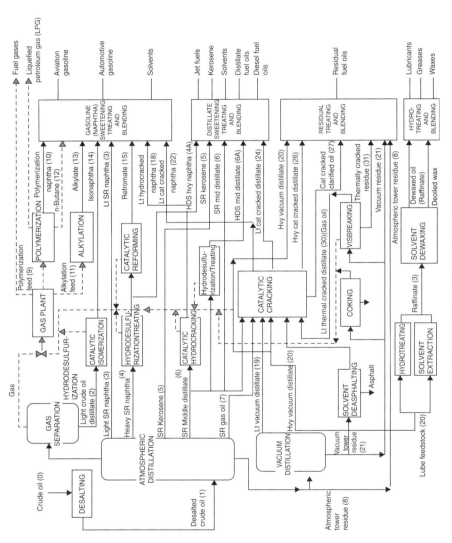

Figure 3.1. General layout of a petroleum refinery.

Table 3.1. Crude Petroleum: A Mixture of Compounds that Can Be Separated into Different Generic Boiling Fractions

Fraction	Boiling Range[a]	
	°C	°F
Light naphtha	−1–150	30–300
Gasoline	−1–180	30–355
Heavy naphtha	150–205	300–400
Kerosene	205–260	400–500[b]
Light gas oil	260–315	400–600
Heavy gas oil	315–425	600–800
Lubricating oil	>400	>750
Vacuum gas oil	425–600	800–1100
Residuum	>510	>950

[a] For convenience, boiling ranges are converted to the nearest 5°.
[b] Kerosene destined for conversion to diesel fuel; crude kerosene is often quoted as having a wider boiling range [150–350°C (302–662°F)] before separation into various fuel and/or solvent products.

of lighter products from the higher-boiling fractions. The means by which a refinery operates in terms of producing the relevant products depends not only on the nature of the petroleum feedstock but also on its configuration (i.e., the number of types of processes that are employed to produce the product slate desired), and the refinery configuration is therefore influenced by the specific demands of a market. Therefore, refineries need to be adapted and upgraded constantly to remain fiscally sound and responsive to ever-changing patterns of crude supply and product market demands. As a result, refineries have been introducing increasingly complex and expensive processes to gain more and more lighter products from the heavier and residual ends of a barrel.

Petroleum products (in contrast to *petrochemicals*) are those bulk fractions (Table 3.1) that are derived from petroleum and have commercial value as a bulk product (Mushrush and Speight, 1995). Petroleum products are, generally speaking, hydrocarbon compounds that have various combinations of hydrogen and carbon and can take many molecular forms. Many of the combinations exist naturally in the original raw materials, but other combinations are created by an ever-growing number of commercial processes for altering one combination to another. The specifications for petroleum products are based on properties such as density and boiling range (to mention only two), used to assure that a petroleum product can perform the task for which it is intended. In addition, each product has its own unique set of chemical and physical properties. As a consequence, each product has a different effect on the environment (EPA, 2004).

In the strictest sense, *petrochemicals* are also petroleum products, but they are individual chemicals that are used as the basic building blocks of the chemical industry. Many petrochemicals are substitutes for earlier products from nonpetroleum sources. Often, a new use imposes additional specifications on the new product, and product specifications evolve to stay abreast of advances in both product application and manufacturing methods. Like petroleum products, petrochemicals cover such a vast range of chemical and physical properties that they also have a wide range of effects on the environment.

Because a critical aspect of assessing the toxic effects of the release of petroleum and petroleum products is the measurement of compounds in the environment, the first approach is to understand the origin and properties of the various fractions.

Therefore, in this chapter we describe major refinery operations and the products therefrom and focus on their composition, properties, and uses. This presents to the reader the essence of petroleum processes, the types of feedstocks employed, and the product produced, as well as warning of the types of the chemicals that can be released to the environment when an accident occurs. Being forewarned offers an environmental analyst the ability to design the necessary test methods to examine the chemical(s) released. It offers environmental scientists and engineers the ability to start forming opinions and predictions about the nature of the chemical(s) released, the potential effects of the chemical(s) on the environment, and the possible methods of cleanup.

3.1. REFINERY PRODUCTS

Refinery processes for crude oil are generally divided into three categories: (1) separation processes, of which atmospheric distillation (Figure 3.2) and vacuum distillation (Table 3.2; Figures 3.2 and 3.3) are the prime examples; (2) conversion processes (Table 3.3), of which delayed coking (Figure 3.4), fluid coking (Figure 3.5), catalytic cracking (Figure 3.6), and hydrocracking (Figure 3.7) are prime examples; (3) finishing processes (Table 3.4), of which hydrotreating (Figure 3.8) to remove sulfur is a prime example; and (4) skeletal alteration processes (Table 3.5), of which butane isomerization (Figure 3.9), alkylation (Figure 3.10), polymerization (Figure 3.11), and catalytic reforming (Figure 3.12) are prime examples. However, before separation of petroleum into its various constituents can proceed, there is a need to clean the petroleum. This is often referred to as

Table 3.2. Separation Processes

Process Name	Action	Method	Purpose	Feedstock(s)	Product(s)
Atmospheric distillation	Separation	Thermal	Separate fractions without cracking	Desalted crude oil	Gas, gas oil, distillate, residual
Vacuum distillation	Separation	Thermal	Separate fractions without cracking	Atmospheric tower residual	Gas oil, lube stock, residual

REFINERY PRODUCTS 61

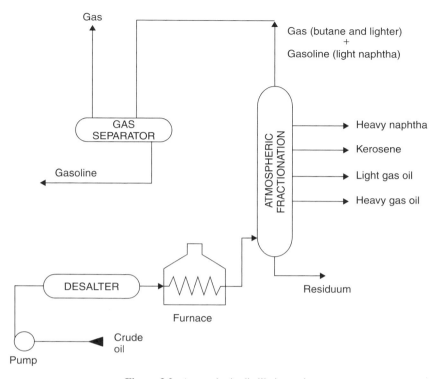

Figure 3.2. Atmospheric distillation unit.

Table 3.3. Conversion Processes

Process Name	Action	Method	Purpose	Feedstock(s)	Product(s)
Catalytic cracking	Alteration	Catalytic	Upgrade gasoline	Gas oil, coke, distillate	Gasoline, petrochemical feedstock
Coking	Polymerize	Thermal	Convert vacuum residuals	Gas oil, coke distillate	Gasoline, petrochemical feedstock
Hydro-cracking	Hydrogenate	Catalytic	Convert to lighter hydro-carbons	Gas oil, cracked oil, residual	Lighter, higher-quality products
Visbreaking	Decompose	Thermal	Reduce viscosity	Atmospheric tower residual	Distillate, tar

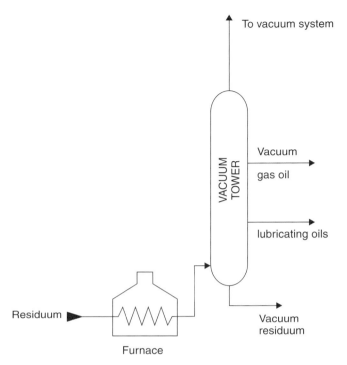

Figure 3.3. Vacuum distillation unit.

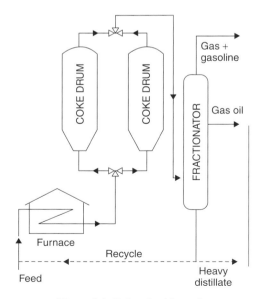

Figure 3.4. Delayed coking unit.

REFINERY PRODUCTS 63

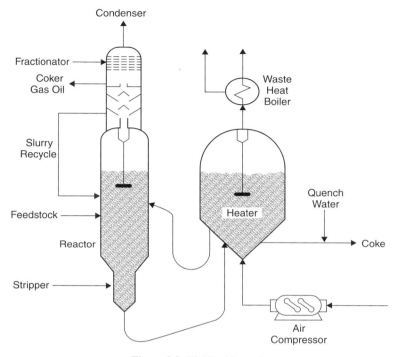

Figure 3.5. Fluid coking unit.

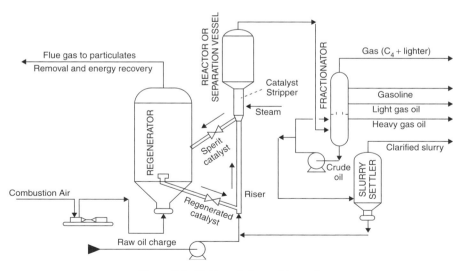

Figure 3.6. Fluid catalytic cracking unit.

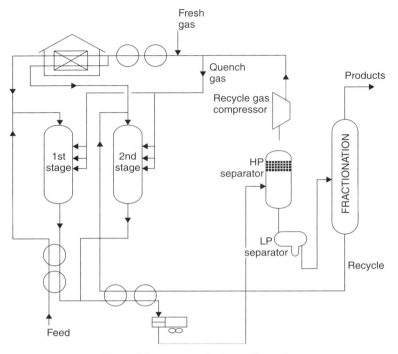

Figure 3.7. Two-stage hydrocracking unit.

desalting (Chapter 4), in which the goal is to remove the constituents of the brine that accompany the crude oil from the reservoir to the wellhead during recovery operations.

3.1.1. Liquefied Petroleum Gas

Fuels with four or fewer carbon atoms in the hydrogen–carbon combination have boiling points lower than room temperature, and these products are gases at ambient temperature and pressure.

Natural gas (predominantly *methane*), denoted by the chemical structure CH_4, is the lowest-boiling and least complex of all hydrocarbons. Natural gas from an underground reservoir, when brought to the surface, can contain other higher-boiling hydrocarbons and is often referred to as *wet gas*. Wet gas is usually processed to remove the entrained hydrocarbons that are higher boiling than methane, and when isolated, the higher-boiling hydrocarbons sometimes liquefy and are called *natural gas condensate*.

Still gas is broad terminology for low-boiling hydrocarbon mixtures and is the lowest-boiling fraction isolated from a distillation (*still*) unit in the refinery. If the distillation unit is separating light hydrocarbon fractions, the still gas will be almost entirely methane, with only traces of ethane (CH_3CH_3) and ethylene

Table 3.4. Finishing Processes

Process Name	Action	Method	Purpose	Feedstock(s)	Product(s)
Amine treating	Treatment	Absorption	Remove acidic contaminants	Sour gas, hydrocarbons with CO_2 and H_2S	Acid-free gases and liquid hydrocarbons
Desalting	Dehydration	Absorption	Remove contaminants	Crude oil	Desalted crude oil
Drying and sweetening	Treatment	Absorption/thermal	Remove H_2O and sulfur compounds	Liquid hydrocarbons, LPG, alkylation feedstock	Sweet and dry hydrocarbons
Furfural extraction	Solvent extraction	Absorption	Upgrade middle distillate and lubes	Cycle oils and lube feedstocks	High-quality diesel and lube oil
Hydrodesulfurization	Treatment	Catalytic	Remove sulfur, contaminants	High-sulfur residual/gas oil	Desulfurized olefins
Hydrotreating	Hydrogenation	Catalytic	Remove impurities, saturate HCs	Residuals, cracked HCs	Cracker feed, distillate, lube
Phenol extraction	Solvent extraction	Absorption/thermal	Improve viscosity index, color	Lube oil base stocks	High-quality lube oils
Solvent deasphalting	Treatment	Absorption	Remove asphalt	Vacuum tower residual, propane	Heavy lube oil, asphalt
Solvent dewaxing	Treatment	Cool/filter	Remove wax from lube stocks	Vacuum tower lube oils	Dewaxed lube base stock
Solvent extraction	Solvent extraction	Absorption/ precipitation	Separate unsaturated oils	Gas oil, reformate, distillate	High-octane gasoline
Sweetening	Treatment	Catalytic	Remove H_2S, convert mercaptan	Untreated distillate/gasoline	High-quality distillate/gasoline

Table 3.5. Skeletal Alteration Processes

Process Name	Action	Method	Purpose	Feedstock(s)	Product(s)
Alkylation	Combining	Catalytic	Unite olefins and isoparaffins	Tower isobutane/cracker olefin	Isooctane (alkylate)
Polymerization	Polymerize	Catalytic	Unite two or more olefins	Cracked olefins	High-octane naphtha, petrochemical stocks

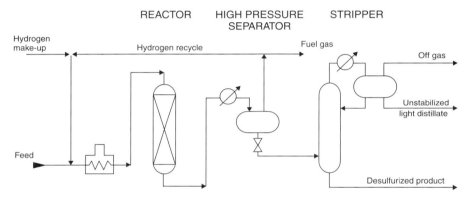

Figure 3.8. Distillate hydrotreater for hydrodesulfurization.

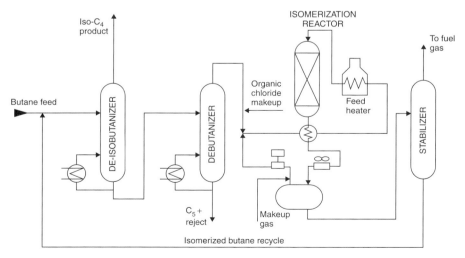

Figure 3.9. Butane isomerization unit.

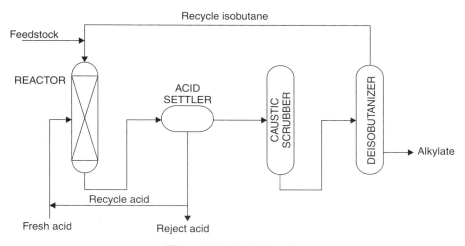

Figure 3.10. Alkylation unit.

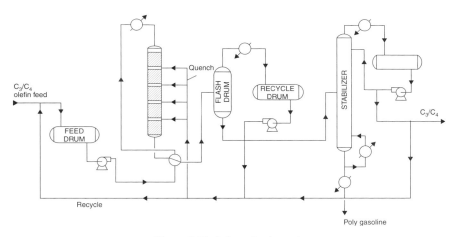

Figure 3.11. Polymerization unit.

($CH_2=CH_2$). If the distillation unit is handling higher-boiling fractions, the still gas might also contain propane ($CH_3CH_2CH_3$), butane ($CH_3CH_2CH_2CH_3$), and their respective isomers. The terms *fuel gas* and *still gas* are often used interchangeably, but the term *fuel gas* is intended to denote the product's destination: to be used as a fuel for boilers, furnaces, or heaters.

Liquefied petroleum gas (LPG) is composed of propane (C_3H_8) and butane (C_4H_{10}) and is stored under pressure in order to keep these hydrocarbons liquefied at normal atmospheric temperatures. Before liquefied petroleum gas is burned, it passes through a pressure relief valve that causes a reduction in pressure and the liquid vaporizes (gasifies). Winter-grade liquefied petroleum gas is mostly

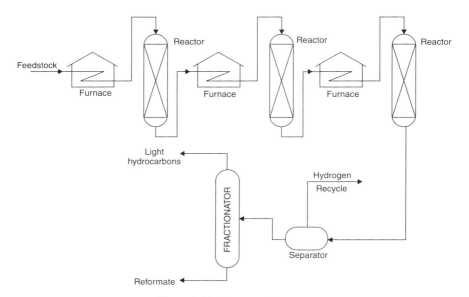

Figure 3.12. Catalytic reforming.

propane, the lower boiling of the two gases, which is easier to vaporize at lower temperatures. Summer-grade liquefied petroleum gas is mostly butane.

Fuel gas or *refinery gas* is produced in considerable quantities during the various refining processes and is used as fuel for the refinery itself and as an important feedstock for the petrochemical industry. Liquefied petroleum gas (LPG) is frequently used as domestic bottled gas for cooking and heating, and forms an important feedstock for the petrochemical industry. It is also used in industry for cutting metals.

3.1.2. Naphtha, Gasoline, and Solvents

Naphtha is the general term that is applied to refined, partly refined, or unrefined low-boiling petroleum products. Naphtha is prepared by any one of several methods, including (1) fractionation of distillates or even crude petroleum, (2) solvent extraction, (3) hydrogenation of distillates, (4) polymerization of unsaturated (olefinic) compounds, and (5) alkylation processes. Naphtha may also be a combination of product streams from more than one of these processes.

The main uses of petroleum naphtha fall into the general areas of (1) precursor to gasoline and other liquid fuels, (2) solvents (diluents) for paints, (3) dry-cleaning solvents, (4) solvents for cutback asphalts, (5) solvents in the rubber industry, and (6) solvents for industrial extraction processes. Turpentine, the older and more conventional solvent for paints has now been replaced almost completely by the cheaper and more abundant petroleum naphtha.

The term *aliphatic naphtha* refers to naphtha containing less than 0.1% benzene and with carbon numbers from C_3 through C_{16}. Aromatic naphtha has

Table 3.6. Component Streams for Gasoline

Stream	Producing Process	Boiling Range °C	Boiling Range °F
Paraffinic			
Butane	Distillation	0	32
	Conversion		
Isopentane	Distillation	27	81
	Conversion		
	Isomerization[a]		
Alkylate	Alkylation[b]	40–150	105–300
Isomerate	Isomerization	40–70	105–160
Naphtha	Distillation	30–100	85–212
Hydrocrackate	Hydrocracking	40–200	105–390
Olefinic			
Catalytic naphtha	Catalytic cracking	40–200	105–390
Cracked naphtha	Steam cracking	40–200	105–390
Polymer	Polymerization[c]	60–200	140–390
Aromatic			
Catalytic reformate	Catalytic reforming[d]	40–200	105–390

[a] Figure 3.9.
[b] Figure 3.10.
[c] Figure 3.11.
[d] Figure 3.12.

carbon numbers from C_6 through C_{16} and contains significant quantities of aromatic hydrocarbons such as benzene (>0.1%), toluene, and xylene. The final gasoline product as a transport fuel is a carefully blended mixture having a predetermined octane value (Table 3.6). Thus, gasoline is a complex mixture of hydrocarbons that boils below 200°C (390°F). The hydrocarbon constituents in this boiling range are those that have four to 12 carbon atoms in their molecular structure.

Gasoline varies widely in composition, and even those with the same octane number may be quite different. The variation in aromatics content as well as the variation in the content of normal paraffins, branched paraffins, cyclopentane derivatives, and cyclohexane derivatives all involve characteristics of any one individual crude oil and influence the octane number of a gasoline.

Automotive gasoline is a mixture of low-boiling hydrocarbon compounds suitable for use in spark-ignited internal combustion engines and having an octane rating of at least 60. Additives that have been used in gasoline include alkyl tertiary butyl ethers (e.g., MTBE), ethanol (ethyl alcohol), methanol (methyl alcohol), tetramethyllead, tetraethyllead, ethylene dichloride, and ethylene dibromide. Other categories of compounds that may be added to gasoline include antiknock agents, antioxidants, metal deactivators, lead scavengers, antirust agents, anti-icing agents, upper-cylinder lubricants, detergents, and dyes.

Automotive gasoline contains 150 or more different chemical compounds and the relative concentrations of the compounds vary considerably, depending on the source of crude oil, refinery process, and product specifications. Typical hydrocarbon constituents are (volume basis) alkanes (4 to 8%), alkenes (2 to 5%), isoalkanes (25 to 40%), cycloalkanes (3 to 7%), cycloalkenes (1 to 4%), and aromatics (20 to 50%). However, these proportions vary greatly.

The reduction in the lead content of gasoline and the introduction of reformulated gasoline have been very successful in reducing automobile emissions, due to changes in gasoline composition, with further improvements in fuel quality proposed for the early decades of this new millennium. These projections are accompanied by a noticeable and measurable decrease in crude oil quality, and the reformulated gasoline will help meet environmental regulations for emissions for liquid fuels but will be subject to continuous review because of the potential for environmental impact.

Aviation fuel comes in two types: aviation gasoline and jet fuel. Aviation gasoline, now usually found in use in light aircraft and older civil aircraft, has a narrower boiling range, 38 to 170°C (100 to 340°F), than that of conventional (automobile) gasoline, −1 to 200°C (30 to 390°F). The narrower boiling range ensures better distribution of the vaporized fuel through the more complicated induction systems of aircraft engines. Since aircraft operate at altitudes where the prevailing pressure is less than the pressure at the surface of the Earth (the pressure at 17,500 ft is 7.5 psi (0.5 atm) compared to 14.8 psi (1.0 atm) at the surface of the Earth), the vapor pressure of aviation gasoline must be limited, to reduce boiling in tanks, fuel lines, and carburetors.

Aviation gasoline consists primarily of straight and branched alkanes and cycloalkanes. Aromatic hydrocarbons are limited to 20 to 25% of the total mixture because they produce smoke when burned. A maximum of 5% alkenes is allowed in JP-4. The approximate distribution by chemical class is 32% straight alkanes, 31% branched alkanes, 16% cycloalkanes, and 21% aromatic hydrocarbons.

Jet fuel is classified as *aviation turbine fuel*, and in the specifications, ratings relative to octane number are replaced with properties concerned with the ability of the fuel to burn cleanly. Jet fuel is a light petroleum distillate that is available in several forms suitable for use in various types of jet engines. The exact composition of jet fuel is established by the U.S. Air Force using specifications that yield maximum performance from the aircraft. The major jet fuels used by the military are JP-4, JP-5, JP-6, JP-7, and JP-8.

Briefly, JP-4 is a wide-cut fuel developed for broad availability in times of need. JP-6 is a higher cut than JP-4 and is characterized by fewer impurities. JP-5 is specially blended kerosene, and JP-7 is a high-flash-point special kerosene used in advanced supersonic aircraft. JP-8 is a kerosene fraction that is modeled on jet A-l fuel (used in civilian aircraft). For this profile, JP-4 will be used as the prototype jet fuel, due to its broad availability and extensive use.

Volatility is an important property of all types of gasoline since it is related to performance and requires sufficient low-boiling hydrocarbons to vaporize easily in cold weather. The gasoline must also contain sufficient high-boiling hydrocarbons to remain a liquid in an engine's fuel supply system during hotter periods.

Petroleum solvents (also called *naphtha*) are valuable because of their good dissolving power. The wide range of naphtha available and the varying degree of volatility possible offer products suitable for many uses.

Stoddard solvent is a petroleum distillate widely used as a dry-cleaning solvent and as a general cleaner and degreaser. It may also be used as paint thinner, as a solvent in some types of photocopier toners, in some types of printing inks, and in some adhesives. Stoddard solvent is considered to be a form of mineral spirits, white spirits, and naphtha; however, not all forms of mineral spirits, white spirits, and naphtha are considered to be Stoddard solvent.

Stoddard solvent consists (volume basis) of linear and branched alkanes (30 to 50%), cycloalkanes (30 to 40%), and aromatic hydrocarbons (10 to 20%). Alcohols, glycols, and ketones are not included in the composition, as few, if any, of these types of compounds are expected to be present in Stoddard solvent. Possible contaminants may include lead (<1 ppm) and sulfur (3.5 ppm).

3.1.3. Kerosene and Diesel Fuel

Kerosene was the major refinery product before the onset of the *automobile age*, but now kerosene might be termed as one of several other petroleum products after gasoline. Kerosene originated as a straight-run (distilled) petroleum fraction that boiled between approximately 150 and 350°C (300 to 660°F). In the early days of petroleum refining some crude oils contained kerosene fractions of very high quality, but other crude oils, such as those having a high proportion of asphaltic materials, must be refined thoroughly to remove aromatics and sulfur compounds before a satisfactory kerosene fraction can be obtained. Apart from the removal of excessive quantities of aromatics, kerosene fractions may need only a lye (alkali) wash if hydrogen sulfide is present.

Diesel fuel also forms part of the kerosene boiling range (or middle distillate group of products). Diesel fuels come in two broad groups, for high-speed engines in cars and trucks requiring a high-quality product, and heavier, lower-quality diesel fuel for slower engines, such as in marine engines or for stationary power plants. An important property of diesel is the cetane number (analogous to the gasoline octane number), which determines the ease of ignition under compression.

The quality of diesel fuel is measured using the cetane number, a measure of the tendency of a diesel fuel to knock in a diesel engine, and the scale, from which the cetane number is derived, is based on the ignition characteristics of two hydrocarbons: (1) n-hexadecane (cetane) and (2) 2,3,4,5,6,7,8-heptamethylnonane.

3.1.4. Fuel Oil

Fuel oil is classified in several ways, but generally, may be divided into two main types: (1) *distillate fuel oil* and (2) *residual fuel oil*. These classifications

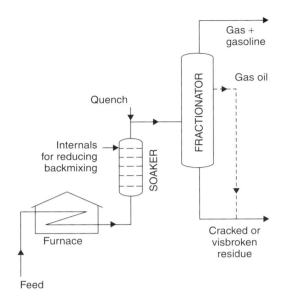

Figure 3.13. Visbreaking unit.

are still employed, but of late, the terms *distillate fuel oil* and *residual fuel oil* have lost some of their significance because fuel oils are now made for specific uses and may be distillates, residuals, or mixtures of the two. The terms domestic *fuel oils, diesel fuel oils*, and *heavy fuel oils* are more indicative of the uses of fuel oils. More often than not, fuel oil is prepared by using a visbreaker unity (Figure 3.13) to perform mild thermal cracking on a residuum or a high-boiling distillate so that the product meets specifications.

Distillate fuel oil is vaporized and condensed during a distillation process; they have a definite boiling range and do not contain high-boiling oils or asphaltic components. A fuel oil that contains any amount of the residue from crude distillation or thermal cracking is a residual fuel oil. *Domestic fuel oil* is fuel oil that is used primarily in the home and includes kerosene, stove oil, and furnace fuel oil. *Diesel fuel oil* is also a distillate fuel oil, but residual oil has been used successfully to power marine diesel engines, and mixtures of distillates and residuals have been used on locomotive diesels. *Furnace fuel oil* is similar to diesel fuel, but the proportion of cracked gas oil in diesel fuel is usually less since the high aromatic content of the cracked gas oil reduces the cetane number of the diesel fuel.

Stove oil is a straight-run (distilled) fraction from crude oil, whereas other fuel oils are usually blends of two or more fractions. The straight-run fractions available for blending into fuel oils are heavy naphtha, light and heavy gas oils, and residua. Cracked fractions such as light and heavy gas oils from catalytic cracking, cracking coal tar, and fractionator bottoms from catalytic cracking may also be used as blends to meet the specifications of the various fuel oils.

Heavy fuel oil includes a variety of oils, ranging from distillates to residual oils, that must be heated to 260°C (500°F) or higher before they can be used. In general, heavy fuel oils consist of residual oils blended with distillates to suit specific needs. Included among heavy fuel oils are various industrial oils; when used to fuel ships, heavy fuel oil is called *bunker oil*.

Fuel oil that is used for heating is graded from No. 1 fuel oil to No. 6 fuel oil, which covers light distillate oils, medium distillate, heavy distillate, a blend of distillate and residue, and residue oil. For example, No. 2 and No. 3 fuel oils medium to light distillate grades used in domestic central heating (Table 3.7).

No. 1 fuel oil is a petroleum distillate that is one of the most widely used fuel oil types. It is used in atomizing burners that spray fuel into a combustion chamber, where the tiny droplets burn while in suspension. It is also used as a carrier for pesticides, as a weed killer, as a mold release agent in the ceramic and pottery industry, and in the cleaning industry. It is found in asphalt coatings, enamels, paints, thinners, and varnishes. No. 1 fuel oil is a light petroleum distillate (straight-run kerosene) consisting primarily of hydrocarbons in the range C_9 to C_{16} and is similar in composition to diesel fuel; the primary difference is in the additives.

No. 2 fuel oil is a petroleum distillate that may be referred to as domestic fuel oil or industrial fuel oil. Domestic fuel oil is usually lighter and straight-run refined; it is used primarily for home heating and to produce diesel fuel. Industrial distillate is the cracked type, or a blend of straight-run and cracked. It is used in smelting furnaces, ceramic kilns, and packaged boilers.

No. 2 fuel oil is characterized by hydrocarbons in the range C_{11} to C_{20}, whereas diesel fuels predominantly contain a mixture of C_{10} to C_{19} hydrocarbons.

Table 3.7. Properties of Various Fuel Oil Grades

Fuel Oil	Properties
No. 1	Similar to kerosene or range oil (fuel used in stoves for cooking).
	Defined as a distillate intended for vaporizing in pot-type burners and other burners where a clean flame is required.
No. 2	Often called *domestic heating oil.*
	Has properties similar to those of diesel and higher-boiling jet fuels.
	Defined as a distillate for general-purpose heating in which the burners do not require the fuel to be vaporized before completely burning.
No. 4	A light industrial heating oil that is intended where preheating is not required for handling or burning.
	Two grades that differ primarily in safety (flash) and flow (viscosity) properties.
No. 5	A heavy industrial oil that often requires preheating for burning and, in cold climates, for handling.
No. 6	A heavy residuum oil.
	Commonly referred to as *bunker C oil* when it is used to fuel ocean-going vessels.
	Preheating is required for both handling and burning this grade of oil.

The composition consists of approximately 64% aliphatic hydrocarbons (straight-chain alkanes and cycloalkanes), 1 to 2% unsaturated hydrocarbons (alkenes), and 35% aromatic hydrocarbons (including alkylbenzenes and two- and three-ring aromatics). No 2 fuel oil contains less than 5% polycyclic aromatic hydrocarbons.

No. 6 fuel oil (also called *bunker C* or *residual fuel oil*) is the residual from crude oil after the light oil, gasoline, naphtha, No. 1 fuel oil, and No. 2 fuel oil have been distilled. No. 6 fuel oil can be blended directly to heavy fuel oil or made into asphalt. It is limited to commercial and industrial uses where sufficient heat is available to fluidize the oil for pumping and combustion.

Residual fuel oil is generally more complex than distillate fuels in composition and impurities. Limited data are available, but there are indications that the composition of No. 6 fuel oil includes (volume basis) aromatics (25%), paraffins (15%), naphthenes (45%), and nonhydrocarbon compounds (15%). Polynuclear aromatic hydrocarbons and their alkyl derivatives and metals are important hazardous and persistent components of No. 6 fuel oil.

3.1.5. Lubricating Oil

Lubricating oil is distinguished from other fractions of crude oil by a usually high [$>400°C$ ($>750°F$)] boiling point, as well as high viscosity. The development of vacuum distillation provided the means of separating lubricating oil fractions with predetermined viscosity ranges and removed the limit on the maximum viscosity that might be obtained in distillate oil. Vacuum distillation prevented residual asphaltic material from contaminating lubricating oils but did not remove other undesirable materials, such as acidic components or components, which caused the oil to thicken excessively when cold and become very thin when hot.

Lubricating oil may be divided into many categories according to the types of service it is intended to perform. However, there are two main groups: (1) oils used in intermittent service, such as motor and aviation oils, and (2) oils designed for continuous service, such as turbine oils.

The mineral-based oils are produced from heavy-end crude oil distillates. Distillate streams may be treated in several ways, such as vacuum-, solvent-, acid-, or hydrotreated, to produce oils with commercial properties. Hydrocarbon types ranging from C_{15} to C_{50} are found in the various types of oils, with the heavier distillates having higher percentages of the higher-carbon-number compounds.

Crankcase oil or motor oil may be either petroleum-based or synthetic. The petroleum-based oils are more widely used than the synthetic oils and may be used in automotive engines, railroad and truck diesel engines, marine equipment, jet and other aircraft engines, and most small two- and four-stroke engines.

The petroleum-based oils contain hundreds to thousands of hydrocarbon compounds, including a substantial fraction of nitrogen- and sulfur-containing compounds. The hydrocarbons are mainly mixtures of straight- and branched-chain hydrocarbons (alkanes), cycloalkanes, and aromatic hydrocarbons. Polynuclear aromatic hydrocarbons, alkyl polynuclear aromatic hydrocarbons, and metals are important components of motor oils and crankcase oils, with the used oils typically having higher concentrations than the new unused oils. Typical carbon numbers range from C_{15} to C_{50}.

Because of the wide range of uses and the potential for close contact with the engine to alter oil composition, the exact composition of crankcase oil/motor oil has not been specifically defined.

3.1.6. White Oil, Insulating Oil, and Insecticides

The term *white distillate* is applied to all the refinery streams with a distillation range between approximately 80 and 360°C (175 to 680°F) at atmospheric pressure and with properties similar to the corresponding straight-run distillate from atmospheric crude distillation. Light distillate products (i.e., naphtha, kerosene, jet fuel, diesel fuel, and heating oil) are all manufactured by appropriate blending of white distillate streams.

There is also a category of petroleum products known as *white oil* that generally falls into two classes: (1) *technical white oil*, which is employed for cosmetics, textile lubrication, insecticide vehicles, and paper impregnation, and (2) *pharmaceutical white oil*, which may is employed medicinally (e.g., as a laxative) or for the lubrication of food-handling machinery.

Insulating oil falls into two general classes: (1) oil used in transformers, circuit breakers, and oil-filled cables, and (2) oil employed for impregnating the paper covering of wrapped cables. The first are highly refined fractions of low viscosity and comparatively high boiling range and resemble heavy-burning oils, such as mineral seal oil, or the very light lubricating fractions known as nonviscous neutral oils. The second are usually highly viscous products and are often naphthenic distillate, which is not usually highly refined.

Insecticides are derived from petroleum oil, which can usually be applied in water-emulsion form and which have marked killing power for certain species of insects. For many applications for which their own effectiveness is too slight, the oils serve as carriers for active poisons, as in the household and livestock sprays.

3.1.7. Grease

Grease is lubricating oil to which a thickening agent has been added for the purpose of holding the oil to surfaces that must be lubricated. The most widely used thickening agents are soaps of various kinds, and grease manufacture is essentially the mixing of soaps with lubricating oils.

Soap is made by chemically combining a metal hydroxide with a fat or fatty acid:

$$\underset{\text{fatty acid}}{R-CO_2H} + NaOH \rightarrow \underset{\text{soap}}{R-CO_2^-Na^+} + H_2O$$

The most common metal hydroxides used for this purpose are calcium hydroxide, lye, lithium hydroxide, and barium hydroxide. Fats are chemical combinations of fatty acids and glycerin. If a metal hydroxide is reacted with a fat, a soap

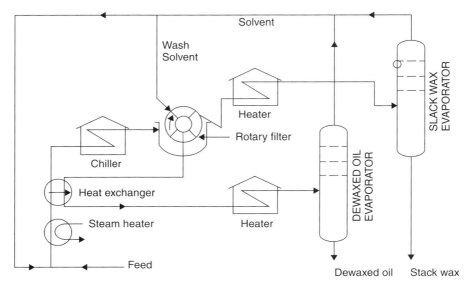

Figure 3.14. Solvent dewaxing unit.

containing glycerin is formed. Frequently, a fat is separated into its fatty acid and glycerin components and only the fatty acid portion is used to make soap. Commonly used fats for grease-making soaps are cottonseed oil, tallow, and lard. Among the fatty acids used are stearic acid (from tallow), oleic acid (from cottonseed oil), and animal fatty acids (from lard).

3.1.8. Wax

Wax is of two general types: (1) paraffin wax in petroleum distillates, and (2) microcrystalline wax in petroleum residua.

Paraffin wax is a solid crystalline mixture of straight-chain (normal) hydrocarbons ranging from 20 to 30 carbon atoms per molecule and even higher. Wax constituents are solid at ordinary temperatures [25°C (77°F)], whereas *petrolatum* (*petroleum jelly*) does contain both solid and liquid hydrocarbons. The melting point of wax is not always directly related to its boiling point, because wax contains hydrocarbons of different chemical structure.

Wax recrystallization, such as wax sweating, separates wax into fractions, but instead of relying on differences in melting points, the process makes use of the different solubility of the wax fractions in a solvent such as a ketone.

More generally, three main methods are used in modern refinery technology:

1. *Solvent dewaxing* (Figure 3.14), in which the feedstock is mixed with one or more solvents, then the mixture is cooled down to allow the formation of wax crystals, and the solid phase is separated from the liquid phase by filtration.

2. *Urea dewaxing,* in which urea forms adducts with straight-chain paraffins that separated by filtration from the dewaxed oil.
3. *Catalytic dewaxing,* in which straight-chain paraffin hydrocarbons are selectively cracked on zeolite-type catalysts, and the lower-boiling reaction products are separated from the dewaxed lubricating oil by fractionation.

3.1.9. Asphalt

Asphalt (referred to as *bitumen* in some parts of the world) is produced from the distillation residuum. In addition to road asphalt, a variety of asphalt grades for roofing and waterproofing are also produced. Asphalt has complex chemical and physical compositions, which usually vary with the source of the crude oil, and it is produced to certain standards of hardness or softness in controlled vacuum distillation processes (Barth, 1962; Bland and Davidson, 1967; Speight, 1999, and references cited therein; Speight and Ozum, 2002).

There are wide variations in refinery operations and in the types of crude oils, so different asphalts will be produced that have different environmental effects (EPA, 1996). Blending with higher- and lower-softening-point asphalts may make asphalts of intermediate softening points. If lubricating oils are not required, the reduced crude may be distilled in a flash drum that is similar to a distillation tower but has few, if any, trays. Asphalt descends to the base of the flasher as the volatile components pass out of the top. Asphalt is also produced by propane deasphalting and can be made softer by blending the hard asphalt with the extract obtained in the solvent treatment of lubricating oils. On the other hand, soft asphalts can be converted into harder asphalts by oxidation (air blowing).

Road oil is liquid asphalt material intended for easy application to earthen roads and provides a strong base or a hard surface and will maintain a satisfactory passage for light traffic. *Cutback asphalt* is a mixture in which hard asphalt has been diluted with a lower-boiling oil to permit application as a liquid without drastic heating. *Asphalt emulsions* are usually oil-in-water emulsions that break on application to a stone or earthen surface, so that the oil clings to the stone and the water disappears. In addition to their usefulness in road and soil stabilization, asphalt emulsions are also used for paper impregnation and waterproofing.

3.1.10. Coke

Petroleum coke is the residue left by the destructive distillation (thermal cracking or coking) of petroleum residua. The coke formed in catalytic cracking operations is usually nonrecoverable because of adherence to the catalyst, as it is often employed as fuel for the process. The composition of coke varies with the source of the crude oil, but in general, is insoluble on organic solvents and has a honeycomb-type appearance.

The use of coke as a fuel must proceed with some caution with the acceptance by refiners of the heavier crude oils as refinery feedstocks. The higher contents of sulfur and nitrogen in these oils results in a coke containing substantial amounts of sulfur and nitrogen. Both of these elements will produce unacceptable pollutants (e.g., sulfur oxides and nitrogen oxides) during combustion. These elements must also be regarded with caution in any coke that is scheduled for electrode manufacture, and removal procedures for these elements are continually under development.

3.2. PETROCHEMICALS

A *petrochemical* is any chemical (as distinct from fuels and petroleum products) manufactured from petroleum (and natural gas) and used for a variety of commercial purposes (Table 3.8). The definition, however, has been broadened to include the entire range of aliphatic, aromatic, and naphthenic organic chemicals, as well as carbon black and inorganic materials such as sulfur and ammonia. Petroleum and natural gas are made up of hydrocarbon molecules, which comprise one or more carbon atoms to which hydrogen atoms are attached. Currently,

Table 3.8. Hydrocarbon Intermediates Used in the Petrochemical Industry

Carbon Number	Hydrocarbon Type		
	Saturated	Unsaturated	Aromatic
1	Methane		
2	Ethane	Ethylene	
		Acetylene	
3	Propane	Propylene	
4	Butanes	n-Butenes	
		Isobutene	
		Butadiene	
5	Pentanes	Isopentenes	
		(Isoamylenes)	
		Isoprene	
6	Hexanes	Methylpentenes	Benzene
7	Cyclohexane	Mixed heptenes	Toluene
8		Diisobutylene	Xylenes
			Ethyl-benzene
			Styrene
9			Cumene
12		Propylene tetramer	
		Triisobutylene	
18			Dodecyl-benzene
6–18		n-Olefins	
11–18	n-Paraffins		

Table 3.9. Sources of Petrochemical Intermediates

Hydrocarbon	Source
Methane	Natural gas
Ethane	Natural gas
Ethylene	Cracking processes
Propane	Natural gas, catalytic reforming, cracking processes
Propylene	Cracking processes
Butane	Natural gas, reforming and cracking processes
Butene(s)	Cracking processes
Cyclohexane	Distillation
Benzene	Catalytic reforming
Toluene	Catalytic reforming
Xylene(s)	Catalytic reforming
Ethylbenzene	Catalytic reforming
Alkylbenzenes	Alkylation
$>C_9$	Polymerization

oil and gas are the principal sources of the raw materials because they are the least expensive, most readily available, and can be processed most easily into the primary petrochemicals (Table 3.9). Primary petrochemicals include olefins (ethylene, propene, and butadiene), aromatics (benzene, toluene, and the isomers of xylene), and methanol.

The petrochemical industry is a large and complex source category that is very difficult to define because its operations are intertwined functionally or physically with (in addition to the petroleum refining industry) other industries. Petrochemical feedstocks can be classified into three general groups: (1) olefins, (2) aromatics, and (3) methanol. A fourth group includes inorganic compounds and synthesis gas (mixtures of carbon monoxide and hydrogen). Thence, the petrochemical industry is involved in the production of several chemicals that fit into one or more of the following three categories: (1) basic raw materials, (2) chemical intermediates, and (3) end products.

Petrochemical intermediates are generally produced by chemical conversion of primary petrochemicals to form more complicated derivative products. Petrochemical derivative products can be made in a variety of ways: directly from primary petrochemicals, through intermediate products that still contain only carbon and hydrogen, and through intermediates that incorporate chlorine, nitrogen, or oxygen in the finished derivative. In some cases, they are finished products; in others, more steps are needed to arrive at the desired composition.

The petrochemical industry also includes the treatment of hydrocarbon streams from the petroleum refining industry and natural gas liquids from the oil and gas production industry. Some of the raw materials used in the petrochemical industry include petroleum, natural gas, ethane, hydrocarbons, naphtha, heavy fractions, kerosene, and gas-oil. Natural gas and petroleum are the main feedstocks for the petrochemical industry. That is why about 65% of petrochemical facilities are located at or near refineries.

The petrochemical industry produces solvents and chemicals of various grades or specifications which are used to produce industrial organic chemicals, including alcohols and aldehydes, butylene, butadiene, ethylene, propylene, toluene, styrene, acetylene, benzene, ethylene oxide, ethylene glycol, acrylonitrile, acetone, acetic acid, acetic anhydride, and ammonia. Petrochemicals are widely used in agriculture, in the manufacture of plastics, synthetic fibers, and explosives, and in the aircraft and automobile industries. The industrial organic chemicals produced from petrochemicals are employed in downstream industries, including plastics and resins, synthetic fibers, elastomers, plasticizers, explosives, surface-active agents, dyes, surface coatings, pharmaceuticals, and pesticides.

Excluding accidental spills, there are four main sources of pollutant emissions in the production of petrochemicals: (1) process vent discharges, (2) equipment leaks, (3) secondary sources, and (4) storage. Process vent discharges can be from reactor vessels, recovery columns, and other process vessels. Equipment leaks include pump seals, process valves, compressors, safety relief valves, flanges, open-ended lines, and sampling connections. Secondary sources include process and other waste streams. However, little information is available regarding amounts of pollutant emissions from the entire petrochemical industry. Many petrochemical processes are located at or near petroleum refining operations; therefore, many of the air pollutants and hazardous wastes generated by the petroleum industry are also present at petrochemical facilities and most reports refer to facility-wide pollution.

In general, waste streams from the petrochemical industry are quite similar to those of the petroleum refining industry. Limited data are available, but almost all assume that waste management operations and facilities are probably of the same degree of sophistication as those of the petroleum refining industry.

Wastewater, which is a basic source of emissions, can be categorized in five ways: (1) wastes containing a principal raw material or product; (2) by-products produced during reactions, (3) spills, leaks, wash downs, vessel cleanouts, or point overflows, (4) cooling tower and boiler blowdown, steam condensate, water treatment wastes, and general washing water; and (5) surface runoff.

Disposal of solid wastes is a significant problem for the petrochemical industry. Waste solids include water treatment sludge, ashes, fly ash and incinerator residue, plastics, ferrous and nonferrous metals, catalysts, organic chemicals, inorganic chemicals, filter cakes, and viscous solids.

3.3. REFINERY CHEMICALS

3.3.1. Alkalis

Several processes use the principle of an alkali (caustic) wash, which is a process in which distillate is treated with sodium hydroxide to remove acidic contaminants that contribute to poor odor and stability. Caustic streams used in various refining, petrochemical, and chemical operations often end up in a waste caustic tank for disposal. This waste caustic can produce a unique disposal challenge, depending on the components to be removed. A common method of remediation is a wet air oxidation (WAO) process that is designed to convert the reactive sulfide components such as sodium sulfide and sodium mercaptide to sodium sulfate. Organic contaminants such as phenols can also be reduced significantly. WAO treatment allows spent caustic to be routed to a standard biological wastewater treatment plant without creating operational or odor problems.

Light and heavy hydrocarbons (oils) will adversely affect the operation of a wet air oxidation process by one or more of the following: (1) foul the system heat exchangers, causing frequent heat-exchanger cleaning, system shutdown, and localized corrosion problems; (2) exceed the supply for of available oxygen, reducing system capacity or resulting in system shutdown due to oxygen deficiency; and (3) cause the system to be oversized in design, which increases the capital and operating cost.

3.3.2. Acids

Naphthenic acids, complex carboxylic acids that are believed to have a cyclopentane ring or cyclohexane ring in the molecule, occur in petroleum. They seem to be of little consequence environmentally since thermal decarboxylation can occur during the distillation process. During this process, the temperature of the crude oil in the distillation column can reach as high as 395°C (740°F). Hence decarboxylation is possible (Speight and Francisco, 1990):

$$R-CO_2H \rightarrow R-H + CO_2$$

However, inorganic acids are used in various processes to treat unfinished petroleum products such as gasoline and kerosene, and lubricating oil stocks are treated with sulfuric acid for improvement of color, odor, and other properties.

For example, treating petroleum distillates with sulfuric acid is generally applied to dissolve unstable or colored substances and sulfur compounds as well as to precipitate asphaltic materials. When drastic conditions are employed, as in the treatment of lubricating fractions with large amounts of concentrated acid or when fuming acid is used in the manufacture of white oils, considerable quantities of petroleum sulfonic acids are formed:

$$\underset{\text{paraffin}}{R-H} + H_2SO_4 \rightarrow \underset{\text{sulfonic acid}}{R-SO_3H} + H_2O$$

Two general methods are applied for the recovery of sulfonic acids from sulfonated oils and their sludge: (1) the acids are removed selectively by adsorbents or by solvents (generally low-molecular-weight alcohols), and (2) the acids are obtained by salting out with organic salts or bases.

Petroleum sulfonic acids may be divided roughly into those soluble in hydrocarbons and those soluble in water. Because of their color, hydrocarbon-soluble acids are referred to as *mahogany acids*, and the water-soluble acids are referred to as *green acids*. The composition of each type varies with the nature of the oil sulfonated and the concentration of acids produced. In general, those formed during light acid treatment are water soluble; oil-soluble acids result from more drastic sulfonation.

Sulfonic acids are used as detergents made by the sulfonation of alkylated benzene. The number, size, and structure of the alkyl side chains are important in determining the performance of the finished detergent. The salts of mixed petroleum sulfonic acids have many other commercial applications. They find use as anticorrosion agents, leather softeners, and flotation agents and have been used in place of red oil (sulfonated castor oil) in the textile industry. Lead salts of the acids have been employed in greases as extreme pressure agents, and alkyl esters have been used as alkylating agents. The alkaline earth metal (magnesium, calcium, and barium) salts are used in detergent compositions for motor oils, and the alkali metal (potassium and sodium) salts are used as detergents in aqueous systems.

The *sulfuric acid sludge* from sulfuric acid treatment is used frequently as a source (through thermal decomposition) to produce sulfur dioxide (SO_2, which is returned to the sulfuric acid plant) and *sludge acid coke*. The coke, in the form of small pellets, is used as a substitute for charcoal in the manufacture of carbon disulfide. Sulfuric acid coke is different from other petroleum coke in that it is pyrophoric in air and also reacts directly with sulfur vapor to form carbon disulfide.

In recent years the petroleum refining industry has placed an increasing emphasis on the safety of the use of hydrogen fluoride (HF) in petroleum refineries. Refineries use the acid in a manufacturing process called *alkylation*, which is increasingly important in producing a high-quality gasoline. Hydrofluoric acid is hazardous and corrosive and if released accidentally can form a vapor cloud. If the vapor cloud is concentrated enough, it can be toxic until sufficiently dispersed. In the past five years there have been a number of accidental releases of this acid from alkylation units at major petroleum refineries in the United States.

Pure hydrogen fluoride is a clear, colorless, corrosive liquid that has roughly the same weight as water (comparing equal volumes). It boils at 67°F and, depending on the release conditions, can form a vapor cloud if released to the atmosphere. It has a sharp, penetrating odor that human beings can detect at very low concentrations (0.04 to 0.13 ppm) in the air. It is completely soluble in water, in which it forms HF, which in concentrated solutions vaporizes in air to form a noticeable cloud.

To protect against adverse effects from exposure to hydrofluoric acid in the workplace, the Occupational Safety and Health Administration has established a permissible exposure limit (PEL) of 3 ppm averaged over an 8-hour work shift. The National Institute for Occupational Safety and Health has found that the concentration of HF that is immediately dangerous to life or health is 30 ppm.

3.3.3. Catalysts

Many processes involve the use of a *catalyst*, a material that aids or promotes a chemical reaction between other substances but does not react itself. Catalysts increase reaction speeds and can provide control by increasing desirable reactions and decreasing undesirable reactions.

The cracking of crude oil fractions occurs over many types of catalytic materials, and cracking catalysts can differ markedly in both activity to promote the cracking reaction and in the quality of the products obtained from cracking the feedstocks (Speight, 1999, and references cited therein; Speight and Ozum, 2002, and references cited therein). Activity can be related directly to the total number of active (acid) sites per unit weight of catalyst and also to the acidic strength of these sites. Differences in activity and acidity regulate the extent of various secondary reactions occurring and thus the product quality differences. The acidic sites are considered to be Lewis- or Brønsted-type acid sites, but there is much controversy as to which type of site predominates

The first cracking catalysts were acid-leached *montmorillonite clays*. The acid leach was to remove various metal impurities, principally iron, copper, and nickel, that could exert adverse effects on the cracking performance of a catalyst. The catalysts were first used in fixed- and moving-bed reactor systems in the form of shaped pellets. Later, with the development of the fluid catalytic cracking process, clay catalysts were made in the form of a ground, sized powder. Clay catalysts are relatively inexpensive and have been used extensively for many years.

The desire to have catalysts that were uniform in composition and catalytic performance led to the development of *synthetic catalysts*. The first synthetic cracking catalyst, consisting of 87% silica (SiO_2) and 13% alumina (Al_2O_3), was used in pellet form and used in fixed-bed units in 1940. Catalysts of this composition were ground and sized for use in fluid catalytic cracking units. In 1944, catalysts in the form of beads about 2.5 to 5.0 mm in diameter were introduced and comprised about 90% silica and 10% alumina and were extremely durable. One version of these catalysts contained a minor amount of chromia (Cr_2O_3) to act as an oxidation promoter.

Neither silica (SiO_2) nor alumina (Al_2O_3) alone is effective in promoting catalytic cracking reactions. In fact, they (and also activated carbon) promote hydrocarbon decompositions of the thermal type. Mixtures of anhydrous silica and alumina ($SiO_2 \cdot Al_2O_3$) or anhydrous silica with hydrated alumina ($2SiO_2 \cdot 2Al_2O_3 \cdot 6H_2O$) are also essentially noneffective. A catalyst having appreciable cracking activity is obtained only when prepared from hydrous oxides followed by partial dehydration (*calcining*). The small amount of water remaining is necessary for proper functioning.

Commercial synthetic catalysts are amorphous and contain more silica than is called for by the preceding formulas; they are generally composed of 10 to 15% alumina (Al_2O_3) and 85 to 90% silica (SiO_2). The natural materials—montmorillonite, a nonswelling bentonite, and halloysite—are hydrosilicates of aluminum, with a well-defined crystal structure and approximate composition of $Al_2O_3 \cdot 4SiO_2 \cdot xH_2O$. Some of the newer catalysts contain up to 25% alumina and are reputed to have a longer active life.

Catalysts are porous and highly adsorptive, and their performance is affected markedly by the method of preparation. Two catalysts that are chemically identical but have pores of different size and distribution may have different activity, selectivity, temperature coefficient of reaction rate, and response to poisons. The intrinsic chemistry and catalytic action of a surface may be independent of pore size, but small pores appear to produce different effects because of the manner and time in which hydrocarbon vapors are transported into and out of the interstices.

In addition to synthetic catalysts comprising silica–alumina, other combinations of *mixed oxides* were found to be catalytically active and were developed during the 1940s. These systems included silica (SiO_2), magnesia (MgO), silica–zirconia ($SiO_2 \cdot ZrO$), silica–alumina–magnesia, silica–alumina–zirconia, and alumina–boria ($Al_2O_3 \cdot B_2O_3$). Of these, only silica–magnesia was used in commercial units, but operating difficulties developed with the regeneration of the catalyst, which at the time demanded a switch to another catalyst. Further improvements in silica–magnesia catalysts have since been made. High yields of desirable products are obtained with hydrated aluminum silicates. These may be either activated (acid-treated natural clays of the bentonite type) or synthesized silica–alumina or silica–magnesia preparations. Both natural and synthetic catalysts can be used as pellets or beads, and also in the form of powder; in either case, replacements are necessary because of attrition and gradual loss of efficiency (Speight, 1999, and references cited therein).

During the period 1940–1962, the cracking catalysts used most widely commercially were the aforementioned acid-leached clays and silica–alumina. The latter was made in two versions; *low alumina* (about 13% Al_2O_3) and *high alumina* (about 25% Al_2O_3) contents. High-alumina-content catalysts showed a higher equilibrium activity level and surface area.

During the 1958–1960 period, *semisynthetic catalysts* of silica–alumina catalyst were used in which approximately 25 to 35% kaolin was dispersed throughout the silica–alumina gel. These catalysts could be offered at a lower price and therefore were disposable, but they were marked by lower catalytic activity and greater stack losses because of increased attrition rates. One virtue of the semisynthetic catalysts was that smaller amounts of adsorbed, unconverted, high-molecular-weight products on the catalyst were carried over to the stripper zone and regenerator. This resulted in a higher yield of more valuable products and also smoother operation of the regenerator as local hot spots were minimized.

The catalysts must be stable to physical impact loading and thermal shocks and must withstand the action of carbon dioxide, air, nitrogen compounds, and steam. They should also be resistant to sulfur compounds; the synthetic catalysts and certain selected clays appear to be better in this regard than average untreated natural catalysts.

Commercially used cracking catalysts are *insulator catalysts* possessing strong protonic (acidic) properties. They function as catalyst by altering the cracking process mechanisms through an alternative mechanism involving *chemisorption* by *proton donation* and *desorption*, resulting in cracked oil and theoretically restored catalyst. Thus, it is not surprising that all cracking catalysts are poisoned by proton-accepting vanadium.

The catalyst–oil volume ratios range from 5:1 to 30:1 for the various processes, although most processes are operated at 10:1. However, for moving-bed processes the catalyst–oil volume ratios may be substantially lower than 10:1.

Crystalline *zeolite catalysts* having molecular sieve properties were introduced as selective adsorbents in the 1955–1959 period. In a relatively short period all of the cracking catalyst manufacturers were offering their versions of zeolite catalysts to refiners. The intrinsically higher activity of the crystalline zeolites vis-à-vis conventional amorphous silica–alumina catalysts coupled with the much higher yields of gasoline and decreased coke and light ends yields served to revitalize research and development in the mature refinery process of catalytic cracking.

A number of *zeolite catalysts* have been mentioned as having catalytic cracking properties, such as synthetic faujasite (X and Y types), offretite, mordenite, and erionite. Of these, the faujasites have been most widely used commercially. While faujasite is synthesized in the sodium form, base exchange removes the sodium with other metal ions, which for cracking catalysts include magnesium, calcium, rare earths (mixed or individual), and ammonium. In particular, mixed rare earths alone or in combination with ammonium ions have been the most commonly used forms of faujasite in cracking catalyst formulations. Empirically, X-type faujasite has a stoichiometric formula of $Na_2O \cdot Al_2O_3 \cdot 2.5SiO_2$ and Y-type faujasite, $Na_2O \cdot Al_2O_3 \cdot 4.8SiO_2$. Slight variations in the silica/alumina (SiO_2/Al_2O_3) ratio exist for each type. Rare earth–exchanged Y-type faujasite retains much of its crystallinity after steaming at 825°C (1520°F) with steam for 12 hours; rare earth–form X-faujasite, which is thermally stable in dry air, will lose its crystallinity at these temperatures in the presence of steam.

The latest technique developed by the refining industry to increase gasoline yield and quality is to treat the catalysts from the cracking units to remove metal poisons that accumulate on the catalyst. Nickel, vanadium, iron, and copper compounds contained in catalytic cracking feedstocks are deposited on the catalyst during the cracking operation, thereby adversely affecting both catalyst activity and selectivity. Increased catalyst metal contents affect catalytic cracking yields by increasing coke formation, decreasing gasoline and butane and butylene production, and increasing hydrogen production. Recent commercial development and adoption of cracking catalyst–treating processes definitely improve the overall catalytic cracking process economics.

Demet

A cracking catalyst is subjected to two pretreatment steps. The first step effects vanadium removal; the second, nickel removal, to prepare the metals on the catalyst for chemical conversion to compounds (chemical treatment step) that can be removed readily through water washing (catalyst wash step). The treatment steps include use of a sulfurous compound followed by chlorination with an anhydrous chlorinating agent (e.g., chlorine gas) and washing with an aqueous solution of a chelating agent (e.g., citric acid). The catalyst is then dried and treated further before being returned to the cracking unit.

Met-X

This process consists of cooling, mixing, and ion-exchange separation, filtration, and resin regeneration. Moist catalyst from the filter is dispersed in oil and returned to the cracking reactor as a slurry. On a continuous basis, the catalyst from a cracking unit is cooled and then transported to a stirred reactor and mixed with an ion-exchange resin (introduced as slurry). The catalyst–resin slurry then flows to an elutriator for separation. The catalyst slurry is taken overhead to a filter, and the wet filter cake is slurried with oil and pumped into the catalytic cracked feed system. The resin leaves the bottom of the elutriator and is regenerated before returning to the reactor.

REFERENCES

Barth, E. J. 1962. *Asphalt Science and Technology*. Gordon and Breach, New York.

Bland, W. F., and Davidson, R. L. 1967. *Petroleum Processing Handbook*. McGraw-Hill, New York.

EPA. 1996. *Study of Selected Petroleum Refining Residuals*. Office of Solid Waste Management, Environmental Protection Agency, Washington, DC.

EPA. 2004. Environmental Protection Agency, Washington, DC. Web site: http://www.epa.gov.

Meyers, R. A. 1997. *Handbook of Petroleum Refining Processes*, 2nd ed. McGraw-Hill, New York.

Mushrush, G. W., and Speight, J. G. 1995. *Petroleum Products: Instability and Incompatibility*. Taylor & Francis, Philadelphia.

Speight, J. G. 1999. *The Chemistry and Technology of Petroleum*, 3rd ed. Marcel Dekker, New York.

Speight, J. G., and Francisco, M. A. 1990. *Rev. Inst. Fr. Pet.*, 45: 733.

Speight, J. G., and Ozum, B. 2002. *Petroleum Refining Processes*. Marcel Dekker, New York.

CHAPTER

4

REFINERY WASTES

The chemicals in petroleum vary from (chemically speaking) simple hydrocarbons of low-to-medium molecular weight to organic compounds containing sulfur, oxygen, and nitrogen, as well as compounds containing metallic constituents, particularly vanadium, nickel, iron, and copper. Many of the latter compounds are of indeterminate molecular weight (Speight, 2001).

Residua produced by distillation, which is a concentration process, contains significantly fewer hydrocarbon constituents than those in original crude oil. The constituents of residua may, depending on the crude oil, be molecular entities of which the majority contain at least one heteroatom.

Typical refinery products include (1) natural gas and liquefied petroleum gas (LPG), (2) solvent naphtha, (3) kerosene, (4) diesel fuel, (5) jet fuel, (6) lubricating oil, (7) various fuel oils, (8) wax, (9) residua, and (10) asphalt (Chapter 3). A single refinery does not necessarily produce all of these products. Some refineries are dedicated to particular products (e.g., the production of gasoline or the production of lubricating oil or the production of asphalt). However, the issue is that refineries also produce a variety of waste products (Table 4.1) that must be disposed of in an environmentally acceptable manner.

Waste treatment processes also account for a significant area of the refinery, particularly sulfur compounds in gaseous emissions, together with various solid and liquid extracts and wastes generated during the refining process. The refinery is therefore composed of a complex system of stills, cracking units, processing and blending units, and vessels in which the various reactions take place, as well as packaging units for products for immediate distribution to the retailer (e.g., lubricating oils). Bulk storage tanks, usually grouped together in tank farms, are used for storage of both crude and refined products. Other tanks are used in the processes outlined (e.g., treating, blending, and mixing), and others are used for spill and fire control systems. A boiler and an electrical generating system usually operate for the refinery as a whole.

There are several hundred individual hydrocarbon chemicals defined as petroleum based. Furthermore, each petroleum product has its own mix of constituents because (Chapter 2) petroleum varies in composition from one reservoir to another, and this variation may be reflected in the finished product(s).

Petroleum hydrocarbons are environmental contaminants, but they are not usually classified as hazardous wastes (Irwin, 1997). Soil and groundwater petroleum

---

Environmental Analysis and Technology for the Refining Industry, by James G. Speight
Copyright © 2005 John Wiley & Sons, Inc.

Table 4.1. Emissions and Waste from Refinery Processes

Process	Air Emissions	Residual Wastes Generated
Crude oil desalting	Heater stack gas (CO, SO_x, NO_x, hydrocarbons, and particulates), fugitive emissions (hydrocarbons)	Crude oil/desalter sludge (iron rust, clay, sand, water, emulsified oil and wax, metals)
Atmospheric distillation	Heater stack gas (CO, SO_x, NO_x, hydrocarbons, and particulates), vents and fugitive emissions (hydrocarbons)	Typically, little or no residual waste
Vacuum distillation	Steam ejector emissions (hydrocarbons), heater stack gas (CO, SO_x, NO_x, hydrocarbons, and particulates), vents and fugitive emissions (hydrocarbons)	
Thermal cracking/visbreaking	Heater stack gas (CO, SO_x, NO_x, hydrocarbons, and particulates), vents and fugitive emissions (hydrocarbons)	Typically, little or no residual waste generated
Coking	Heater stack gas (CO, SO_x, NO_x, hydrocarbons, and particulates), vents and fugitive emissions (hydrocarbons), and decoking emissions (hydrocarbons and particulates)	Coke dust (carbon particles and hydrocarbons)
Catalytic cracking	Heater stack gas (CO, SO_x, NO_x, hydrocarbons, and particulates), fugitive emissions (hydrocarbons), and catalyst regeneration (CO, NO_x, SO_x, and particulates)	Spent catalysts (metals from crude oil and hydrocarbons), spent catalyst fines from electrostatic precipitators (aluminum silicate and metals)
Catalytic hydrocracking	Heater stack gas (CO, SO_x, NO_x, hydrocarbons, and particulates), fugitive emissions (hydrocarbons), and catalyst regeneration (CO, NO_x, SO_x, and catalyst dust)	Spent catalyst fines
Hydrotreating/hydroprocessing	Heater stack gas (CO, SO_x, NO_x, hydrocarbons, and particulates), vents and fugitive emissions (hydrocarbons), and catalyst regeneration (CO, NO_x, SO_x)	Spent catalyst fines (aluminum silicate and metals)
Alkylation	Heater stack gas (CO, SO_x, NO_x, hydrocarbons, and particulates), vents and fugitive emissions (hydrocarbons)	Neutralized alkylation sludge (sulfuric acid or calcium fluoride, hydrocarbons)

Table 4.1. (*continued*)

Process	Air Emissions	Residual Wastes Generated
Isomerization	Heater stack gas (CO, SO_x, NO_x, hydrocarbons, and particulates), HCl (potentially in light ends), vents and fugitive emissions (hydrocarbons)	Calcium chloride sludge from neutralized HCl gas
Polymerization	H_2S from caustic washing	Spent catalyst containing phosphoric acid
Catalytic reforming	Heater stack gas (CO, SO_x, NO_x, hydrocarbons, and particulates), fugitive emissions (hydrocarbons), and catalyst regeneration (CO, NO_x, SO_x)	Spent catalyst fines from electrostatic precipitators (alumina silicate and metals)
Solvent extraction	Fugitive solvents	Little or no residual waste generated
Dewaxing	Fugitive solvents, heaters	Little or no residual waste generated
Propane deasphalting	Heater stack gas (CO, SO_x, NO_x, hydrocarbons, and particulates), fugitive propane	Little or no residual waste generated
Wastewater treatment	Fugitive emissions (H_2S, NH_3, and hydrocarbons)	API separator sludge (phenols, metals, and oil), chemical precipitation sludge (chemical coagulants, oil), DAF floats, biological sludge (metals, oil, suspended solids), spent lime

Source: EPA, 1995, 2004.

hydrocarbon contamination has long been of concern and has spurred various analytical and site remediation developments (e.g., risk-based corrective actions). In some instances, it may appear that such cleanup operations were initiated with incomplete knowledge of the charter and behavior of the contaminants. The most appropriate first assumption is that the spilled constituents are toxic to the ecosystem. The second issue is an investigation of the products of the spilled material to determine an appropriate cleanup method. The third issue is whether or not the chemical nature of the constituents has changed during the time since the material was released into the environment. If it has, a determination must be made of the effect of any such changes on the potential cleanup method.

Despite the large number of hydrocarbons found in petroleum products and the widespread nature of petroleum use and contamination, many of the lower-boiling constituents are well characterized in terms of physical properties, but only a relatively small number of the compounds are well characterized for toxicity. The health effects of some fractions can be well characterized, based on their components or representative compounds (e.g., light aromatic fraction benzene–toluene–ethylbenzene–xylenes). However, higher-molecular-weight (higher-boiling) fractions have far fewer well-characterized compounds.

This chapter deals with the toxicity of petroleum and petroleum products, the effects of petroleum constituents on the environment, individual process wastes, and the means by which petroleum, petroleum products, and process wastes are introduced into the environment. The processes are restricted to those processes by which the common products are produced (Chapter 3).

4.1. PROCESS WASTES

Petroleum refineries are complex but integrated unit process operations that produce a variety of products from various feedstock blends (Figure 4.1). During petroleum refining, refineries use and generate an enormous amount of chemicals, some of which are present in air emissions, wastewater, or solid wastes. Emissions are also created through the combustion of fuels and as by-products of chemical reactions occurring when petroleum fractions are upgraded. A large source of air emissions is, generally, process heaters and boilers, which produce carbon monoxide, sulfur oxides, and nitrogen oxides, leading to pollution and the formation of acid rain.

$$CO_2 + H_2O \rightarrow H_2CO_3 \quad \text{(carbonic acid)}$$
$$SO_2 + H_2O \rightarrow H_2SO_3 \quad \text{(sulfurous acid)}$$
$$2SO_2 + O_2 \rightarrow 2SO_3$$
$$SO_3 + H_2O \rightarrow H_2SO_4 \quad \text{(sulfuric acid)}$$
$$NO + H_2O \rightarrow HNO_2 \quad \text{(nitrous acid)}$$
$$2NO + O_2 \rightarrow NO_2$$
$$NO_2 + H_2O \rightarrow HNO_3 \quad \text{(nitric acid)}$$

Hence, there is a need for gas-cleaning operations on a refinery site so that such gases are cleaned from the gas stream prior to entry into the atmosphere.

In addition, some processes create considerable amounts of particulate matter and other emissions from catalyst regeneration or decoking processes. Volatile chemicals and hydrocarbons are also released from equipment leaks, storage tanks, and wastewaters. Other cleaning units, such as the installation of filters, electrostatic precipitators, and cyclones, can mitigate part of the problem.

Process wastewater is also a significant effluent from a number of refinery processes. Atmospheric and vacuum distillation create the largest volumes of process wastewater, about 26 gallons per barrel of oil processed. Fluid catalytic cracking and catalytic reforming also generate considerable amounts of wastewater (15 and 6 gallons per barrel of feedstock, respectively). A large portion of

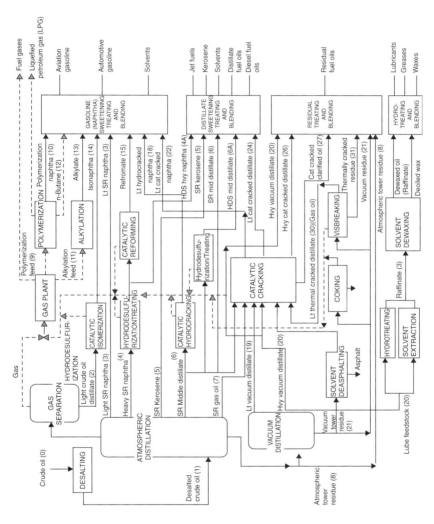

Figure 4.1. General layout of a petroleum refinery (Meyers, 1997; Speight, 1999, and references cited therein; Speight and Ozum, 2002, and references cited therein).

wastewater from these three processes is contaminated with oil and other impurities and must be subjected to primary, secondary, and sometimes tertiary water treatment processes, some of which also create hazardous waste.

Wastes, residua, and by-products are produced by a number of processes. Residuals produced during refining can be but are not necessarily wastes. They can be recycled or regenerated, and in many cases do not become part of the waste stream but are useful products. For example, processes utilizing caustics for the neutralization of acidic gases or solvent (e.g., alkylation, sweetening/chemical treating, lubricating oil manufacture) create the largest source of residuals in the form of spent caustic solutions. However, nearly all of these caustics are recycled.

The treatment of oily wastewater from distillation, catalytic reforming, and other processes generates the next-largest source of residuals, in the form of biomass sludge from biological treatment and pond sediments. Water treatment of oily wastewater also produces a number of sludge materials associated with oil–water separation processes. Such sludge is often recycled in the refining process and is not considered waste.

Catalytic processes (fluid catalytic cracking, catalytic hydrocracking, hydrotreating, isomerization, ether manufacture) also create some residuals in the form of spent catalysts and catalyst fines or particulates. The latter are sometimes separated from exiting gases by electrostatic precipitators or filters. These are collected and disposed of in landfills or may be recovered by off-site facilities. The potential for waste generation and hence leakage of emissions is discussed below for individual processes.

4.1.1. Desalting

As already noted (Chapter 3), petroleum oil often contains water, inorganic salts, suspended solids, and water-soluble trace metals. As a first step in the refining process, to reduce corrosion, plugging, and fouling of equipment and to prevent poisoning the catalysts in processing units, these contaminants must be removed by desalting (dehydration).

The two most typical methods of petroleum desalting, chemical separation and electrostatic separation, use hot water as the extraction agent. In chemical desalting, water and chemical surfactant (demulsifiers) are added to the petroleum, heated so that salts and other impurities dissolve into the water or attach to the water, and are then held in a tank, where they settle out. Electrical desalting is the application of high-voltage electrostatic charges to concentrate suspended water globules in the bottom of the settling tank. Surfactants are added only when the crude has a large amount of suspended solids. Both methods of desalting are continuous. A third and less common process involves filtering heated petroleum using diatomaceous earth.

The feedstock crude oil is heated to between 65 and 177°C (150 to 350°F) to reduce viscosity and surface tension for easier mixing and separation of the water. The temperature is limited by the vapor pressure of the petroleum constituents. In both methods, other chemicals may be added. Ammonia is often

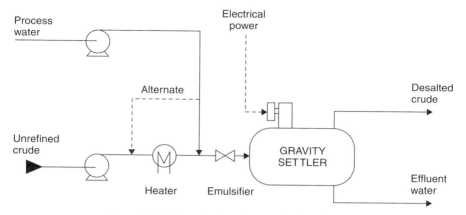

Figure 4.2. Schematic of an electrostatic desalting unit.

used to reduce corrosion. Caustic or acid may be added to adjust the pH of the water wash. Wastewater and contaminants are discharged from the bottom of the settling tank to the wastewater treatment facility. The desalted crude is drawn continuously from the top of the settling tanks and sent to the crude distillation (fractionating) tower.

Since desalting is a closed process, there is little potential for exposure to the feedstock unless a leak or release occurs. However, whenever elevated temperatures are used when desalting sour (sulfur-containing) petroleum, hydrogen sulfide will be present. Depending on the crude feedstock and the treatment chemicals used, the wastewater will contain varying amounts of chlorides, sulfides, bicarbonates, ammonia, hydrocarbons, phenol, and suspended solids. If diatomaceous earth is used in filtration, exposures should be minimized or controlled.

Desalting (Figure 4.2) creates an oily desalter sludge that may be a hazardous waste and a high-temperature salt wastewater stream (treated along with other refinery wastewaters). The primary polluting constituents in desalter wastewater include hydrogen sulfide, ammonia, phenol, high levels of suspended solids, and dissolved solids, with a high biochemical oxygen demand (BOD). In some cases it is possible to recycle the desalter effluent water back into the desalting process, depending on the type of crude being processed.

4.1.2. Distillation

Atmospheric and vacuum distillation units (Figures 4.3 and 4.4) are closed processes, and exposures are expected to be minimal. Both atmospheric distillation units and vacuum distillation units produce refinery fuel gas streams containing a mixture of light hydrocarbons, hydrogen sulfide, and ammonia. These streams are processed through gas treatment and sulfur recovery units to recover fuel gas and sulfur. Sulfur recovery creates emissions of ammonia, hydrogen sulfide, sulfur oxides, and nitrogen oxides.

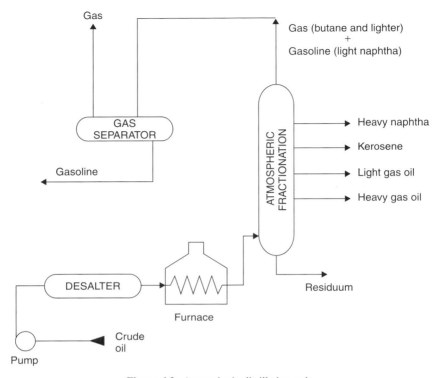

Figure 4.3. Atmospheric distillation unit.

When sour (high-sulfur) petroleum is processed, there is potential for exposure to hydrogen sulfide in the preheat exchanger and furnace, tower flash zone and overhead system, vacuum furnace and tower, and bottoms exchanger. Hydrogen chloride may be present in the preheat exchanger, tower top zones, and overheads. Wastewater may contain water-soluble sulfides in high concentrations and other water-soluble compounds, such as ammonia, chlorides, phenol, and mercaptans, depending on the crude feedstock and the treatment chemicals. Safe work practices and/or the use of appropriate personal protective equipment may be needed for exposures to chemicals and other hazards, such as heat and noise, and during sampling, inspection, maintenance, and turnaround activities.

Air emissions from a petroleum distillation unit include emissions from the combustion of fuels in process heaters and boilers, fugitive emissions of volatile constituents in the crude oil and fractions, and emissions from process vents. The primary source of emissions is combustion of fuels in the crude preheat furnace and in boilers that produce steam for process heat and stripping. When operating in an optimum condition and burning cleaner fuels (e.g., natural gas, refinery gas), these heating units create relatively low emissions of sulfur oxides, (SO_x), nitrogen oxides (NO_x), carbon monoxide (CO), hydrogen sulfide (H_2S), particulate

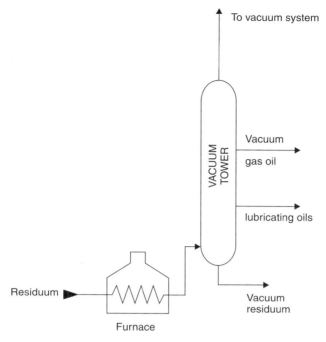

Figure 4.4. Vacuum distillation unit.

matter, and volatile hydrocarbons. If fired with lower-grade fuels (e.g., refinery fuel pitch, coke) or operated inefficiently (incomplete combustion), heaters can be a significant source of emissions.

Fugitive emissions of volatile hydrocarbons arise from leaks in valves, pumps, flanges, and similar sources where crude and its fractions flow through the system. Although individual leaks may be minor, the combination of fugitive emissions from various sources can be substantial. Those potentially released during crude distillation include ammonia, benzene, toluene, and xylenes, among others. These emissions are controlled primarily through leak detection and repair programs and occasionally through the use of special leak-resistant equipment.

Petroleum distillation units generate considerable wastewater. The process water used in distillation often comes in direct contact with oil and can be highly contaminated. Both atmospheric distillation and vacuum distillation produce an oily, sour wastewater (condensed steam containing hydrogen sulfide and ammonia) from side-stripping fractionators and reflux drums.

Many refineries now use vacuum pumps and surface condensers in place of barometric condensers to eliminate generation of the wastewater stream and reduce energy consumption. Reboiled side-stripping towers rather than open steam stripping can also be utilized on the atmospheric tower to reduce the quantity of sour-water condensate.

Typical constituents of sour wastewater streams from crude distillation include hydrogen sulfide, ammonia, suspended solids, chlorides, mercaptans, and phenol, characterized by a high pH. Combined flows from atmospheric and vacuum distillation are about 26.0 gallons per barrel of oil, and represent one of the largest sources of wastewater in a refinery.

4.1.3. Visbreaking and Coking

Like many thermal cracking processes, *visbreaking* (Figure 4.5) tends to produce a relatively small amount of fugitive emissions and sour wastewater. Usually, some wastewater is produced from steam strippers and the fractionator. Wastewater is also generated during unit cleanup and cooling operations and from the steam injection process to remove organic deposits from the soaker or from the coil. Combined wastewater flows from thermal cracking and coking processes are about 3.0 gallons per barrel of process feed.

Coking processes (Figures 4.6 and 4.7) produce a relatively small amount of sour wastewater from steam strippers and fractionators. Wastewater is generated during coke removal and cooling operations and from the steam injection process to cut coke from the coke drums. Combined wastewater flows from thermal cracking and coking processes are about 3.0 gallons per barrel of process feed.

Like most separation processes in the refinery, the process water used in coker fractionators (as is also the case in other product fractionators) often comes in direct contact with oil and can have a high oil content (much of that oil can be recovered through wastewater oil recovery processes). Thus, the main constituents

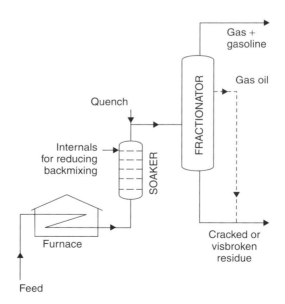

Figure 4.5. Soaker visbreaking unit.

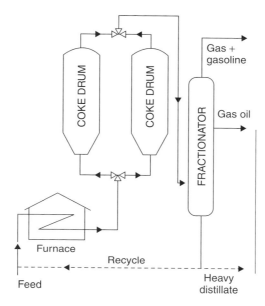

Figure 4.6. Delayed coking unit.

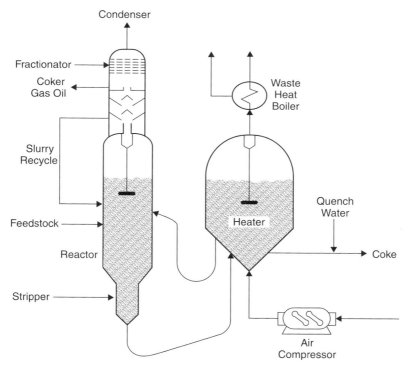

Figure 4.7. Fluid coking unit.

of sour water from catalytic cracking include high levels of oil, suspended solids, phenols, cyanides, hydrogen sulfate, and ammonia. Typical wastewater flow from catalytic cracking is about 15.0 gallons per barrel of feed processed (more than one-third of a gallon of wastewater for every gallon of feed processed) and represents the second largest source of wastewater in the refinery.

Particulate emissions from decoking can also be considerable. Coke-laden water from decoking operations in delayed cokers (hydrogen sulfide, ammonia, suspended solids) and coke dust (carbon particles and hydrocarbons) occur.

4.1.4. Fluid Catalytic Cracking

Fluid catalytic cracking (Figure 4.8) is one of the largest sources of air emission in refineries. Air emissions are released in process heater flue gas, as fugitive emissions from leaking valves and pipes and during regeneration of the cracking catalyst. If not controlled, catalytic cracking is one of the most substantial sources of carbon monoxide and particulate emissions in the refinery. In nonattainment areas where carbon monoxide and particulates are above acceptable levels, carbon monoxide waste heat boilers (CO boiler) and particulate controls are employed. Carbon monoxide produced during regeneration of the catalyst is converted to carbon dioxide either in the regenerator or farther downstream in a carbon monoxide waste heat boiler (CO boiler). Catalytic crackers are also significant sources of sulfur oxides and nitrogen oxides. The nitrogen oxides produced by catalytic crackers is expected to be a major target of emissions reduction in the future.

Like coking units, catalytic cracking units usually include some form of fractionation or steam stripping as part of the process configuration. These units all

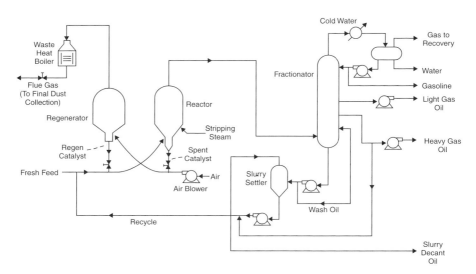

Figure 4.8. Schematic of a fluid catalytic cracking unit.

produce sour waters and sour gases containing some hydrogen sulfide and ammonia. Like crude oil distillation, some of the toxic releases reported by the refining industry are generated through sour water and gases, notably ammonia. Gaseous ammonia often leaves fractionating and treating processes in the sour gas along with hydrogen sulfide and fuel gases.

Catalytic cracking produces large volumes of wastewater and spent catalysts. Catalytic cracking (primarily fluid catalytic cracking) generates considerable sour wastewater from fractionators used for product separation, from steam strippers used to strip oil from catalysts, and in some cases from scrubber water. The steam stripping process used to purge and regenerate catalysts may contain metal impurities from the feed in addition to oil and other contaminants. Sour wastewater from the fractionator/gas concentration units and steam strippers contain oil, suspended solids, phenols, cyanides, hydrogen sulfide, ammonia, and spent catalysts (metals from crude oil and hydrocarbons).

Catalytic cracking generates significant quantities of spent process catalysts (containing metals from crude oils and hydrocarbons), which are often sent off-site for disposal or recovery or recycling. Management options can include land filling, treatment, or separation and recovery of the metals. Metals deposited on catalysts are often recovered by third-party recovery facilities. Spent catalyst fines (containing aluminum silicate and metals) from electrostatic precipitators are also sent off-site for disposal and/or recovery options.

Catalytic crackers also produce a significant amount of fine catalyst dust, which results from the constant movement of catalyst grains against each other. This dust contains primarily alumina (Al_2O_3) and small amounts of nickel (Ni) and vanadium (V) and is generally carried along with the carbon monoxide stream to the carbon monoxide waste heat boiler. The dust is separated from the carbon dioxide stream exiting the boiler through the use of cyclones, flue gas scrubbing, or electrostatic precipitators, and may be disposed of at an off-site facility.

4.1.5. Hydrocracking and Hydrotreating

Hydrocracking (Figure 4.9) generates air emissions through process heater flue gas, vents, and fugitive emissions. Unlike fluid catalytic cracking catalysts, hydrocracking catalysts are usually regenerated off-site after months or years of operations, and little or no emissions or dust is generated. However, the use of heavy oil as feedstock to the unit can change this balance.

Hydrocracking produces less sour wastewater than does catalytic cracking. Like catalytic cracking, hydrocracking produces sour wastewater at the fractionator. These processes include processing in a separator (API separator, corrugated plate interceptor) that creates a sludge. Physical or chemical methods are then used to separate the remaining emulsified oils from the wastewater. Treated wastewater may be discharged to public wastewater treatment, to a refinery secondary treatment plant for ultimate discharge to public wastewater treatment, or may be recycled and used as process water. The separation process permits recovery of usable oil and creates a sludge that may be recycled or treated as a hazardous waste.

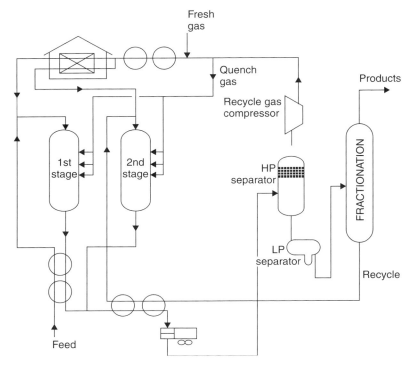

Figure 4.9. Two-stage hydrocracking unit.

In addition, oily sludge from a wastewater treatment facility that results from treating sour wastewaters may be a hazardous waste (unless recycled in the refining process). These include API separator sludge, primary treatment sludge, sludge from various gravitational separation units, and float from dissolved air flotation units.

Propylene, another source of toxic releases from refineries, is produced as a light end during cracking and coking processes. It is volatile as well as soluble in water, which increases its potential for release to both air and water during processing.

Like catalytic cracking, hydrocracking processes generate toxic metal compounds, many of which are present in spent catalyst sludge and catalyst fines generated from catalytic cracking and hydrocracking. These include metals such as nickel, cobalt, and molybdenum.

Hydrotreating (Figure 4.10) generates air emissions through process heater flue gas, vents, and fugitive emissions. Unlike fluid catalytic cracking catalysts, hydrotreating catalysts are usually regenerated off-site after months or years of operations, and little or no emission or dust is generated from the catalyst regeneration process at the refinery. Section 4.2.2 provides air emissions factors for emissions from process heaters and boilers used throughout the refinery.

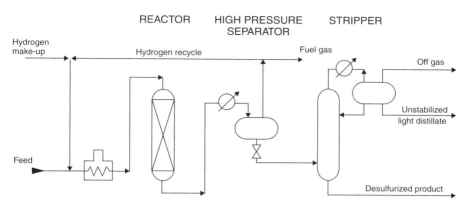

Figure 4.10. Distillate hydrotreating unit.

The off-gas stream from hydrotreating is usually very rich in hydrogen sulfide and light fuel gas. This gas is usually sent to a sour gas treatment and sulfur recovery unit along with other refinery sour gases.

Fugitive air emissions of volatile components released during hydrotreating may also be toxic components. These include toluene, benzene, xylenes, and other volatiles that are reported as toxic chemical releases under the EPA Toxics Release Inventory.

Hydrotreating generates sour wastewater from fractionators used for product separation. Like most separation processes in a refinery, the process water used in fractionators often comes in direct contact with oil and thus can be highly contaminated. It also contains hydrogen sulfide and ammonia and must be treated along with other refinery sour waters. In hydrotreating, sour wastewater from fractionators is produced at the rate of about 1.0 gallon per barrel of feed.

Oily sludge from a wastewater treatment facility that results from treating oily and/or sour wastewaters from hydrotreating and other refinery processes may be hazardous waste, depending on how it is managed. Oily sludge may be API separator sludge, primary treatment sludge, sludge from various gravitational separation units, and float from dissolved air flotation units.

Hydrotreating also produces some residuals in the form of spent catalyst fines, usually consisting of aluminum silicate and some metals (e.g., cobalt, molybdenum, nickel, tungsten). Spent hydrotreating catalyst is now listed as a hazardous waste (K171) (except for most support material). Hazardous constituents of this waste include benzene and arsenia (arsenic oxide, As_2O_3). The support material for these catalysts is usually an inert ceramic (e.g., alumina, Al_2O_3).

4.1.6. Alkylation and Polymerization

Alkylation (Figure 4.11) combines low-molecular-weight olefins (primarily a mixture of propylene and butylene) with isobutene in the presence of a catalyst, either sulfuric acid or hydrofluoric acid. The product, called *alkylate*, consists of

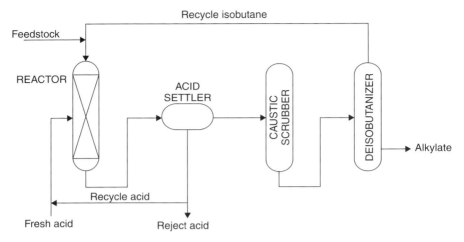

Figure 4.11. Alkylation unit (sulfuric acid catalyst).

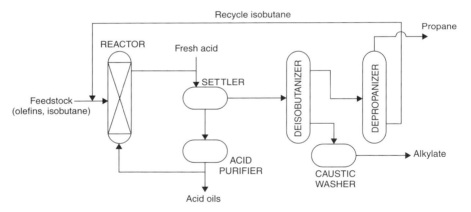

Figure 4.12. Alkylation unit (hydrogen fluoride catalyst).

a mixture of high-octane branched-chain paraffinic hydrocarbons. Alkylate is a premium blending stock because it has exceptional antiknock properties and is clean burning. The octane number of alkylate depends primarily on the type of olefin used and on operating conditions.

Emissions from alkylation processes (Figures 4.11 and 4.12) and polymerization processes (Figure 4.13) include fugitive emissions of volatile constituents in the feed and emissions that arise from process vents during processing. These can take the form of acidic hydrocarbon gases, nonacidic hydrocarbon gases, and fumes that may have a strong odor (from sulfonated organic compounds and organic acids, even at low concentrations). To prevent releases of hydrofluoric acid, refineries install a variety of mitigation and control technologies (e.g., acid

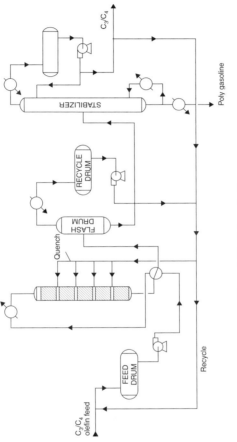

Figure 4.13. Polymerization process.

inventory reduction, hydrogen fluoride detection systems, isolation valves, rapid acid transfer systems, and water spray systems).

In hydrofluoric acid alkylation processes, acidic hydrocarbon gases can originate anywhere that hydrogen fluoride is present (e.g., during a unit upset, unit shutdown, or maintenance). Hydrofluoric acid alkylation units are designed to pipe these gases from acid vents and valves to a separate closed-relief system where the acid is neutralized. The basins are tightly covered and equipped with a gas scrubbing system to remove odors, using either water or activated charcoal as the scrubbing agent. Another source of emissions is a combustion of fuels in process boilers to produce steam for strippers. As with all process heaters in a refinery, these boilers produce significant emissions of sulfur oxides, nitrogen oxides, carbon monoxide, particulate matter, and volatile hydrocarbons.

Alkylation generates relatively low volumes of wastewater, primarily from water washing of the liquid reactor products. Wastewater is also generated from steam strippers, depropanizers, and debutanizers, and can be contaminated with oil and other impurities. Liquid process waters (hydrocarbons and acid) originate from minor undesirable side reactions and from feed contaminants, and usually exit as a bottoms stream from the acid regeneration column. The bottoms stream is an acid–water mixture that is sent to the neutralizing drum. The acid in this liquid eventually ends up as insoluble calcium fluoride.

Sulfuric acid alkylation generates considerable quantities of spent acid that must be removed and regenerated. Nearly all the spent acid generated at refineries is regenerated and recycled, and although technology for on-site regeneration of spent sulfuric acid is available, the supplier of the acid may perform this task off-site. If sulfuric acid production capacity is limited, acid regeneration is often done on-site. The development of internal acid regeneration for hydrofluoric acid units has virtually eliminated the need for external regeneration, although most operations retain one for startups or during periods of high feed contamination.

Both sulfuric acid and hydrofluoric acid alkylation units generate neutralization sludge from treatment of acid-laden streams with caustic solutions in neutralization or wash systems. Sludge from hydrofluoric acid alkylation neutralization systems consists largely of calcium fluoride and unreacted lime and is usually disposed of in a landfill. It can also be directed to steel manufacturing facilities, where the calcium fluoride can be used as a neutral flux to lower the slag-melting temperature and improve slag fluidity. Calcium fluoride can also be routed back to a hydrofluoric acid manufacturer.

A basic step in hydrofluoric acid manufacture is the reaction of sulfuric acid with fluorspar (calcium fluoride) to produce hydrogen fluoride and calcium sulfate. Spent alumina is also generated by the defluorination of some hydrofluoric acid alkylation products over alumina. It is disposed of or sent to the alumina supplier for recovery. Other solid residuals from hydrofluoric acid alkylation include any porous materials that may have come in contact with the hydrofluoric acid.

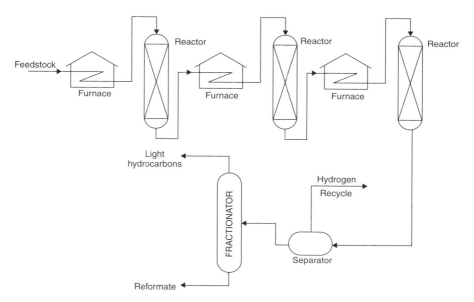

Figure 4.14. Catalytic reforming unit.

4.1.7. Catalytic Reforming

Emissions from catalytic reforming (Figure 4.14) include fugitive emissions of volatile constituents in the feed and emissions from process heaters and boilers. As with all process heaters in the refinery, combustion of fossil fuels produces emissions of sulfur oxides, nitrogen oxides, carbon monoxide, particulate matter, and volatile hydrocarbons.

Toluene, xylene, and benzene are toxic aromatic chemicals that are produced during the catalytic reforming process and used as feedstocks in chemical manufacturing. Due to their highly volatile nature, fugitive emissions of these chemicals are a source of their release to the environment during the reforming process. Point air sources may also arise during the process of separating these chemicals.

In a continuous reformer, some particulate and dust matter can be generated as the catalyst moves from reactor to reactor and is subject to attrition. However, due to catalyst design little attrition occurs, and the only outlet to the atmosphere is the regeneration vent, which is most often scrubbed with a caustic to prevent emission of hydrochloric acid (this also removes particulate matter). Emissions of carbon monoxide and hydrogen sulfide may occur during regeneration of catalyst.

4.1.8. Isomerization

Isomerization (Figure 4.15) converts n-butane, n-pentane and n-hexane into their respective isoparaffins of substantially higher octane number. The straight-chain

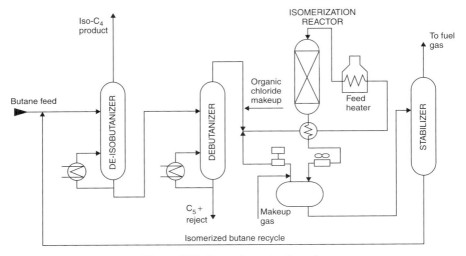

Figure 4.15. Butane isomerization unit.

paraffins are converted to their branched-chain counterparts, whose component atoms are the same but are arranged in a different geometric structure. Isomerization is important for the conversion of n-butane into isobutane, to provide additional feedstock for alkylation units, and the conversion of normal pentanes and hexanes into higher-branched isomers for gasoline blending.

Isomerization processes produce sour water and caustic wastewater. The ether manufacturing process utilizes a water wash to extract methanol or ethanol from the reactor effluent stream. After the alcohol is separated, this water is recycled back to the system and is not released. In those cases where chloride catalyst activation agents are added, a caustic wash is used to neutralize any entrained hydrogen chloride. This process generates a caustic wash water that must be treated before being released. This process also produces a calcium chloride neutralization sludge that must be disposed of off-site.

4.1.9. Deasphalting and Dewaxing

Propane deasphalting (Figure 4.16) produces lubricating oil base stocks by extracting asphaltenes and resins from vacuum distillation residua. Propane is the usual solvent of choice, due to its unique solvent properties. At lower temperatures (38 to 60°C, 100 to 140°F), paraffins are very soluble in propane, and at higher temperatures (about 93°C, 200°F), hydrocarbons are almost insoluble in propane. The propane deasphalting process is similar to solvent extraction in that a packed or baffled extraction tower or rotating disk contactor is used to mix the oil feedstocks with the solvent. In the tower method, four to eight volumes of propane are fed to the bottom of the tower for every volume of feed flowing down from the top of the tower. The oil, which is more soluble in

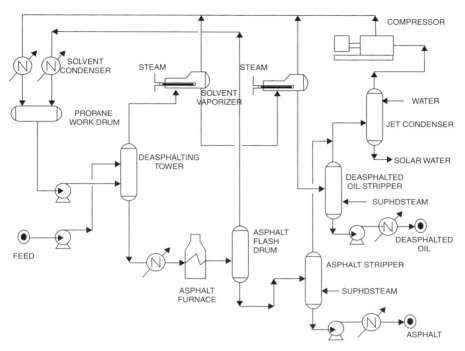

Figure 4.16. Deasphalting unit.

the propane, dissolves and flows to the top. The higher-molecular-weight polar asphalt constituents flow to the bottom of the tower, where they are removed in a propane mix. Propane is recovered from the two streams through two-stage flash systems followed by steam stripping, in which propane is condensed and removed by cooling at high pressure in the first stage and at low pressure in the second stage. The asphalt recovered can be blended with other asphalt, or heavy fuel oil, or can be used as feed to the coker. The propane recovery stage results in propane-contaminated water, which typically is sent to the wastewater treatment plant.

Air emissions may arise from fugitive propane emissions and process vents. These include heater stack gas (carbon monoxide, sulfur oxides, nitrogen oxides, and particulate matter) as well as hydrocarbon emission, such as fugitive propane and fugitive solvents. Steam stripping wastewater (oil and solvents) and solvent recovery wastewater (oil and propane) are also produced.

Dewaxing (Figure 4.17) processes also produce heater stack gas (carbon monoxide, sulfur oxides, nitrogen oxides, and particulate matter) as well as hydrocarbon emission such as fugitive propane and fugitive solvents. Steam stripping wastewater (oil and solvents) and solvent recovery wastewater (oil and propane) are also produced. The fugitive solvent emissions may be toxic (toluene, methyl ethyl ketone, methyl isobutyl ketone).

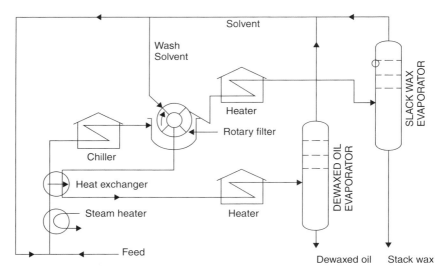

Figure 4.17. Solvent dewaxing unit.

4.2. ENTRY INTO THE ENVIRONMENT

It is almost impossible to transport, store, and refine crude oil without spills and losses. It is difficult to prevent spills resulting from failure of or damage to pipelines. It is also impossible to install control devices for controlling the ecological properties of water and the soil along the length of all pipelines. The soil suffers the most ecological damage in the damage areas of pipelines. Crude oil spills from pipelines lead to irreversible changes in the soil properties. The most affected soil properties by crude oil losses from pipelines are filtration, physical, and mechanical properties. These properties of the soil are important for maintaining the ecological equilibrium in the damaged area.

Principal sources of releases to air from refineries include (1) combustion plants, emitting sulfur dioxide, oxides of nitrogen, and particulate mater; (2) refining operations, emitting sulfur dioxide, oxides of nitrogen, carbon monoxide, particulate matter, volatile organic compounds, hydrogen sulfide, mercaptans, and other sulfurous compounds; and (3) bulk storage operations and handling of volatile organic compounds (various hydrocarbons). In light of this, it is necessary to consider (1) regulatory requirements (air emission permits stipulating limits for specific pollutants) and possibly health and hygiene permit requirements, (2) requirements for monitoring programs, and (3) requirements to upgrade pollution abatement equipment.

4.2.1. Storage and Handling of Petroleum Products

Large quantities of environmentally sensitive petroleum products are stored in (1) tank farms (multiple tanks), (2) single aboveground storage tanks (ASTs),

(3) semiunderground or underground storage tanks (USTs). Smaller quantities of materials may be stored in drums and containers of assorted compounds (such as lubricating oil, engine oil, other products for domestic supply).

In light of this, it is also necessary to consider (1) secondary containment of tanks and other storage areas and the integrity of hard standing (without cracks, an impervious surface) to prevent spills reaching the wider environment: also secondary containment of pipelines where appropriate; (2) age, construction details, and testing program of tanks; (3) labeling and environmentally secure storage of drums (including waste storage); (4) accident/fire precautions, emergency procedures; and (5) disposal/recycling of waste or "out of spec" oils and other materials.

There is a potential for significant soil and groundwater contamination to have arisen at petroleum refineries. Such contamination consists of (1) petroleum hydrocarbons, including lower-boiling, very mobile fractions (paraffins, cycloparaffins, and volatile aromatics such as benzene, toluene, ethylbenzene, and xylenes) typically associated with gasoline and similar boiling-range distillates, (2) middle distillate fractions (paraffins, cycloparaffins, and some polynuclear aromatics) associated with diesel, kerosene, and lower-boiling fuel oil, which are also of significant mobility; (3) higher-boiling distillates (long-chain paraffins, cycloparaffins, and polynuclear aromatics) that are associated with lubricating oil and heavy fuel oil; (4) various organic compounds associated with petroleum hydrocarbons or produced during the refining process (e.g., phenols, amines, amides, alcohols, organic acids, nitrogen- and sulfur-containing compounds); (5) other organic additives [e.g., antifreeze (glycols), alcohols, detergents, and various proprietary compounds]; (6) organic lead, associated with leaded gasoline and other heavy metals.

Key sources of such contamination at petroleum refineries are at (1) transfer and distribution points in tankage and process areas and in general loading and unloading areas; (2) land farm areas; (3) tank farms; (4) individual aboveground storage tanks, particularly individual underground storage tanks; (5) additive compounds; and (6) pipelines, drainage areas as well as on-site waste treatment facilities, impounding basins, and lagoons, especially if unlined.

Although contamination may be associated with specific facilities, the contaminants are relatively highly mobile in nature and have the potential to migrate significant distances from the source in soil and groundwater. Petroleum hydrocarbon contamination can take several forms: free-phase product, dissolved phase, emulsified phase, or vapor phase. Each form will require different methods of remediation, so that cleanup may be complex and expensive. In addition, petroleum hydrocarbons include a number of compounds of significant toxicity (e.g., benzene and some polyaromatics are known carcinogens). Vapor-phase contamination can be of significance in terms of odor issues.

Due to the obvious risk of fire, refineries are equipped with sprinkler or spray systems that may draw on the main supply of water, or water held in lagoons, or from reservoirs or neighboring water courses. Such water will be polluting and require containment.

Refining facilities require significant volumes of water for on-site processes (e.g., coolants, blowdowns) as well as for sanitary and potable use. Wastewater will derive from these sources (process water) and from stormwater runoff. The latter could contain significant concentrations of petroleum product.

Petroleum hydrocarbons, either dissolved, emulsified, or occurring as free-phase, will be the key constituents, although wastewater may also contain significant concentrations of phenols, amines, amides, alcohols, ammonia, sulfide, heavy metals, and suspended solids.

Wastewaters may be collected in separate drainage systems (for process, sanitary, and storm water) although industrial and stormwater systems may in some cases be combined. In addition, ballast water from bulk crude tankers may be pumped to receiving facilities at the refinery site prior to removal of floating oil in an interceptor and treatment as for other wastewater streams.

On-site treatment facilities may exist for wastewater, or treatment may take place at a public wastewater treatment plant. Stormwater/process water is generally passed to a separator or interceptor prior to leaving the site, which takes out free-phase oil (i.e., floating product) from the water prior to discharge or prior to further treatment (e.g., in settling lagoons). Discharge from wastewater treatment plants is usually passed to a nearby watercourse.

Other wastes that are typical of a refinery include (1) waste oils, process chemicals, and still resides; (2) nonspecification chemicals and/or products; (3) waste alkali (sodium hydroxide); (4) waste oil sludge (from interceptors, tanks, and lagoons); and (5) solid wastes (cartons, rags, catalysts, and coke).

4.2.2. Release into the Environment

Petroleum products released into the environment undergo weathering processes with time. These processes include evaporation; leaching (transfer to the aqueous phase), through solution and entrainment (physical transport along with the aqueous phase); chemical oxidation; and microbial degradation. The rate of weathering is highly dependent on environmental conditions. For example, gasoline, a volatile product, will evaporate readily in a surface spill, whereas gasoline released below 10 ft of clay topped with asphalt will tend to evaporate slowly (weathering processes may not be detectable for years).

An understanding of weathering processes is valuable to environmental test laboratories. Weathering changes product composition and may affect testing results, the ability to bioremediate, and the toxicity of the spilled product. Unfortunately, the database available on the composition of weathered products is limited. However, biodegradation processes, which influence the presence and the analysis of petroleum hydrocarbon at a particular site, can be very complex. The extent of biodegradation is dependent on many factors, including the type of microorganisms present, environmental conditions (e.g., temperature, oxygen levels, and moisture), and the predominant hydrocarbon types. In fact, the primary factor controlling the extent of biodegradation is the molecular composition of the petroleum contaminant. Multiple ring cycloalkanes are difficult to degrade, whereas polynuclear aromatic hydrocarbons display varying degrees of degradation. Straight-chain alkanes biodegrade rapidly with branched alkanes and single saturated ring compounds degrading more slowly.

The primary processes determining the fate of crude oils and oil products after a spill are (1) dispersion, (2) dissolution, (3) emulsification, (4) evaporation, (5) leaching, (6) sedimentation, (7) spreading, and (8) wind. These processes are influenced by the spill characteristics, environmental conditions, and physicochemical properties of the material spilled.

Dispersion

The physical transport of oil droplets into the water column, called *dispersion*, is often a result of water surface turbulence but may also result from the application of chemical agents (dispersants). These droplets may remain in the water column or coalesce with other droplets and gain enough buoyancy to resurface. Dispersed oil tends to biodegrade and dissolve more rapidly than floating slicks because of high surface area relative to volume. Most of this process occurs from about half an hour to half a day after the spill.

Dissolution

Dissolution is the loss of individual oil compounds into the water. Many of the acutely toxic components of oils, such as benzene, toluene, and xylene, will dissolve readily in water. This process also occurs quickly after a discharge but tends to be less important than evaporation. In a typical marine discharge, generally less than 5% of the benzene is lost to dissolution while more than 95% is lost to evaporation. For alkylated polynuclear aromatic compounds, solubility is inversely proportional to the number of rings and the extent of alkylation. The dissolution process is thought to be much more important in rivers because natural containment may prevent spreading, reducing the surface area of the slick and thus retarding evaporation. At the same time, river turbulence increases the potential for mixing and dissolution. Most of this process occurs within the first hour of the spill.

Aromatics, especially BTEX, tend to be the most water-soluble fraction of petroleum. Petroleum-contaminated groundwater tends to be enriched in aromatics relative to other petroleum constituents. Relatively insoluble hydrocarbons may be entrained in water through adsorption into kaolinite particles suspended in the water or as an agglomeration of oil droplets (microemulsion). In cases where groundwater contains only dissolved hydrocarbons, it may not be possible to identify the original petroleum product because only a portion of the free product will be present in the dissolved phase. As whole product floats on groundwater, the free product will gradually lose the water-soluble compounds. Groundwater containing entrained product will have a gas chromatographic fingerprint that is a combination of the free product chromatogram plus enhanced amounts of the soluble aromatics.

Generally, dissolved aromatics may be found quite far from the origin of a spill, but entrained hydrocarbons may be found in water close to the petroleum source. Oxygenates such as methyl-t-butyl ether (MTBE) are even more water soluble than aromatics and are highly mobile in the environment.

Emulsification

Certain oils tend to form water-in-oil emulsions (where water is incorporated into oil) or "mousse" as weathering occurs. This process is significant because, for example, the apparent volume of the oil may increase dramatically, and the emulsification will slow the other weathering processes, especially evaporation. Under certain conditions, these emulsions may separate and release relatively fresh oil. Most of this process occurs from about half a day to two days after the spill.

Evaporation

Evaporative processes are very important in the weathering of volatile petroleum products and may be the dominant weathering process for gasoline. Automotive gasoline, aviation gasoline, and some grades of jet fuel (e.g., JP-4) contain 20 to 99% highly volatile constituents (i.e., constituents with fewer than nine carbon atoms).

Evaporative processes begin immediately after oil is discharged into the environment. Some light products (e.g., one- to two-ring aromatic hydrocarbons and/or low-molecular-weight alkanes less than n-C_{15}) may evaporate entirely; a significant fraction of heavy refined oils also may evaporate. For crude oils, the amount lost to evaporation can typically range from approximately 20 to 60%. The primary factors that control evaporation are the composition of the oil, slick thickness, temperature and solar radiation, wind speed, and wave height. While evaporation rates increase with temperature, this process is not restricted to warm climates. For the *Exxon Valdez* incident, which occurred in cold conditions (March 1989), it has been estimated that appreciable evaporation occurred even before all the oil escaped from the ship, and that evaporation ultimately accounted for 20% of the oil. Most of this process occurs within the first few days after the spill.

It is not unusual for evaporative processes, however, to be working simultaneously with other processes to remove volatile aromatics such as benzene and toluene.

Leaching

Leaching processes introduce hydrocarbon into the water phase by solubility and entrainment. Leaching processes of petroleum products in soils can have a variety of potential scenarios. Part of the aromatic fraction of a petroleum spill in soil may partition into water that has been in contact with the contamination.

Sedimentation or Adsorption

As mentioned above, most oils are buoyant in water. However, in areas with high levels of suspended sediment, petroleum constituents may be transported to the river, lake, or ocean floor through the process of sedimentation. Oil may adsorb to sediments and sink or be ingested by zooplankton and excreted in fecal pellets that may settle to the bottom. Oil stranded on shorelines also may pick up sediments, float with the tide, and then sink. Most of this process occurs from about 2 to 7 days after the spill.

Spreading

As oil enters the environment, it begins to spread immediately. The viscosity of the oil, its pour point, and the ambient temperature will determine how rapidly the oil will spread, but light oils typically spread more rapidly than heavy oils. The rate of spreading and ultimate thickness of the oil slick will affect the rates of the other weathering processes. For example, discharges that occur in geographically contained areas (such as a pond or slow-moving stream) will evaporate more slowly than if the oil were allowed to spread. Most of this process occurs within the first week after the spill.

Wind

Wind (aeolian) transport (relocation by wind) can also occur and is particularly relevant when catalyst dust and coke dust are considered. Dust becomes airborne when winds traversing arid land with little vegetation cover pick up small particles such as catalyst dust, coke dust, and other refinery debris and send them skyward. Wind transport may occur through *suspension*, *saltation*, or *creep* of the particles.

4.3. TOXICITY

With few exceptions, the constituents of petroleum, petroleum products, and the various emissions are hazardous to the health. There always exceptions that will be cited in opposition to such a statement, the most common exception being the liquid paraffin that is used medicinally to lubricate the alimentary tract. The use of such medication is common among miners who breathe and swallow coal dust every day during their work shifts.

Another approach is to consider petroleum constituents in terms of transportable materials, the character of which is determined by several chemical and physical properties (i.e., solubility, vapor pressure, and propensity to bind with soil and organic particles). These properties are the basis of measures of leachability and volatility of individual hydrocarbons. Thus, petroleum transport fractions can be considered by equivalent carbon number to be grouped into 13 different fractions. The analytical fractions are then set to match these transport fractions, using specific n-alkanes to mark the analytical results for aliphatic compounds and selected aromatic compounds to delineate hydrocarbons containing benzene rings.

Although chemicals grouped by transport fraction generally have similar toxicological properties, this is not always the case. For example, benzene is a carcinogen, but many alkyl-substituted benzenes do not fall under this classification. However, it is more appropriate to group benzene with compounds that have similar environmental transport properties than to group it with other carcinogens, such as benzo[a]pyrene, that have very different environmental transport properties.

Nevertheless, consultation of any reference work that lists the properties of chemicals will show the properties and hazardous nature of the types of chemicals that are found in petroleum. In addition, petroleum is used to make petroleum products, which can contaminate the environment. The range of chemicals in petroleum and petroleum products is so vast that summarizing the properties and/or the toxicity or general hazard of petroleum in general or even for a specific crude oil is a difficult task. However, because of the hydrocarbon content, petroleum and some petroleum products are at least theoretically biodegradable, but large-scale spills can overwhelm the ability of the ecosystem to break the oil down. The toxicological implications from petroleum occur primarily from exposure to or biological metabolism of aromatic structures. These implications change as an oil spill ages or is weathered.

4.3.1. Lower-Boiling Constituents

Many of the gaseous and liquid constituents of the lower-boiling fractions of petroleum and also those in petroleum products fall into the class of chemicals that have one or more of the following characteristics, considered to be hazardous by the EPA.

1. *Ignitability–flammability*. A liquid that has a flash point of less than 60°C (140°F) is considered ignitable. Some examples are benzene, hexane, heptane, benzene, pentane, petroleum ether (low boiling), toluene, and xylene(s).

2. *Corrosivity*. An aqueous solution that has a pH less than or equal to 2, or greater than or equal to 12.5, is considered corrosive. Most petroleum constituents and petroleum products are not corrosive, but many of the chemicals used in refineries are corrosive. Corrosive materials also include substances such as sodium hydroxide and some other acids or bases.

3. *Reactivity*. Chemicals that react violently with air or water are considered hazardous; examples are sodium metal, potassium metal, and phosphorus. Reactive materials also include strong oxidizers such as perchloric acid, and chemicals capable of detonation when subjected to an initiating source, such as solid, dry < 10% H_2O picric acid, benzoyl peroxide, or sodium borohydride. Solutions of certain cyanide or sulfides that could generate toxic gases are also classified as reactive. The potential for finding such chemicals in a refinery is

subject to the function and product slate of the refinery and/or the petrochemical complex.

4. *Hazardous chemicals.* Many chemicals have been shown in scientific studies to have toxic, carcinogenic, mutagenic, or teratogenic effects on humans or other life-forms and are designated either as *acutely hazardous waste* or *toxic waste* by the EPA. Substances found to be fatal to humans in low doses, or in the absence of data on human toxicity, have been shown in studies to have an oral LD_{50} toxicity (rat) of less than 2 mg/L or a dermal LD_{50} toxicity (rabbit) of less than 200 mg/kg or is otherwise capable of causing or significantly contributing to an increase in serious irreversible or incapacitating reversible illness are designated as *acute hazardous waste*. Materials containing any of the toxic constituents so listed are to be considered hazardous waste, unless after considering the following factors it can reasonably be concluded (by the Department of Environmental Health and Safety) that the waste is not capable of posing a substantial present or potential hazard to public health or the environment when improperly treated, stored, transported, or disposed of, or otherwise managed.

The issues to be held in consideration are (1) the nature of the toxicity presented by the constituent, (2) the concentration of the constituent in the waste, (3) the potential of the constituent or any toxic degradation product of the constituent to migrate from the waste into the environment under the types of improper management considered in item 7 below, (4) the persistence of the constituent or any toxic degradation product of the constituent, (5) the potential for the constituent or any toxic degradation product of the constituent to degrade into nonharmful constituents and the rate of degradation, (6) the degree to which the constituent or any degradation product of the constituent accumulates in an ecosystem, (7) the plausible types of improper management to which the waste could be subjected, (8) the quantities of the waste generated at individual generation sites or on a regional or national basis, (9) the nature and severity of the public health threat and environmental damage that has occurred as a result of the improper management of wastes containing the constituent, and (10) actions taken by other governmental agencies or regulatory programs based on the health or environmental hazard posed by the waste or waste constituent. Other factors that may be appropriate may also be considered.

For the analysts, laboratories wishing to dispose of materials containing dilute concentrations of these constituents should contact the Department of Environmental Health and Safety for advice regarding the proper disposition of the materials. In addition, the list of such materials is not included here, as it is subject to periodic updates. Furthermore, the list is not meant to be complete and may not include substances that have the hazardous characteristics as defined above. Omission of a chemical from this list does not mean that it is without toxic properties or any other hazard.

More specifically regarding petroleum and petroleum products, the alkanes in gasoline and some other petroleum products are central nervous system depressants. In fact, gasoline was once evaluated as an anesthetic agent. However, sudden deaths, possibly as a result of irregular heartbeats, have been attributed to those inhaling vapors of hydrocarbons such as those in gasoline.

Alkanes of various types of crude oils and various petroleum products were biodegraded faster than the *unresolved fractions*. Different types of crude oils and products biodegraded at different rates in the same environments. An oil product is a complex mixture of organic chemicals and contains within it less persistent and more persistent fractions. The range between these two extremes is greatest for crude oils. Since the many different substances in petroleum have different physical and chemical properties, summarizing the fate of petroleum in general (or even a particular oil) is very difficult. Solubility-fate relationships must be considered.

The relative proportion of hazardous constituents present in petroleum is typically quite variable. Therefore, contamination will vary from one site to another. In addition, the farther one progresses from lighter toward heavier constituents (the general progression from lower-to higher-molecular-weight constituents), the greater the percentage of polynuclear aromatic hydrocarbons and other semivolatile constituents or nonvolatile constituents (many of which are not so immediately toxic as the volatiles but which can result in long-term/chronic impacts). These higher-molecular-weight constituents thus need to be analyzed for the semivolatile compounds that typically pose the greatest long-term risk.

In addition to large oil spills, petroleum hydrocarbons are released into the aquatic environments from natural seeps as well as non-point-source urban runoffs. Acute impacts from massive one-time spills are obvious and substantial. The impacts from small spills and chronic releases are the subject of much speculation and continued research. Clearly, these inputs of petroleum hydrocarbons have the potential for significant environmental impacts, but the effects of chronic low-level discharges can be minimized by the net assimilative capacities of many ecosystems, resulting in little detectable environmental harm.

Short-term (acute) hazards of lighter, more volatile, and water-soluble aromatic compounds (such as benzenes, toluene, and xylenes) include potential acute toxicity to aquatic life in the water column (especially in relatively confined areas) as well as potential inhalation hazards. However, the compounds that pass through the water column often tend to do so in small concentrations and/or for short periods of time, and fish and other pelagic or generally mobile species can often swim away to avoid impacts from spilled oil in open waters. Most fish are mobile and it is not known whether or not they can sense, and thus avoid, toxic concentrations of oil.

However, there are some potential effects of spilled oil on fish. The impacts on fish are primarily to the eggs and larvae, with limited effects on the adults. The sensitivity varies by species; pink salmon fry are affected by exposure to water-soluble fractions of crude oil, and pink salmon eggs are very tolerant to benzene and water-soluble petroleum. The general effects are difficult to assess and document quantitatively, due to the seasonal and natural variability of the species. Fish rapidly metabolize aromatic hydrocarbons, due to their enzyme system.

Long-term (chronic) potential hazards of lighter, more volatile, and water-soluble aromatic compounds include contamination of groundwater. Chronic effects of benzene, toluene, and xylene include changes in the liver and harmful effects on the kidneys, heart, lungs, and nervous system.

At the initial stages of a release, when the benzene-derived compounds are present at their highest concentrations, acute toxic effects are more common than they are later. These noncarcinogenic effects include subtle changes in detoxifying enzymes and liver damage. Generally, the relative aquatic acute toxicity of petroleum will be the result of the fractional toxicities of the different hydrocarbons present in the aqueous phase. Tests indicate that naphthalene-derived chemicals have a similar effect.

Except for short-term hazards from concentrated spills, BTEX compounds (benzene, toluene, ethylbenzene, and xylenes) have been more frequently associated with risk to humans than with risk to nonhuman species such as fish and wildlife. This is partly because plants, fish, and birds take up only very small amounts, and because this volatile compound tends to evaporate into the atmosphere rather than persisting in surface waters or soils. However, volatiles such as this compound can pose a drinking water hazard when they accumulate in groundwater. See also BTEX entry, and entries for benzene, toluene, ethylbenzene, and xylenes.

Petroleum is naturally weathered according to its physical and chemical properties, but during this process, living species within the local environment may be affected via one or more routes of exposure, including ingestion, inhalation, dermal contact, and to a much lesser extent, bioconcentration through the food chain. Aromatic compounds of concern include alkylbenzenes, toluene, naphthalenes, and polynuclear aromatic hydrocarbons (PNAs). Moreover, both atmospheric and hydrospheric impacts must be assessed when considering toxic implications from a petroleum release containing significant quantities of these single-ring aromatic compounds.

4.3.2. Higher-Boiling Constituents

Naphthalene and its homologs are less acutely toxic than benzene but are more prevalent for a longer period during oil spills. The toxicity of different crude oils and refined oils depends not only on the total concentration of hydrocarbons but also the hydrocarbon composition in the water-soluble fraction (WSF) of petroleum, water solubility, concentrations of individual components, and toxicity of the components. The water-soluble fractions prepared from different oils will vary in these parameters. Water-soluble fractions (WSFs) of refined oils (e.g., No. 2 fuel oil and bunker C oil) are more toxic than water-soluble fraction of crude oil to several species of fish (killifish and salmon). Compounds with either more rings or methyl substitutions are more toxic than less substituted compounds, but tend to be less water soluble and thus less plentiful in the water-soluble fraction.

Among the polynuclear aromatic hydrocarbons, the toxicity of petroleum is a function of its di- and triaromatic hydrocarbon content. Like the single aromatic ring variations, including benzene, toluene, and the xylenes, all are relatively volatile compounds with varying degrees of water solubility.

There are indications that pure naphthalene (a constituent of mothballs, which are, by definition, toxic to moths) and alkylnaphthalenes are from three to 10 times more toxic to test animals than are benzene and alkylbenzenes. In addition, and because of the low water solubility of tricyclic and polycyclic (polynuclear) aromatic hydrocarbons (i.e., those aromatic hydrocarbons heavier than naphthalene), these compounds are generally present at very low concentrations in the water-soluble fraction of oil. Therefore, the results of this study and others conclude that the soluble aromatics of crude oil (such as benzene, toluene, ethylbenzene, xylenes, and naphthalenes) produce the majority of its toxic effects in the environment.

Once the acutely toxic lighter compounds have left the aquatic environment through volatilization or degradation, the main concern is the chronic effects from heavier and more alkylated PAHs. Bird species with water habitats are the species most commonly affected by oil spills and releases. Oil itself breaks down the protective waxes and oils in the feathers and fur of birds and animals and disrupts the fine strand structure of the feathers, resulting in a loss of heat retention and buoyancy and possible hypothermia and death. Oiled birds often ingest petroleum while attempting to remove the petroleum from their feathers. The effects of ingested petroleum include anemia, pneumonia, kidney, and liver damage, decreased growth, altered blood chemistry, and decreased egg production and viability. Ingestion of oil can also kill animals by interfering with their ability to digest food. Chicks may be exposed to petroleum by ingesting food regurgitated by affected adults.

The dynamics of oil-in-water dispersion (OWD) are complex and have relevance related to potential toxicity or hazard. In comparing the toxicities to marine animals of oil-in-water dispersions prepared from different oils, not only the amount of oil added but also the concentrations of oil in the aqueous phase and the composition and dispersion-forming characteristics of the parent oil must be taken into consideration. In comparing the potential impacts of spills of different oils on the marine biotic community, the amount of oil per unit water volume required to cause mortality is of greater importance than any other aspect of the crude oil behavior.

Several compounds in petroleum products are carcinogenic. The larger and higher-molecular-weight aromatic structures (with four to five aromatic rings), which are more persistent in the environment, have the potential for chronic toxicological effects. Since these compounds are nonvolatile and are relatively insoluble in water, their principal routes of exposure are through ingestion and epidermal contact. Some of the compounds in this classification are considered possible human carcinogens; these include benzo[*a,e*]pyrene, benzo[*a*]anthracene, benzo[*b,j,k*]fluorene, benzo[*ghi*]perylene, chrysene, dibenzo[*ah*]anthracene, and pyrene.

Mixtures of polynuclear aromatic hydrocarbons are often carcinogenic and possibly phototoxic. One way to approach site-specific risk assessments would be to collect the complex mixture of polynuclear aromatic hydrocarbons and other lipophilic contaminants in a semipermeable membrane device (SPMD, also known as a *fat bag*), then test the mixture for carcinogenicity, toxicity, and phototoxicity.

The solubility of hydrocarbon components in petroleum products is an important property when assessing toxicity. The water solubility of a substance determines the routes of exposure that are possible. Solubility is approximately inversely proportional to molecular weight; lighter hydrocarbons are more soluble in water than higher-molecular-weight compounds. Lower-molecular-weight hydrocarbons (C_4 to C_8, including the aromatic compounds) are relatively soluble, up to about 2000 ppm, while the higher-molecular-weight hydrocarbons are nearly insoluble. Usually, the most soluble components are also the most toxic.

Finally, the toxicity of crude oil may be affected by factors such as "weathering" time or the addition of oil dispersants. *Weathered* crude oil and *fresh* crude oil may have different toxicities, depending on oil type and weathering time.

4.3.3. Total Petroleum Hydrocarbons

As a result of the wide variety of chemicals in petroleum and in petroleum products, it is not practical to measure each one separately. After an incident, it is more usual to measure the amount of total petroleum hydrocarbons at the site. The term *total petroleum hydrocarbons* (TPHs) is used environmentally to describe the family of several hundred chemical compounds that originally come from petroleum (Weisman, 1998).

Chemicals that may be included in the total petroleum hydrocarbons are hexane, heptane (and higher-molecular-weight homologs), benzene, toluene, xylenes (and the higher-molecular-weight homologs), and naphthalene as well as the constituents of other petroleum products, such as gasoline and diesel fuel. It is likely that samples of the total petroleum hydrocarbons collected at a specific site will contain only some, or a mixture, of these chemicals.

Petroleum hydrocarbons may enter the environment through accidents, from industrial releases, or as by-products from commercial or private uses such as direct release into water through spills or leaks. When release into water occurs, some of the hydrocarbons float on the water and form surface films, while others may sink and form bottom sediments. Bacteria and microorganisms in the water have the potential to break down some of the hydrocarbons over varying periods of time that are dependent on the ambient conditions. On the other hand, hydrocarbons that are spilled onto the soil may remain for a long time.

In addition, the amount of total petroleum hydrocarbons is the measurable parameter of petroleum-based hydrocarbon in an environmental medium, whether it is air, water, or land. It is thus dependent on analysis of the medium in which it is found, and since it is a measured, gross quantity without identification of its constituents, the total petroleum hydrocarbons data still represent a mixture. Thus, the data derived from measurement of the petroleum hydrocarbons in a particular environment is not a direct indicator of risk to humans or to the environment.

The data may even be the results from one of several analytical methods, some of which have been used for decades and others developed in the past several years.

Petroleum products themselves are the source of the many components but do not adequately define *total petroleum hydrocarbons*. However, the composition of petroleum products assist in understanding the hydrocarbons that become environmental contaminants, but any ultimate exposure is also determined by how the product changes with use, by the nature of the release, and by the hydrocarbon's environmental fate. When petroleum products are released into the environment, changes occur that affect their potential effects significantly. Physical, chemical, and biological processes change the location and concentration of hydrocarbons at any particular site.

Analysis for total petroleum hydrocarbons (EPA Method 418.1) provides a *one-number* value of the petroleum hydrocarbons in a given environmental medium. It does not, however, provide information on the composition (i.e., individual constituents) of the hydrocarbon mixture. The amount of hydrocarbon contaminants measured by this method depends on the ability of the solvent used to extract the hydrocarbon from the environmental media and the absorption of infrared light (infrared spectroscopy) by the hydrocarbons in the solvent extract. The method is not specific to hydrocarbons and does not always indicate petroleum contamination, since humic acid, a nonpetroleum material and a constituents of many soils, can be detected by this method.

Another analytical method commonly used for total petroleum hydrocarbons (EPA Method 8015 Modified) gives the concentration of purgeable and extractable hydrocarbons. These are sometimes referred to as gasoline range organics (GROs) and diesel range organics (DROs) because the boiling-point ranges of the hydrocarbon in each correspond roughly to those of gasoline (C_6 to C_{10-12}) and diesel fuel (C_{8-12} to C_{24-26}), respectively.

Purgeable hydrocarbons are measured by purge-and-trap gas chromatography (GC) analysis using a flame ionization detector, whereas the extractable hydrocarbons are extracted and concentrated prior to analysis. The results are most frequently reported as single numbers for purgeable and extractable hydrocarbons. Another method (based on EPA Method 801.5 Modified) gives a measure of the aromatic and aliphatic content of the hydrocarbon in each of several carbon number ranges (fractions).

An important feature of the analytical methods for the total petroleum hydrocarbons is the use of an *equivalent carbon number index* (EC). This index represents equivalent boiling points for hydrocarbons and is the physical characteristic that is the basis for separating petroleum (and other) components in chemical analysis.

4.3.4. Wastewater

A number of wastewater issues face the refining industry, including chemicals in waste process waters. However, efforts by the industry are being continued to eliminate any water contamination that may occur, whether it be from inadvertent leakage of petroleum or petroleum products or leakage of contaminated water from one or more processes. In addition to monitoring organics in the water, metals concentration must be monitored continually since heavy metals tend to concentrate in the body tissues of fish and animals and increase in concentration as they go up the food chain. General sewage problems face every municipal sewage treatment facility, regardless of size.

Primary treatment (solid settling and removal) is required, secondary treatment (use of bacteria and aeration to enhance organic degradation) is becoming more routine, and tertiary treatment (filtration through activated carbon, applications of ozone, and chlorination) has been, or is being, implemented by all refineries.

Wastewater pretreaters that discharge water into sewer systems have new requirements. Pollutant standards for sewage sludge have been set. Toxics in the water must be identified and plans must be developed to alleviate any problems. In addition, regulators have established, and continue to establish, water-quality standards for priority toxic pollutants.

REFERENCES

EPA. 1995. *Profile of the Petroleum Refining Industry*. Environmental Protection Agency, Washington, DC.

EPA. 2004. Environmental Protection Agency, Washington, DC. Web site: http://www.epa.gov.

Irwin, R. J. 1997. Petroleum. In *Environmental Contaminants Encyclopedia*. National Park Service, Water Resources Division, Water Operations Branch, Fort Collins, CO.

Meyers, R. A. 1997. *Handbook of Petroleum Refining Processes*, 2nd ed. McGraw-Hill, New York.

Speight, J. G. 1999. *The Chemistry and Technology of Petroleum*, 3rd ed. Marcel Dekker, New York.

Speight, J. G. 2001. *Handbook of Petroleum Analysis*. Wiley, New York.

Speight, J. G., and Ozum, B. 2002. *Petroleum Refining Processes*. Marcel Dekker, New York.

Weisman, W. (Ed.). 1998. *Analysis of Petroleum Hydrocarbons*, Environmental Media Total Petroleum Hydrocarbon Criteria Working Group Series. Amherst Scientific Publishers, Amherst, MA.

PART
II

ENVIRONMENTAL TECHNOLOGY AND ANALYSIS

CHAPTER 5

ENVIRONMENTAL REGULATIONS

Although it may be felt that environmental analysts must, by definition, focus on analysis of refinery products and wastes, knowledge of the various environmental regulations (Table 5.1) is always helpful in determining the analyses that must be performed.

The definitions of hazardous substances (40 CFR 300.5; CFR, 2004) and pollutants or contaminants (40 CFR 300.5; CFR, 2004) specifically exclude *petroleum, including crude oil or any fraction thereof* unless specifically listed. Although there is no definition of *petroleum* in Superfund, the Environmental Protection Agency interprets the *petroleum exclusion* provision to include petroleum and fractions of petroleum, including the hazardous substances, such as benzene, that are indigenous in petroleum and are therefore included in the term *petroleum*. The term also includes hazardous substances that are normally mixed with or added to crude oil or crude oil fractions during the refining process, including hazardous substances whose levels are increased during refining. These substances are also part of *petroleum* because their addition is part of the normal oil separation and processing operations at refineries that produce the product commonly understood to be petroleum. However, hazardous substances that are added to petroleum (e.g., mixing of solvents with used oil) or that increase in concentration solely as a result of contamination of the petroleum during use are not part of the petroleum and thus are not excluded from Superfund (Wagner, 1999; EPA, 2004).

Despite this exclusion, the refining industry has come under considerable strain because of several important factors and changes in the industry. Over the years, there has been an increased demand for petroleum products and a decrease in domestic production. However, there has been no new major refinery construction in the United States in the last three decades. This lack of infrastructure growth has caused a strain on the industry in meeting existing demand and has resulted in an increase in the amount of petroleum imports to meet the increasing need for liquid fuel.

Furthermore, as a result of the evolving environmental awareness, petroleum refinery operators face more stringent regulation of the treatment, storage, and disposal of hazardous wastes. Under recent regulations, a larger number of compounds have been, and are being, studied. Long-time methods of disposal, such as land farming of refinery waste, are being phased out. New regulations are becoming even more stringent, and they encompass a broader range of chemical constituents and processes.

Environmental Analysis and Technology for the Refining Industry, by James G. Speight
Copyright © 2005 John Wiley & Sons, Inc.

Table 5.1. Federal Regulations Relevant to Petroleum Refineries

Name	Code of Federal Regulation (CFR) Cite
CLEAN AIR ACT (CAA)	
New Source Performance Standards (NSPSs)	40 CFR Part 60
Subpart A: General Provisions	40 CFR Part 60
Subpart Cb: Designated Facilities—Existing Sulfuric Acid Units	40 CFR Part 60
Subpart D: Fossil-Fuel-Fired Steam Generators Constructed After 8/17/71	40 CFR Part 60
Subpart Da: Electric Utility Steam Generating Units Constructed After 9/18/78	40 CFR Part 60
Subpart Db: Industrial–Commercial–Institutional Steam Generating Units	40 CFR Part 60
Subpart Dc: Small Industrial–Commercial–Institutional Steam Generating Units	40 CFR Part 60
Subpart H: Sulfuric Acid Units	40 CFR Part 60
Subpart J: Petroleum Refineries	40 CFR Part 60
Subpart K: Storage Vessels for Petroleum Liquids Constructed, Reconstructed, or Modified between 6/11/73 and 5/19/78	40 CFR Part 60
Subpart Ka: Storage Vessels for Petroleum Liquids Constructed, Reconstructed, or Modified between 5/18/78 and 7/23/84	40 CFR Part 60
Subpart Kb: Volatile Organic Liquid Storage	40 CFR Part 60
Subpart GG: Stationary Gas Turbines	40 CFR Part 60
Subpart UU: Asphalt Processing and Roofing Manufacturing	40 CFR Part 60
Subpart VV: Equipment Leaks of VOC in the Synthetic Organic Chemicals Manufacturing Industry (SOCMI)	40 CFR Part 60
Subpart XX: Bulk Gasoline Terminals	40 CFR Part 60
Subpart GGG: Equipment Leaks of VOC in Petroleum Refineries	40 CFR Part 60
Subpart III: VOC Emissions for SOCMI Air Oxidation Unit Processes	40 CFR Part 60
Subpart NNN: VOC Emissions for SOCMI Distillation Processes	40 CFR Part 60
Subpart QQQ: VOC Emissions for Petroleum Refinery Wastewater Systems	40 CFR Part 60
Subpart RRR: SOCMI Reactor Processes	40 CFR Part 60
National Emission Standards for Hazardous Air Pollutants (NESHAPs)	
Subpart A: General Provisions	40 CFR Part 61
Subpart J/V: Equipment Leaks (Fugitive Emission Sources) of Benzene	40 CFR Part 61
Subpart M: Asbestos	40 CFR Part 61
Subpart Y: Benzene Emissions from Benzene Storage Vessels	40 CFR Part 61
Subpart BB: Benzene Emissions from Benzene Transfer Operations	40 CFR Part 61
Subpart FF: Benzene Waste Operations	40 CFR Part 61

Table 5.1. (*continued*)

Name	Code of Federal Regulation (CFR) Cite
NESHAPs for Source Categories	
Subpart A: General Provisions	40 CFR Part 63
Subpart B: Control Technology Determination	40 CFR Part 63
Subpart F: SOCMI	40 CFR Part 63
Subpart G: SOCMI Process Vents, Storage Vessels, Transfer Operations, and Wastewater	40 CFR Part 63
Subpart H: Equipment Leaks	40 CFR Part 63
Subpart I: NESHAP for Organic Hazardous Air Pollutants (HON); Certain Processes Subject to the Negotiated Regulation for Equipment Leaks	40 CFR Part 63
NESHAP for HON (partially under stay pending reconsideration for compressors, surge control vessels, and bottom receivers)	40 CFR Part 63
Subpart Q: Industrial Cooling Towers	40 CFR Part 63
Subpart R: Stage I Gasoline Distribution Facilities	40 CFR Part 63
Subpart T: Halogenated Solvent Cleansing (MACT)	40 CFR Part 63
Subpart Y: NESHAP for Marine Tank Vessel Loading and Unloading Operations (MACT)	40 CFR Part 9, 63
Subpart CC: NESHAP for Petroleum Refining, Phase I (MACT)	40 CFR Parts 9, 60, 63
Stack Height Provisions	40 CFR Part 51, Subpart G
Control Technology Guidelines (CTGs)	
Petroleum Liquid Storage in External Floating Roof Tanks	40 CFR Part 52
Petroleum Liquid Storage in Fixed Roof Tanks	40 CFR Part 52
Petroleum Refinery Equipment Leaks	40 CFR Part 52
Refinery Vacuum Producing Systems, Wastewater Separators, and Process Unit Turnarounds	40 CFR Part 52
SOCMI Air Oxidation Processes	40 CFR Part 52
SOCMI Distillation Operations and Reactor Processes	40 CFR Part 52
Tank Truck Gasoline Loading Terminals	40 CFR Part 52
Fuels	
Fuel and Fuel Additives	
Registration Requirements	40 CFR Part 79
Interim Requirements for Deposit Control Gasoline Additives	40 CFR Part 80
Reid Vapor Pressure Limitation	40 CFR Part 80
Oxygenated Fuel Requirement	40 CFR Part 80
Lead Phaseout	40 CFR Part 80
Reformulated Gasoline	40 CFR Part 80
Low-Sulfur Diesel	40 CFR Part 85
Permits	
State Operating Permit Program, Title V (Revised 8/29/94)	40 CFR Part 70
Prevention of Significant Deterioration (new sources in attainment areas) and New Source Review (new sources in nonattainment areas); LAER requirements (existing source)	40 CFR Part 52

(*continued overleaf*)

Table 5.1. (*continued*)

Name	Code of Federal Regulation (CFR) Cite
Stratospheric Ozone	40 CFR Part 82
Acid Rain Provisions	40 CFR Parts 72, 73, 75, 77, 78
Nitrogen Oxides Emission Reduction Program	40 CFR Part 76
CLEAN WATER ACT (CWA)	
Discharge of Oil: Notification Requirements	40 CFR Part 110
Designation of Hazardous Substances	40 CFR Part 116
Notice of Discharge of a Reportable Quantity	40 CFR Part 117
Spill Prevention, Control, and Countermeasures (SPCC) Requirements for Oil Storage	40 CFR Part 112
General Provisions for Effluent Guidelines and Standards	40 CFR Part 401
Toxic Pollutant Effluent Standards	40 CFR Part 129
Effluent Guidelines and Categorical Pretreatment Standards	40 CFR Part 419
Water Quality Standards for Toxic Pollutants	40 CFR Part 131
General National Pretreatment Standards	40 CFR Part 403
Great Lakes Water Quality Guidance	40 CFR Parts 9, 122, 123, 131, 132
NPDES	
Stormwater Application, Permit, and Reporting Requirements Associated with Industrial Activities	40 CFR Part 122
Permit	40 CFR Part 121–125
OIL POLLUTION ACT (OPA)	
Natural Resource Damage Assessments (NRDA) under National Oceanic and Atmospheric Administration	15 CFR Part 990
Response Plans for Marine Transportation-Related Facilities (interim final rule)	33 CFR Parts 150, 154
Oil Pollution Prevention; Non-Transportation-Related Onshore Facilities	40 CFR Part 9, 112
RESOURCE CONSERVATION AND RECOVERY ACT (RCRA)	
Nonhazardous Waste Requirements (Subtitle D)	40 CFR Parts 256, 257 (federal guidelines for state/local requirements)
Subtitle C Requirements	
General Requirements for Hazardous Waste Management	40 CFR Part 260
Identification and Listing of Hazardous Wastes and Toxicity Characteristics	40 CFR Part 261

Table 5.1. (*continued*)

Name	Code of Federal Regulation (CFR) Cite
Standards Applicable to Generators of Hazardous Wastes	
Subpart A: General Provisions	40 CFR Part 262
Subpart B: Shipping Manifest	40 CFR Part 262
Subpart C: Packaging, Labeling, Marking, and Placarding	40 CFR Part 262
Subpart D: Recordkeeping and Reporting	40 CFR Part 262
Subparts E & F: Exports and Imports	40 CFR Part 262
Subparts A & B: General Provisions and Facility Standards	40 CFR Part 264 (265)
Subparts C & D: Preparedness, Prevention and Emergency Plans	40 CFR Part 264 (265)
Subpart E: Recordkeeping/Reporting Requirements	40 CFR Part 264 (265)
Subpart F: Releases from Units	40 CFR Part 264
Subpart F: Groundwater Monitoring Requirements (interim status only)	40 CFR Part 265
Subpart G: Closure and Post-closure Requirements	40 CFR Part 264 (265)
Subpart H: Financial Responsibility Requirements	40 CFR Part 264 (265)
Subparts I, J, K, and L: Use and Management of Containers, Tank Systems, Surface Impoundments, and Waste Piles	40 CFR Part 264 (265)
Liners and Leak Detection for Hazardous Waste Land Disposal Units	40 CFR Part 264 (265)
Double Liners and Leachate Collection Systems for Hazardous Waste Disposal Units	40 CFR Parts 144, 264 (265)
Subparts M, N, and O: Land Treatment, Landfills, and Incinerators	40 CFR Part 264 (265)
Subpart S: Corrective Action	40 CFR Part 264 (265)
Subparts AA, BB, and CC: Air Emission Standards for Process Vents; Equipment Leaks; and Tanks, Surface Impoundments, and Containers	40 CFR Part 264 (265)
Phase I	40 CFR Part 264 (265)
Phase II	40 CFR Part 264 (265)
Standards for the Management of Specific Hazardous Wastes	40 CFR Part 266
Land Disposal Restrictions	40 CFR Part 268
Phase I: Contaminated Debris and Newly Identified Wastes, F037 and F038 Petroleum	40 CFR Part 148, 268
Phase II: Set Treatment Standards (BDAT) for TC Wastes and Establish Universal Treatment Standards	40 CFR Part 148, 268

(*continued overleaf*)

Table 5.1. (*continued*)

Name	Code of Federal Regulation (CFR) Cite
Permits	40 CFR Parts 270, 271, 272
Standards for the Management of Used Oil: Used Oil Destined for Recycling	40 CFR Part 279
Underground Storage Tanks: Technical Standards and Corrective Action	40 CFR Part 280
SAFE DRINKING WATER ACT (SDWA)	
Underground Injection Control Regulations	40 CFR Parts 144, 146
SUPERFUND (CERCLA)	
Natural Resource Damage Assessments (also under CWA)	43 CFR Part 11
Reportable Quantities Releases (Notification to National Response Center)	40 CFR Part 302
Extremely Hazardous Substances (EHSs) Emergency Planning	40 CFR Part 355
EHS Release Notification (Notification to State Emergency Response Commission, Local Emergency Response Commission) and Follow-up	60 CFR Part 355
Community Right-to-Know	
Hazardous Chemicals (Material Safety Data Sheet Chemicals) Inventory Reporting	40 CFR Part 370
Toxic Chemical Release Reporting	40 CFR Part 372
Expansion of TRI List	40 CFR Part 372
TOXIC SUBSTANCES CONTROL ACT (TSCA)	
General Provisions	40 CFR Part 702
Reporting and Recordkeeping Requirements	40 CFR Parts 704, 710
Chemical Information Rule	40 CFR Part 712
Health and Safety Data Reporting	40 CFR Part 716
Premanufacture Notification (and Exemptions)	40 CFR Parts 720 (723)
Significant New Uses	40 CFR Part 721
Chromium Comfort D Cooling Towers	40 CFR Part 749
Rules for Controlling Polychlorinated Biphenyls	40 CFR Part 761
Asbestos-Containing Products Labeling Requirements	40 CFR Part 763

However, it is not our intent in this chapter to enter into a political discussion of the levy of fines for infringement of the environmental laws. Our purpose is to provide the reader with an overview of a selection of the many and varied regulations that shows the types of emissions from refinery processes and the laws that regulate these emissions.

5.1. ENVIRONMENTAL IMPACT OF REFINING

Petroleum refining is one of the largest industries in the United States, and potential environmental hazards associated with refineries have caused increased concern for communities in close proximity to them. This update provides a general overview of the processes involved and some of the potential environmental hazards associated with petroleum refineries (see also Chapter 3).

Briefly, petroleum refining involves a series of steps, including separation and blending of petroleum products. The five major processes are described briefly below.

1. *Separation processes.* These processes involve separating the different constituents into fractions based on their boiling-point differences. Additional processing of these fractions is usually needed to produce final products to be sold within the market.
2. *Conversion processes.* Coking and cracking are conversion processes used to break down higher-molecular-weight constituents into lower-molecular-weight product by heating and by use of catalysts.
3. *Treating.* Petroleum-treating processes are used to remove the undesirable components and impurities such as sulfur, nitrogen, and heavy metals from the products. This involves processes such as hydrotreating, deasphalting, acid gas removal, desalting, hydrodesulfurization, and sweetening.
4. *Blending/combination processes.* Refineries use blending/combination processes to create mixtures with the various petroleum fractions to produce a desired final product, such as gasoline with various octane ratings.
5. *Auxiliary processes.* Refineries also have other processes and units that are vital to operations by providing power, waste treatment, and other utility services, such as boilers, wastewater treatment, and cooling towers. Products from these facilities are usually recycled and used in other processes within the refinery and are also important with regard to minimizing water and air pollution.

Refineries are generally considered a major source of pollutants in areas where they are located and are regulated by a number of environmental laws related to air, land, and water (Table 5.1). Thus, refineries are generally considered a major source of pollutants in areas where they are located and are regulated by a number of environmental laws related to air, land, and water.

5.1.1. Air Pollution

Petroleum refineries are a source of hazardous and toxic air pollutants, such as BTEX compounds (benzene, toluene, ethylbenzene, and xylene). They are also a major source of criteria air pollutants: particulate matter (PM), nitrogen oxides (NO_x), carbon monoxide (CO), hydrogen sulfide (H_2S), and sulfur oxides (SO_x).

Refineries also release less toxic hydrocarbons, such as natural gas (methane) and other light volatile fuels and oils.

Air emissions can come from a number of sources within a petroleum refinery, including equipment leaks (from valves or other devices), high-temperature combustion processes in the actual burning of fuels for electricity generation, the heating of steam and process fluids, and the transfer of products. These pollutants are typically emitted into the environment over the course of a year through normal emissions, fugitive releases, accidental releases, or plant upsets. The combination of volatile hydrocarbons and oxides of nitrogen also contribute to ozone formation, one of the most important air pollution problems.

5.1.2. Water Pollution

Refineries are also potential contributors to groundwater and surface water contamination. Some refineries use deep-injection wells to dispose of wastewater generated inside the plants, and some of these wastes end up in aquifers and groundwater. These wastes are then regulated under the Safe Drinking Water Act (SDWA). Wastewater in refineries may be highly contaminated and may arise from various processes (such as wastewaters from desalting, water from cooling towers, stormwater, distillation, or cracking). This water is recycled through many stages during the refining process and goes through several treatment processes, including a wastewater treatment plant, before being released into surface waters.

The wastes discharged into surface waters are subject to state discharge regulations and are regulated under the Clean Water Act (CWA). These discharge guidelines limit the amounts of sulfides, ammonia, suspended solids, and other compounds that may be present in the wastewater. Although these guidelines are in place, contamination from past discharges may remain in surface-water bodies.

5.1.3. Soil Pollution

Contamination of soils from the refining processes is generally a less significant problem than that of contamination of air and water. Past production practices may have led to spills on the refinery property that now need to be cleaned up. Natural bacteria that may use the petroleum products as food are often effective at cleaning up petroleum spills and leaks compared to many other pollutants. Many residuals are produced during the refining processes, and some of them are recycled through other stages in the process. Other residuals are collected and disposed of in landfills, or they may be recovered by other facilities. Soil contamination including some hazardous wastes, spent catalysts or coke dust, tank bottoms, and sludge from the treatment processes can occur from leaks as well as accidents or spills on- or off-site during the transport process.

5.2. ENVIRONMENTAL REGULATIONS IN THE UNITED STATES

The toxic chemicals found within the refining industry are not necessarily unique, and although general air pollution, water pollution, and land pollution controls are affected by petroleum refining, these problems and solutions are not unique to the refining industry. In fact, because the issues are so diverse, the refining industry (because a refinery is an industrial complex consisting of many integrated unit processes) may be looked upon as a series of complex pollution issues, each unique to the unit processes from which the effluent originates. Therefore, there may be many examples of laws and controls that have been enacted by governments with input from refiners that address pollution prevention and control of hazardous materials.

Pollution prevention and control of hazardous materials is an issue not only for the petroleum industry but also for many industries and has been an issue for decades (Table 5.2). In this context, there are specific definitions for terms such as *hazardous substances, toxic substances*, and *hazardous waste* (Chapter 1). These are all terms of art and must be fully understood in the context of their statutory or regulatory meanings and not merely limited to their plain English or dictionary meanings. It is absolutely imperative from a legal sense that each statute or regulation promulgated be read in conjunction with terms defined in that specific statute or regulation (Majumdar, 1993).

To combat any threat to the environment, it is necessary to understand the nature and magnitude of the problems involved. It is in such situations that environmental technology has a major role to play. Environmental issues even arise when outdated laws are taken to task. Thus, the concept of what seemed to be a good idea at the time the action occurred no longer holds when the law influences the environment.

Finally, it is worthy note that regulatory disincentives to voluntary reductions of emissions from petroleum refineries also exist. Many environmental statutes define a baseline period and measure progress in pollution reductions from that baseline. Any reduction in emissions before it is required could lower a facility's baseline emissions. Consequently, future regulations requiring a specified reduction from the baseline could be more difficult (and consequently, more costly) to achieve because the most easily applied and hence the most cost-effective reductions would already have been made. With no credit given for voluntary reductions, those facilities that do the minimum may be in fact be rewarded when emissions reductions are required.

5.2.1. Clean Air Act

The Clean Air Act Amendments (CAAA) of 1990 have made significant changes in the basic Clean Air Act enacted in 1970. The Clean Air Act allowed the establishment of air quality standards and provisions for their implementation and enforcement. This law was strengthened in 1977, and the Clean Air Act Amendments of 1990 imposed many new standards that included controls for industrial pollutants.

The Clean Air Act of 1970 and the 1977 amendments that followed consist of three titles. Title I deals with stationary air emission sources, Title II deals with mobile air emission sources, and Title III includes definitions of appropriate terms, provisions for citizen suits, and applicable standards for judicial review.

Table 5.2. Chronology of Environmental Events and Regulations in the United States (Not Necessarily Related to the Petroleum Refining Industry)

1906	The Pure Food and Drug Act established the Food and Drug Administration (FDA), which now oversees the manufacture and use of all foods, food additives, and drugs; amendments (1938, 1958, and 1962) strengthened the law considerably.
1924	The Oil Pollution Act.
1935	The Chemical Manufacturers' Association (CMA), a private group of people working in the chemical industry and involved especially in the manufacture and selling of chemicals, established a Water Resources Committee to study the effects of their products on water quality.
1948	The Chemical Manufacturers' Association established an Air Quality Committee to study methods of improving the air that could be implemented by chemical manufacturers.
1953	The Delaney Amendment to the Food and Drug Act defined and controlled food additives; any additives showing an increase in cancer tumors in rats, even if extremely large doses were used in the animal studies, had to be outlawed in foods; recent debates have focused on a number of additives, including the artificial sweetener cyclamate.
1959	Just before Thanksgiving the government announced that it had destroyed cranberries contaminated with a chemical, aminotriazole, that produced cancer in rats; the cranberries were from a lot frozen from two years earlier when the chemical was still an approved weed killer.
1960	Diethylstilbestrol (DES), taken in the late 1950s and early 1960s to prevent miscarriages and also used as an animal fattener, was reported to cause vaginal cancer in the daughters of these women as well as premature deliveries, miscarriages, and infertility.
1962	Thalidomide, a prescription drug used as a tranquilizer and flu medicine for pregnant women in Europe to replace dangerous barbiturates that cause 2000 to 3000 deaths per year by overdoses, was found to cause birth defects. Thalidomide had been kept off the market in the United States because of the insistence that more safety data be produced for the drug.
1962	The Kefauver–Harris Amendment to the Food and Drug Act required that drugs be proven safe before put on the market.
1962	The publication of *Silent Spring* (authored by Rachel Carson), which outlined many environmental problems associated with chlorinated pesticides, caused a ban on the use of DDT in 1972.
1965	Nonlinear, nonbiodegradable synthetic detergents made from propylene tetramer were banned after these materials were found in large amounts in rivers, so much as to cause soapy foam in many locations. Phosphates in detergents were banned in detergents by many states in the 1970s.

Table 5.2. (*continued*)

1965	Mercury poisoning from concentration in the food chain was recognized.
1966	Polychlorinated biphenyls (PCBs) were first found in the environment and in contaminated fish; banned in 1978 except in closed systems.
1968	TCDD (a dioxin derivative) tested positive as a teratogen in rats.
1969	The artificial sweetener cyclamate was banned because of its link to bladder cancer in rats fed with large doses; 20 subsequent studies have failed to confirm this result, but cyclamate remains banned.
1970	Earth Day recognized because of concern about the effects of many substances on the environment.
1970	The Clean Air.
1971	TCDD (see above) outlawed by the Environmental Protection Agency.
1971	The Chemical Manufacturers' Association established the Chemical Emergency Transportation System (CHEMTREC) to provide immediate information on chemical transportation emergencies.
1972	Federal Water Pollution Control Act.
1972	The Clean Water Act.
1974	Safe Drinking Water Act.
1974	Vinyl chloride investigated as a possible carcinogen.
1976	The Toxic Substances Control Act (TSCA or TOSCA); Environmental Protection Agency developed rules to limit manufacture and use of PCBs.
1976	The Resource Conservation and Recovery Act (RCRA).
1977	Saccharin found to cause cancer in rats; banned by the FDA temporarily, but Congress placed a moratorium on this ban because of public pressure; saccharin is still available.
1977	Dibromochloropropane (DBCP) investigated for causes leading to sterility; now banned.
1977	Benzene was linked to an abnormally high rate of leukemia; increased concern with benzene use in industry.
1978	Ban on chlorofluorocarbons (CFCs) as aerosol propellants; react with ozone in the stratosphere, causing an increase in the penetration of ultraviolet sunlight and increase the risk of skin cancer.
1978	Love Canal, Niagara Falls, New York.
1980	CHEMTREC (see above) recognized by the Department of Transportation as the central service to provide immediate information on chemical transportation emergencies.
1980	The Comprehensive Environmental Response, Compensation, and Liability Act.
1986	The Safe Drinking Water Act Amendments.

(*continued overleaf*)

Table 5.2. (*continued*)

1986	The Emergency Planning and Community-Right-to-Know Act; companies must also report inventories of specific chemicals kept in the workplace and annual release of hazardous materials into the environment.
1986	The Superfund Amendments and Reauthorization Act.
1989	Pasadena, Texas: explosion caused by leakage of ethylene and isobutane from a pipeline.
1990	Channelview, Texas: explosion in a petrochemical treatment tank of wastewater and chemicals.
1991	Sterlington, Louisiana: explosion at a nitro-paraffin plant.
1991	Charleston, South Carolina: explosion at a plant manufacturing Antiblaze 19, a phosphonate ester and flame retardant used in textiles and polyurethane foam; manufactured from trimethyl phosphite, dimethyl methylphosphonate, and trimethyl phosphate.

Source: Noyes, 1993.

However, in contrast to the previous clean air statutes, the 1990 amendments contained extensive provisions for control of the accidental release of air toxics from storage or transportation (TPG, 1995) as well as the formation of acid rain. At the same time, the 1990 amendments provided new and added requirements for such original ideas as state implementation plans for attainment of the national ambient air quality standards and permitting requirements for the attainment and nonattainment areas. Title III now calls for a vastly expanded program to regulate *hazardous air pollutants* (HAPs) or *air toxics*.

Under the Clean Air Act Amendments of 1990, the mandate is to establish, during the first phase, technology-based maximum achievable control technology (MACT) emission standards that apply to the major categories or subcategories of sources of the listed hazardous air pollutants (EPA, 1997). In addition, Title III provides for health-based standards that address the issue of residual risks due to air toxic emissions from the sources equipped with MACT and to determine whether the MACT standards can protect health with an *ample margin of safety*.

Section 112 of the original Clean Air Act that dealt with hazardous air pollutants has been greatly expanded by the 1990 amendments. The list of hazardous air pollutants has been increased manyfold. In addition, the standards for emission control have been tightened and raised to a very high level, referred to as the *best of the best*, in order to reduce the risk of exposure to various hazardous air pollutants.

Thus, the Clean Air Act Amendments of 1990 aimed to encourage voluntary reductions above the regulatory requirements by allowing facilities to obtain emission credits for voluntary reductions in emissions. These credits would serve as offsets against any potential future facility modifications, resulting in an increase in emissions. Other regulations established by the amendments, however, will require the construction of major new units within existing refineries to produce reformulated fuels. These new operations will require emission offsets in order to be permitted. This will consume many of the credits available for existing facility modifications. A shortage of credits for facility modifications will make it difficult to receive credits for emission reductions through pollution prevention projects.

Thus, under this Clean Air Act, the EPA sets limits on how much of a pollutant can be in the air anywhere in the United States. The law does allow individual states to have stronger pollution controls, but states are not allowed to have weaker pollution controls than those set for the entire country. The law recognizes that it makes sense for states to take the lead in carrying out the Clean Air Act, because pollution control problems often require special understanding of local industries and geography as well as housing developments near to industrial sites.

Individual states are required to develop state implementation plans (SIPs), a collection of the regulations that a state will use to clean up polluted areas. The states must involve the public, through hearings and opportunities to comment, in the development of each state implementation plan. The EPA must approve each plan, and if a state implementation is not acceptable, the EPA can take over, enforcing the Clean Air Act in that state.

Air pollution often travels from its source in one state to another state. In many metropolitan areas, people live in one state and work or shop in another; air pollution from cars and trucks may spread throughout the interstate area. The Clean Air Act Amendments of 1990 provide for interstate commissions on air pollution control, which are to develop regional strategies for cleaning up air pollution. The 1990 amendments also cover pollution that originates in nearby countries, such as Mexico and Canada, and drifts into the United States as well as pollution from the United States that reaches Canada and Mexico.

In the current context, the 1990 amendments provide economic incentives for cleaning up pollution. For instance, refiners can get credits if they produce cleaner gasoline than required, and the credits when the gasoline does not quite meet cleanup requirements.

Furthermore, gasoline (like many industrial products) contains toxic chemicals that can eventually contribute to smog formation. To combat such effects, refiners have started to reformulate gasoline sold in the smoggiest areas. This gasoline contains smaller quantities of volatile organic compounds (VOCs) such as benzene (which is also a hazardous air pollutant that causes cancer and aplastic anemia, a potentially fatal blood disease). Gasoline also contains detergents, which, by preventing the buildup of engine deposits, keep engines working smoothly and burning fuel cleanly.

5.2.2. Resource Conservation and Recovery Act

The hazardous waste regulatory program as we know it today began with the Resource Conservation and Recovery Act (RCRA) in 1976. Since its enactment in 1976, the RCRA has been amended several times, to promote safer solid and hazardous waste management programs (Dennison, 1993). The Used Oil Recycling Act of 1980 and the Hazardous and Solid Waste Amendments of 1984 (HSWA) were the major amendments to the original law.

The 1984 amendments also brought the owners and operators of underground storage tanks under the RCRA umbrella. This can have a significant effect on refineries that store products in underground tanks. Now, in addition to the hazardous waste being controlled, RCRA Subtitle I regulates petroleum products.

The RCRA controls the disposal of solid waste and requires that all wastes destined for land disposal be evaluated for their potential hazard to the environment. Solid waste includes liquids, solids, and containerized gases and is divided into nonhazardous waste and hazardous waste. The various amendments are aimed at preventing the disposal problems that lead to a need for the Comprehensive Environmental Response Compensation and Liability Act (CERCLA), or *Superfund*, as it is known.

Subtitle C of the original RCRA lists the requirements for the management of hazardous waste. This includes criteria for identifying hazardous waste and the standards for generators, transporters, and companies that treat, store, or dispose of the waste. The RCRA regulations also provide standards for design and operation of such facilities. However, before any action under RCRA is planned, it is essential to understand what constitutes a solid waste and what constitutes a hazardous waste. The first step to be taken by a generator of waste is to determine whether that waste is hazardous. Waste may be hazardous by being listed in the regulations or by meeting any of the four characteristics: ignitability, corrosivity, reactivity, and extraction procedure (EP) toxicity.

Section 1004 (27) of RCRA defines *solid waste* as garbage, refuse, or sludge from a waste treatment plant, water supply treatment plant, or air pollution control facility and other discarded material, including solid, liquid, semisolid, or contained gaseous material resulting from industrial, commercial, mining, and agricultural operations and from community activities, but does not include solid or dissolved materials in domestic sewage, or solid or dissolved materials in irrigation return flows or industrial discharges, which are point sources subject to permits under Section 402 of the Federal Water Pollution Control Act, as amended (86 Stat. 880), or source, special nuclear, or by-product material as defined by the Atomic Energy Act of 1954, as amended (68 Stat. 923).

This statutory definition of solid waste is pursuant to the regulations of the EPA insofar as a solid waste is a hazardous waste if it exhibits any one of four specific characteristics: ignitability, reactivity, corrosivity, and toxicity.

1. *Ignitability.* A waste is an ignitable hazardous waste if it has a flash point of less than 140°F (40 CFR 261.21) as determined by the Pensky–Martens closed cup flash point test; readily causes fires and burns so vigorously as to create a hazard; or is an ignitable compressed gas or an oxidizer (as defined by the Department of Transport regulations). A simple method of determining the flash point of a waste is to review the material. Ignitable wastes carry the waste code D1001. Naphtha is an example of an ignitable hazardous waste.

2. *Corrosivity.* A liquid waste that has a pH of less than or equal to 2 or greater than or equal to 12.5 is considered to be a corrosive hazardous waste (40 CFR 261.22). Sodium hydroxide, a caustic solution with a high pH, is often used

by the refining industry in the form of a caustic wash to remove sulfur compounds or acid gases. When these caustic solutions become contaminated and must be disposed of, the waste would be a corrosive hazardous waste. Corrosive wastes carry the waste code D002. Acid solutions also fall under this category.

3. *Reactivity.* A material is considered to be a reactive hazardous waste if it is normally unstable, reacts violently with water, generates toxic gases when exposed to water or corrosive materials, or if it is capable of detonation or explosion when exposed to heat or a flame (40 CFR 261.23). Materials that are defined as forbidden explosives or class A or B explosives by the Department of Transportation are also considered reactive hazardous waste.

4. *Toxicity.* The fourth characteristic that could make a waste a hazardous waste is toxicity (40 CFR 261.24). To determine if a waste is a toxic hazardous waste, a representative sample of the material must be subjected to a test conducted in a certified laboratory using a test procedure [toxicity characteristic leaching procedure, (TCLP)]. Under federal rules (40 CFR 261), all generators are required to use the toxicity characteristic leaching procedure test when evaluating wastes.

Wastes that fail a toxicity characteristic test are considered hazardous under the RCRA. There is less incentive for a refinery to attempt to reduce the toxicity of such waste below the toxicity characteristic levels because even though such toxicity reductions may render the waste nonhazardous, it may still have to comply with new land disposal treatment standards under Subtitle C of the RCRA before disposal. Similarly, there is little positive incentive to reduce the toxicity of listed refinery hazardous wastes because, once listed, the waste is subject to Subtitle C regulations without regard to how much the toxicity levels are reduced.

In addition to the four characteristics of hazardous wastes, the EPA has established three hazardous waste lists: hazardous wastes from nonspecific sources (e.g., spent nonhalogenated solvents), hazardous wastes from specific sources (e.g., bottom sediment sludge from the treatment of wastewaters from wood preserving), and discarded commercial chemical products and off-specification species, containers, and spill residues.

Under regulations of the EPA, certain types of solid wastes (e.g., household waste) are not considered to be hazardous wastes, irrespective of their characteristics. Additionally, the EPA has provided certain regulatory exemptions based on very specific criteria. For example, hazardous waste generated in a product or raw material storage tank, transport vehicle, or manufacturing processes and samples collected for monitoring and testing purposes are exempt from the regulations.

Under the RCRA, the hazardous waste management program is based on a *cradle-to-grave* concept, so that all hazardous wastes can be traced and accounted for fully. Section 3010(a) of the act requires all generators and transporters of hazardous wastes as well as owners and operators of all TSD facilities to file a notification with the EPA within 90 days after the promulgation of the regulations. The notification should state the location of the facility and include a general description of the activities as well as those hazardous wastes being handled that have been identified and listed.

Submission of the Part A permit application for existing facilities prior to November 19, 1980, qualified a refinery for interim status. This meant that the refinery was allowed to continue operation according to certain regulations during the permitting process.

The Hazardous and Solid Waste Amendments (HSWA) represented a strong bias against land disposal of hazardous waste. Some of the provisions that affect refineries are:

1. A ban on the disposal of bulk or noncontainerized liquids in landfills. The prohibition also bans solidification of liquids using absorbent material, including absorbents used for spill cleanup.
2. Five hazardous wastes from specific sources in the refining industry come under schedules for disposal prohibition and/or treatment standards. These five are: dissolved air flotation (DAF) float, slop oil emulsion solids, heat exchanger bundle cleaning sludge, API separator sludge, and leaded tank bottoms (EPA waste numbers K047 to K051).
3. Petroleum refineries must retrofit surface impoundments that are used for hazardous waste management. Retrofitting must involve the use of double liners and leak detection systems.

Under the RCRA, the EPA has the authority to require a refinery to clean up releases of hazardous waste or waste constituents. The regulation provides for cleanup of hazardous waste released from active treatment, storage, and disposal facilities. Superfund was expected to handle contamination that had occurred before that date.

5.2.3. Clean Water Act

The Clean Water Act (CWA; or the Water Pollution Control Act) is the cornerstone of surface water quality protection in the United States and employs a variety of regulatory and nonregulatory tools to sharply reduce direct pollutant discharges into waterways and manage polluted runoff. The objective of the Clean Water Act is to restore and maintain the chemical, physical, and biological integrity of water systems.

The Clean Water Act (CWA) established the basic structure for regulating discharges of pollutants into the waters of the United States and gave the EPA the authority to implement pollution control programs such as setting wastewater standards for industry. The CWA also continued requirements to set water quality standards for all contaminants in surface waters. The act is credited with the first comprehensive program for controlling and abating water pollution.

The federal clean water statute makes a distinction between conventional and toxic pollutants. As a result, two standards of treatment are required prior to their discharge into the navigable waters of the nation. For conventional pollutants that generally include degradable nontoxic organic compounds and inorganic compounds, the applicable treatment standard is the *best conventional technology* (BCT). For toxic pollutants, on the other hand, the required treatment standard is best available technology (BAT), which is a higher standard than BCT.

The statutory provisions of the CWA have five major sections that deal with specific issues: (1) nationwide water quality standards; (2) effluent standards from certain specific industries; (3) permit programs for discharges into receiving water bodies based on the National Pollutant Discharge Elimination System (NPDES); (4) discharge of toxic chemicals, including oil spills; and (5) construction grant program for publicly owned treatment works (POTWs).

In addition, Section 311 of the CWA includes elaborate provisions for regulating intentional or accidental discharges of oil and hazardous substances. Included there are response actions required for oil spills and the release or discharge of toxic and hazardous substances. Pursuant to this, certain elements and compounds are designated as hazardous substances and an appropriate list has been developed (40 CFR 116.4). The person in charge of a vessel or an onshore or offshore facility from which any designated hazardous substances is discharged, in quantities equal to or exceeding its reportable quality, must notify the appropriate federal agency as soon as such knowledge is obtained. Such notice should be provided in accordance with the designated procedures (33 CFR 153.203).

Under the CWA, the discharge of waterborne pollutants is limited by National Pollutant Discharge Elimination System (NPDES) permits. Refineries that easily meet their permit requirements will often have their permit limits changed to lower values. Because occasional system upsets do occur, resulting in significant excursions above the normal performance values, refineries feel that they must maintain a large operating margin below the permit limits to ensure continuous compliance. Those refineries that can significantly reduce waterborne emissions may find the risk of having their permit limits lowered to be a substantial disincentive.

5.2.4. Safe Drinking Water Act

The Safe Drinking Water Act (SDWA), enacted in 1974 to assure high-quality water supplies through public water system. The act is truly the first federal intervention to set the limits of contaminants in drinking water. The 1986 amendments came two years after passage of the Hazardous and Solid Waste Amendments (HSWA) or the RCRA amendments of 1984. As a result, certain statutory provisions were added to these 1986 amendments to reflect the changes made in the underground injection control (UIC) systems.

In addition, the Superfund Amendments and Reauthorization Act (SARA) of 1986 set the groundwater standards the same as the drinking water standards for the purpose of necessary cleanup and remediation of an inactive hazardous waste disposal site. The 1986 amendments of the SDWA included additional elements to establish maximum contaminant-level goals (MCLGs) and national primary drinking water standards. The MCLGs must be set at a level at which no known or anticipated adverse effects on human health occur, thus providing an *adequate margin of safety*. Establishment of a specific MCLG depends on the evidence of carcinogenicity in drinking water or a reference dose that is calculated individually for each contaminant. The MCLG, an enforceable standard, however, must be set to operate as the nation primary drinking water standard (NPDWS).

The Safe Drinking Water Act calls for regulations that (1) apply to public water systems, (2) specify contaminants that may have any adverse effect on the health of persons, and (3) specify contaminant levels. The difference between primary and secondary drinking water regulations are defined as well as other applicable terms. Information concerning national drinking water regulations and the protection of underground sources of drinking water is given.

In the context of the petroleum industry, the priority list of drinking water contaminants is very important since it includes the contaminants known for their adverse effect on public health. Furthermore, most if not all are known or suspected to have hazardous or toxic characteristics that can compromise human health.

5.2.5. Comprehensive Environmental Response, Compensation, and Liability Act

The Comprehensive Environmental Response, Compensation, and Liability Act (CERCLA), commonly known as Superfund, 1980, created a tax on the chemical and petroleum industries and provided broad federal authority to respond directly to releases or threatened releases of hazardous substances that may endanger public health or the environment. The act was amended by the Superfund Amendments and Reauthorization Act (SARA) in 1986 and stressed the importance of permanent remedies and innovative treatment technologies in cleaning up hazardous waste sites.

A CERCLA response or liability will be triggered by an actual release or the threat of a *hazardous substance or pollutant or contaminant* being released into the environment. A hazardous substance [CERCLA 101(14)] is any substance requiring special consideration due to its toxic nature under the Clean Air Act, Clean Water Act, or Toxic Substances Control Act (TSCA) and as defined under the RCRA. Additionally [CERCLA 101(33)], a pollutant or contaminant can be any other substance not necessarily designated or listed that "will or may reasonably" be anticipated to cause any adverse effect in organisms and/or their offspring.

The central purpose of CERCLA is to provide a response mechanism for cleanup of any hazardous substance released, such as an accidental spill, or of a threatened release of a hazardous substance (Nordin et al., 1995). Section 102 of CERCLA is a catchall provision because it requires regulations to establish *that quantity of any hazardous substance the release of which shall be reported pursuant to Section 103* of CERCLA. Thus, under CERCLA, the list of potentially responsible parties (PRPs) can include all direct and indirect culpable parties who have either released a hazardous substance or violated any statutory provision. In addition, responsible private parties are liable for cleanup actions and/or costs as well as for reporting requirements for an actual or potential release of a hazardous substance, pollutant, or contaminant.

CERCLA (Superfund) legislation deals with actual or potential releases of hazardous materials that have the potential to endanger people or the surrounding environment at uncontrolled or abandoned hazardous waste sites. The act requires responsible parties or the government to clean up waste sites. Among CERCLA's major purposes are the following: (1) site identification, (2) evaluation of danger from waste sites, (3) evaluation of damages to natural resources, (4) monitoring of release of hazardous substances from sites, and (5) removal or cleanup of wastes by responsible parties or government.

The Superfund Amendments and Reauthorization Act (SARA) addresses closed hazardous waste disposal sites that may release hazardous substances into any environmental medium. Title III of SARA also requires regular review of emergency systems for monitoring, detecting, and preventing releases of extremely hazardous substances at facilities that produce, use, or store such substances.

The most revolutionary part of SARA is the Emergency Planning and Community Right-to-Know Act (EPCRA), which is covered under Title III of SARA. EPCRA includes three subtitles and four major parts: emergency planning, emergency release notification, hazardous chemical reporting, and toxic chemical release reporting. Subtitle A is a framework for emergency planning and release notification. Subtitle B deals with various reporting requirements for *hazardous chemicals* and *toxic chemicals*. Subtitle C provides various dimensions of civil, criminal, and administrative penalties for violations of specific statutory requirements.

Other provisions of SARA basically reinforce and/or broaden the basic statutory program dealing with the releases of hazardous substances (CERCLA, Section 313). It requires owners and operators of certain facilities that manufacture, process, or otherwise use one of the listed chemicals and chemical categories to report all environmental releases of these chemicals annually. This information about total annual releases of chemicals from the industrial facilities can be made available to the public.

5.2.6. Oil Pollution Act

The Oil Pollution Act of 1924 was the first federal statute prohibiting pollution of waters strictly by oil. As indicated earlier, the Federal Water Pollution Control Act (FWPCA) of 1972 provided a comprehensive plan for the cleanup of waters polluted by oil spills and intentional or accidental release of oil into the water. The subsequent laws, including the Clean Water Act of 1977 and with its later amendments, provide for regulation of pollution of waters by oil spills and other forms of discharges. These legislations also incorporate certain provisions of the Rivers and Harbors Act of 1899, which was intended to prevent any obstruction to the use of navigable waters for interstate commerce.

The Oil Pollution Act (OPA) of 1990 streamlined and strengthened the EPA's ability to prevent and respond to catastrophic oil spills. A trust fund financed by a tax on oil is available to clean up spills when the responsible party is incapable or unwilling to do so. The act requires oil storage facilities and vessels to submit to the federal government plans detailing how they will respond to large discharges. The act also requires the development of area contingency plans to prepare and plan for oil spill response on a regional scale.

5.2.7. Occupational Safety and Health Act

The objective of the OSHA Hazard Communication Standard is to inform workers of potentially dangerous substances in the workplace and to train them on how to protect themselves against potential dangers.

OSHA is entrusted with the major responsibility for workplace safety and workers' health (Wang, 1994). It is responsible for the means by which chemicals are contained (TPG, 1995) through the inspection of workplaces to ensure compliance and enforcement of applicable standards under OSHA. It is also the means by which guidelines have evolved for the destruction of chemicals used in chemical laboratories (Lunn and Sansone, 1994; Studt, 1995).

The goal of OSHA is to ensure that "no employee will suffer material impairment of health or functional capacity," due to a lifetime occupational exposure to chemicals and hazardous substances. The statute imposes a duty on employers to provide employees with a safe workplace environment, free of known hazards that may cause death or serious bodily injury.

The statute covers all employers and their employees in all the states and federal territories, with certain exceptions. Generally, the statute does not cover self-employed persons, farms employing solely family members, and those workplaces covered under other federal statutes. Refiners must evaluate whether the chemicals they manufacture and sell are hazardous. Under the General Duty Clause of OSHA, employers are required to provide an environment that is free from recognized hazards that could cause physical harm or death.

All employers are required to develop, implement, and maintain at the workplace a written hazard communication program. The program must include the following components: (1) a list of hazardous chemicals in the workplace, (2) the methods the employer will use to inform employees of the hazards associated with these chemicals, and (3) a description of how the labeling, material safety data sheet (MSDS), and employee training requirements will be met.

The following information must be included in the program for employers who produce, use, or store hazardous chemicals in the workplace: (1) the means by which MSDSs will be made available to outside contractors for each hazardous chemical, (2) the means by which the employer will inform outside contractors of precautions necessary to protect the contractor's employees both during normal operating conditions and in foreseeable emergencies, and (3) the methods that the employer will use to inform contractors of the labeling system used in the workplace.

5.2.8. Toxic Substances Control Act

The Toxic Substances Control Act (TSCA), enacted in 1976, was designed to understand the use or development of chemicals and to provide controls, if necessary, for those chemicals that may threaten human health or the environment (Ingle, 1983; Sittig, 1991). This act has probably had more effect on the producers of chemicals, and the refining industry, than any other act. It has caused many changes in the industry and may create even further modifications in the future. The basic thrust of the act is (1) to develop data on the effects of chemicals on our health and environment, (2) to grant authority to the EPA to regulate substances presenting an unreasonable risk, and (3) to assure that this authority is exercised so as not to impede technological innovation. In short, the act calls for regulation of *chemical substances* and *chemical mixtures* that present an unreasonable risk or injury to health or the environment. Furthermore, the introduction and evolution of this act has led to a central bank of information on existing commercial chemical substances and chemical mixtures, procedures for further testing of hazardous chemicals, and detailed permit requirements for submission of proposed new commercial chemical substances and chemical mixtures.

As used in the act, the term *chemical substance* means any organic or inorganic substance of a particular molecular identity, including any combination of such substances occurring in whole or in part as a result of a chemical reaction or occurring in nature and any element or uncombined radical. Items not considered *chemical substances* are listed in the definition section of the act. The term *mixture* means any combination of two or more chemical substances if the combination does not occur in nature and is not, in whole or in part, the result of a chemical reaction; except that such term does include any combination that occurs, in whole or in part, as a result of a chemical reaction if none of the chemical substances comprising the combination is a new chemical substance and if the combination could have been manufactured for commercial purposes without a chemical reaction at the time the chemical substances comprising the combination were combined.

For many, familiarity with the TSCA generally stems from its specific reference to polychlorinated biphenyls, which raise a vivid, deadly characterization of the harm caused by them. But the TSCA is not a statute that deals with a single chemical or chemical mixture or product. In fact, under the TSCA, the EPA is authorized to institute testing programs for various chemical substances that may enter the environment. Under the TSCA's broad authorization, data on the production and use of various chemical substances and mixtures may be obtained to protect public health and the environment from the effects of harmful chemicals. In actuality, the TSCA supplements the appropriate sections dealing with toxic substances in other federal statutes, such as the Clean Water Act (Section 307) and the Occupational Safety and Health Act (Section 6).

At the heart of the TSCA is a premanufacture notification (PMN) requirement under which a manufacturer must notify the EPA at least 90 days prior to the production of a new chemical. In this context, a *new chemical* is a chemical that is not listed in the TSCA-based Inventory of Chemical Substances or is an unlisted reaction product of two or more chemicals. For chemicals already on this list, a notification is required if there is a new use that could significantly increase human or environmental exposure. No notification is required for chemicals that are manufactured in small quantities solely for scientific research and experimentation.

The TSCA chemical substances inventory is a comprehensive list of the names of all existing chemical substances, and currently contains over 70,000 existing chemicals. Information in the inventory is updated every four years. A facility must submit a premanufacture notice (PMN) prior to manufacturing or importation for any chemical substances not on the list and not excluded by the TSCA. Examples of regulated chemicals include lubricants, paints, inks, fuels, plastics, and solvents.

5.2.9. Hazardous Materials Transportation Act

The Hazardous Materials Transportation Act, passed in 1975, is the primary law governing transportation of chemicals and hazardous materials. The act includes a comprehensive assessment of the regulations, information systems, container safety, and training for emergency response and enforcement. The regulations apply to "any person who transports, or causes to be transported or shipped, a hazardous material; or who manufactures, fabricates, marks, maintains, reconditions, repairs, or tests a package or container which is represented, marked, certified, or sold by such person for use in the transportation in commerce of certain hazardous materials."

Under this statutory authority, the Secretary of Transportation has broad authority to determine what constitutes a hazardous material, using the dual tools of quantity and type. By this two-part approach, any material that may pose an unreasonable risk to human health or the environment may be declared a hazardous material. Such a designated hazardous material obviously includes both the quantity and the form that make the material hazardous.

The basic purpose of HMTA is to ensure safe transportation of hazardous materials through the nation's highways, railways, and waterways. The basic theme of HMTA is to prevent any person from offering or accepting a hazardous material for transportation anywhere within this nation if that material is not properly classified, described, packaged, marked, labeled, and properly authorized for shipment pursuant to the regulatory requirements.

Under the Department of Transportation (DOT) regulations, a hazardous material is *any substance or material, including a hazardous substance and hazardous waste that is capable of posing an unreasonable risk to health, safety, and property when transported in commerce.* DOT thus has broad authority to regulate the transportation of hazardous materials that, by definition, include hazardous substances as well as hazardous wastes.

5.3. REFINERY OUTLOOK

5.3.1. Hazardous Waste Regulations

Petroleum refinery operators face more stringent regulation of the treatment, storage, and disposal of hazardous wastes. Under recent regulations, a larger number of compounds have been, and are being, studied. Long-time methods of disposal, such as land farming of refinery waste, are being phased out. As a result, many refineries are changing their waste management practices. An informal survey of nine refineries showed that eight were planning to close land treatment units because of the uncertainty of continuing the practice.

New regulations are becoming even more stringent, and they encompass a broader range of chemical constituents and processes. Continued pressure from the U.S. Congress has led to more explicit laws allowing little leeway for industry, the U.S. EPA, or state agencies. A summary of the current regulations and what they mean to refiners follows.

5.3.2. Regulatory Background

As we know it today, the hazardous waste regulatory program began with the Resource Conservation and Recovery Act (RCRA) in 1976. The Used Oil Recycling Act of 1980 and Hazardous and Solid Waste Amendments of 1984 (HSWA) were the major amendments to the original law.

The RCRA provides for the tracking of hazardous waste from the time it is generated, through storage and transportation, to the treatment or disposal sites. The RCRA and its amendments are aimed at preventing disposal problems that lead to a need for the Comprehensive Environmental Response, Compensation, and Liability Act (CERCLA), or Superfund, as it is known. Subtitle C of the original RCRA lists the requirements for the management of hazardous waste. This includes the EPA criteria for identifying hazardous waste, and the standards for generators, transporters, and companies that treat, store, or dispose of waste. The RCRA regulations also provide standards for the design and operation of such facilities.

5.3.3. Requirements

The first step to be taken by a generator of waste is to determine whether that waste is hazardous. Waste may be hazardous by being listed in the regulations, or by meeting any of four characteristics: ignitability, corrosivity, reactivity, and extraction procedure (EP) toxicity.

Generally: (1) if the material has a flash point less than 140°F, it is considered ignitable; (2) if the waste has a pH less than 2.0 or above 12.5, it is considered corrosive. (it may also be considered corrosive if it corrodes stainless steel at a certain rate); (3) a waste is considered reactive if it is unstable and produces toxic materials, or it is a cyanide or sulfide-bearing waste that generates toxic gases or fumes; and (4) a waste that is analyzed for EP toxicity and fails is also considered a hazardous waste. This procedure subjects a sample of the waste to an acidic environment. After an appropriate time has elapsed, the liquid portion of the sample (or the sample itself if the waste is liquid) is analyzed for certain metals and pesticides. Limits for allowable concentrations are given in the regulations. The specific analytical parameters and procedures for these tests are referred to in 40CRF 261.

The 1984 amendments also brought the owners and operators of underground storage tanks into the RCRA fold. This can have a significant effect on refineries that store product in underground tanks. In addition, petroleum products are also regulated by the RCRA Subtitle I.

5.4. MANAGEMENT OF REFINERY WASTE

The refining industry, as well as other industries, will increasingly feel the effects of land bans on their hazardous waste management practices. Current practices of land disposal must change along with management attitudes for waste handling. The way that refineries will handle their waste in the future will depend largely on the ever-changing regulations. Waste management is the focus, and reuse/recycle options must be explored to maintain a balanced waste management program. This requires that a waste be recognized as either *nonhazardous* or *hazardous*.

However, before a refinery can determine if its waste is hazardous, it must determine that the waste is indeed a solid waste. In 40 CFR 261.2, the definition of solid waste can be found. If a waste material is considered a solid waste, it may be a hazardous waste in accordance with 40 CFR 261.3. There are two ways to determine whether a waste is hazardous. These are to see if the waste is listed in the regulations or to test the waste to see if it exhibits one of the characteristics (40 CFR 261).

There are four lists of hazardous wastes in the regulations: wastes from nonspecific sources (F list), wastes from specific sources (K list), acutely toxic wastes (P list), and toxic wastes (U list); there are also the four characteristics mentioned before: ignitability, corrosivity, reactivity, and extraction procedure toxicity. Certain waste materials are excluded from regulation under the RCRA. The various definitions and situations that allow waste to be exempted can be confusing and difficult to interpret. One such case is the interpretation of the mixture and derived-from rules. According to the *mixture rule*, mixtures of solid waste and listed hazardous wastes are, by definition, considered hazardous. Similarly, the *derived-from rule* defines solid waste resulting from the management of hazardous waste to be hazardous (40 CFR 261.3a and 40 CFR 261.1c).

There are five specific listed hazardous wastes (K list) generated in refineries, K048 to K052. Additional listed wastes, those from nonspecific sources (F list) and those from the commercial chemical product lists (P and U), may also be generated at refineries. Because of the mixture and derived-from rules, special care must be taken to ensure that hazardous wastes do not *contaminate* nonhazardous waste. Under the mixture rule, adding one drop of hazardous waste in a container of nonhazardous materials makes the entire container contents a hazardous waste.

As an example of the problems such mixing can cause, consider the case with API separator sludge that is a listed hazardous waste (K051). The wastewater from a properly operating API separator is not hazardous unless it exhibits one of the characteristics of a hazardous waste. That is, the derived-from rule does not apply to the wastewater. However, if the API separator is not functioning properly, solids carryover in the wastewater can occur. In this case, the wastewater contains a listed hazardous waste, the solids from the API sludge, and the wastewater would be considered a hazardous waste because it is a mixture of a nonhazardous waste and a hazardous waste.

This wastewater is often cleaned further by other treatment systems (filters, impoundments, etc.). The solids separating in these systems continue to be API separator sludge, a listed hazardous waste. Therefore, all downstream wastewater treatment systems are receiving and treating a hazardous waste and are considered hazardous waste management units subject to regulation.

Oily wastewater is often treated or stored in unlined wastewater treatment ponds in refineries. These wastes appear to be similar to API separator waste.

REFERENCES

CFR. 2004. *Code of Federal Regulations*, U.S. Government, Washington, DC. *The Code of Federal Regulations* (CFR) is the codification of the general and permanent rules published in the *Federal Register* by the executive departments and agencies of the federal government. It is divided into 50 titles that represent broad areas subject to federal regulation. Each volume of the CFR is updated once each calendar year and is issued on a quarterly basis.

EPA. 1997. *Petroleum Refinery MACT Standard Guidance*. Environmental Protection Agency, Washington, DC.

EPA. 2004. Environmental Protection Agency, Washington, DC. Web site: http://www.epa.gov.

Dennison, M. S. 1993. *RCRA Regulatory Compliance Guide*. Noyes Data Corp., Park Ridge, NJ.

Ingle, G. W. (Ed.). 1983. *TSCA's Impact on Society and the Chemical Industry*, ACS Symposium Series. American Chemical Society, Washington, DC.

Lunn, G., and Sansone, E. B. 1994. *Destruction of Hazardous Chemicals in the Laboratory*, 2nd ed. McGraw-Hill, New York.

Majumdar, S. B. 1993. *Regulatory Requirements for Hazardous Materials*. McGraw-Hill, New York.

Nordin, J. S., Sheesley, D. C., King, S. B., and Routh, T. K. 1995. *Environ. Solut.*, 8(4): 49.

Noyes, R. (Ed.). 1993. *Pollution Prevention Technology Handbook*. Noyes Data Corp., Park Ridge, NJ.

Sittig, M. 1991. *Handbook of Toxic and Hazardous Chemicals and Carcinogens*, 3rd ed. Noyes Data Corp., Park Ridge, NJ.

Studt, T. 1995. *R&D Mag.*, February, p. 69.

TPG. 1995. *Aboveground Storage Tank Guide*, Vol. 1 and 2. Thompson Publishing Group, Washington, DC.

Wagner, T. P. 1999. *The Complete Guide to Hazardous Waste Regulations*. Wiley, New York.

Wang, C. C. K. 1994. *OSHA Compliance and Management Handbook*. Noyes Data Corp., Park Ridge, NJ.

CHAPTER

6

SAMPLE COLLECTION AND PREPARATION

Despite the nature of the environmental regulations and the precautions taken by the refining industry, the accidental release of nonhazardous chemicals and hazardous chemicals into the environment has occurred and, without being unduly pessimistic, will continue to occur (by all industries—not wishing to select the refining industry as the only industry that suffers accidental release of chemicals into the environment). To paraphrase *chaos theory*, no matter how well one prepares, the unexpected is always inevitable.

It is at this point that the environmental analyst has to identify the nature of the chemicals and their potential effects on the ecosystem(s) (Smith, 1999). Although petroleum itself and its various products are complex mixtures of many organic chemicals (Chapters 2 and 3), the predominance of one particular chemical or one particular class of chemicals may offer the environmental analyst or scientist an opportunity for the predictability of behavior of the chemical(s).

6.1. PETROLEUM CHEMICALS

Briefly, for environmental purposes, chemicals are subdivided into two classes: (1) *organic chemicals* and (2) *inorganic chemicals*. Furthermore, classification occurs insofar as organic chemicals are classified as *volatile organic compounds* or *semivolatile organic compounds* (on occasion, the word *chemicals* is substituted for the word *compounds* without affecting the definition).

The first class of organic compounds, the *volatile organic compounds* (VOCs), is subdivided into *regulated compounds* and *unregulated compounds*. Regulated compounds have maximum contaminant levels, but unregulated compounds do not. Regulated compounds generally (but not always) have low-boiling points, or low-boiling ranges, and some are gases. Many of these chemicals can be detected at extremely low levels by a variety of instruments, including the human nose! In the case of the petroleum industry, sources for volatile organic compounds typically are petroleum refineries, fuel stations, naphtha (i.e., dry cleaning solvents, paint thinners, cleaning solvents for auto parts), and in some cases, refrigerants that are manufactured from petrochemicals.

The second class of organic compounds, the *semivolatile compounds*, typically have high boiling points, or high boiling ranges, and are not always easily detected

Environmental Analysis and Technology for the Refining Industry, by James G. Speight
Copyright © 2005 John Wiley & Sons, Inc.

by the instrumentation that may be used to detect volatile organic compounds (including the human nose). Some of the common sources of contamination are high-boiling petroleum products (e.g., lubricating oils), pesticides, herbicides, fungicides, wood preservatives, and a variety of other chemicals that can be linked to the refining industry.

Regulations are in place that set the maximum contamination concentration levels that are designed to ensure public safety. There are primary and secondary standards for inorganic chemicals. Primary standards are those chemicals that cause neurological damage, cancer, or blood disorders. Secondary standards are developed for other environmental reasons. In some instances, the primary standards are referred to as the *inorganic chemical group*. The secondary standards are referred to as the *general mineral group* and *general physical testing group*.

The inorganic chemical group includes aluminum, antimony, arsenic, barium, beryllium, cadmium, chromium, lead, mercury, nickel, selenium, silver, fluoride, nitrate, nitrite, and thallium. The mineral group includes calcium, magnesium, sodium, potassium, bicarbonate, carbonate, chloride, sulfate, pH, alkalinity, hardness, electrical conductivity, total dissolved solids, surfactants, copper, iron, manganese, and zinc. The physical group includes turbidity, color, and odor. Many of these chemicals arise from desalting residues and from other processes where catalysts are used. A high level of any of these three chemicals in the soil or in the water is an indication that one or more specific processes (identified from the chemicals that have been released) or pollution prevention processes are not performing according to operational specifications.

Another source of toxic compounds is combustion (Chapter 4). In fact, some of the greater dangers of fires are from toxic products and by-products of combustion. The most obvious of these is carbon monoxide (CO), which can cause serious illness or death because it forms carboxyhemoglobin with hemoglobin in the blood so that the blood no longer carries oxygen to body tissues. Toxic sulfur dioxide and hydrogen chloride are formed by the combustion of sulfur compounds and organic chlorine compounds, respectively. In addition, a large number of noxious organic compounds such as aldehydes are generated as by-products of combustion. In addition to forming carbon monoxide, combustion under oxygen-deficient conditions produces polynuclear aromatic hydrocarbons consisting of fused-ring structures. Some of these compounds, such as benzo[a]pyrene, are precarcinogenic compounds, insofar as they are acted upon by enzymes in the body to yield cancer-producing metabolites.

Most investigations involving petroleum hydrocarbons are regulated by various agencies that may require methodologies, action levels, and cleanup criteria that are different. Indeed, the complex chemical composition of petroleum and petroleum products can make it extremely difficult to select the most appropriate analytical test methods for evaluating environmental samples and to interpret and use the data accurately.

Accordingly, general methods of environmental analysis (Smith, 1999), that is, analysis for the determination of petroleum or petroleum products that have

Table 6.1. Suggested Items to be Included in a Sampling Log

1.	The precise (geographic or other) location (or site or refinery or process) from which the sample was obtained.
2.	The identification of the location (or site or refinery or process) by name.
3.	The character of the bulk material (solid, liquid, or gas) at the time of sampling.
4.	The means by which the sample was obtained.
5.	The means and protocols that were used to obtain the sample.
6.	The date and the amount of sample that was originally placed into storage.
7.	Any chemical analyses (elemental analyses, fractionation by adsorbents or by liquids, functional type analyses) that have been determined to date.
8.	Any physical analyses (API gravity, viscosity, distillation profile) that have been determined to date.
9.	The date of any such analyses included in items 7 and 8.
10.	The methods used for analyses that were employed in items 7 and 8.
11.	The analysts who carried out the work in items 7 and 8.
12.	A log sheet showing the names of the persons (with the date and the reason for the removal of an aliquot) who removed the samples from storage and the amount of each sample (aliquot) that was removed for testing.

Source: Dean, 2003.

been released, are available. The data determine whether or not a release of such chemicals will be detrimental to the environment and may lead to regulations governing the use and handling of such chemicals. But first, sample collection, preservation, preparation, and handling protocols must be followed to the letter. This, of course, includes *chain of custody* or *sampling handling protocols*, which will be defensible if and when legal issues arise. Thus, an accurate sample handling and storage log should be maintained and should include the basic necessary information (Table 6.1). Attention to factors such as these enables standardized comparisons to be made when subsequent samples are taken.

In summary, many of the specific chemicals in petroleum are hazardous because of their chemical reactivity, fire hazard, toxicity, and other properties. In fact, a simple definition of a hazardous chemical (or hazardous waste) is that it is a chemical substance (or chemical waste) that has been inadvertently released, discarded, abandoned, neglected, or designated as a waste material and has the potential to be detrimental to the environment. Alternatively, a hazardous chemical may be a chemical that may interact with other (chemical) substances to give a product that is hazardous to the environment. Whatever the case, methods of analysis must be available to determine the nurture of the released chemical (waste) and from the data predict the potential hazard to the environment.

6.2. SAMPLE COLLECTION AND PREPARATION

The ability to collect and preserve a sample that is representative of the site is a critically important step (Dean, 2003; Patnaik, 2004). Obtaining representative environmental samples is always a challenge, due to the heterogeneity of different sample matrixes. Additional difficulties are encountered with petroleum hydrocarbons, due to the wide range in volatility, solubility, biodegradation, and adsorption potential of individual constituents. The procedures used for sample collection and preparation must be legally defensible.

There are many recommended sampling protocols (Table 6.2). The sampling methods used for petroleum hydrocarbons are generally thought of as methods for determination of the total petroleum hydrocarbons. In part due to the complexity of the components of the total petroleum hydrocarbons fractions, little is known about their potential for health or environmental impacts. As gross measures of petroleum contamination, the total petroleum hydrocarbons data simply show that petroleum hydrocarbons are present in the sampled media. Measured total petroleum hydrocarbons values suggest the relative potential for human exposure and therefore the relative potential for human health effects.

Although most site investigations to determine the assessment of contamination of an ecosystem by petroleum hydrocarbons are regulated by local or by regional (state) governments, sample collection and preservation recommendations follow strict guidelines (Table 6.1), thereby reducing the potential for sample compromise. Therefore, before a sample is collected, the particular sample collection and preservation requirements must be investigated. Because of holding (storage) time considerations, the laboratory must be selected and notified prior to the collection of the samples.

6.2.1. Sample Collection

The value of any analysis is judged by the characteristics of the sample as determined by laboratory tests. The sample used for the test(s) must be representative of the bulk material, or data will be produced that are not representative of the material and will be incorrect no matter how accurate or precise the test method. In addition, the type and cleanliness of sample containers are important if the container is contaminated or is made of material that either reacts with the product or is a catalyst, as the test results may be wrong.

Thus, the importance of correct sampling of any sample destined for analysis should always be overemphasized. Incorrect sampling protocols can lead to erroneous analytical data, from which decisions about regulatory issues cannot be made accurately. In addition, adequate records of the circumstances and conditions during sampling have to be made; for example, in sampling from storage tanks, the temperatures and pressures of the separation plant and the atmospheric temperature would be noted.

At the other end of the volatility scale, samples that contain, or are composed of, high-molecular-weight paraffin hydrocarbons (wax) that are also in a solid state may require judicious heating (to dissolve the wax) and agitation (homogenized, to ensure thorough mixing) before sampling. If room-temperature sampling is the *modus operandi* and product cooling causes wax to precipitate, homogenization to ensure correct sampling is also necessary.

Table 6.2. Sampling Protocols Recommended by the Environmental Protection Agency

Analytical Parameter	Analytical Method(s)	Medium	Sample Container[a]			Preservatives[b]	Holding Time
			Volume	Type			
Trph[c]	EPA 418.1 (IR); gravimetric	Water	1 liter	Glass jar with Teflon-lined cap		Acid fix pH < 2; cool to 4°C	Extract in 7 days; analyze in 40 days
	GC/FID	Soil	125 mL	Widemouth glass with Teflon-lined cap		Cool to 4°C	Extract in 7 days; analyze in 40 days
Volatile petroleum hydrocarbons[d]	Various	Water	40 mL	Glass vial with Teflon-lined septum		Acid fix pH < 2; cool to 4°C	14 days
		Soil	40 mL	Glass vial with Teflon-lined septum		Cool to 4°C[e]	14 days
Extractable petroleum hydrocarbons[f]	Various	Water	1 liter	Glass jar with Teflon-lined cap		Acid fix pH < 2; cool to 4°C	Extract in 7 days; analyze in 40 days
		Soil	60 mL	Widemouth glass with Teflon-lined cap		Cool to 4°C	Extract in 7 days; analyze in 40 days

(*continued overleaf*)

Table 6.2. (*continued*)

Analytical Parameter	Analytical Method(s)	Sample Container[a] Medium	Volume	Type	Preservatives[b]	Holding Time
BTEX	EPA 8240/8260[g]	Water	40 mL	Glass vial with Teflon-lined septum	Acid fix pH < 2; cool to 4°C	14 days
	EPA 8020/8021[g]					
	EPA 624[h], EPA 602[h], EPA 524[h]	Soil	40 mL	Glass vial with Teflon-lined septum	Cool to 4°C[e]	14 days
PAHs	EPA 8270[g]	Water	1 liter	Glass jar with Teflon-lined cap	Acid fix pH < 2; cool to 4°C	Extract in 7 days; analyze in 40 days
	EPA 8310[g] EPA 8100[g]	Soil	60 mL	Widemouth glass with Teflon-lined cap	Cool to 4°C	Extract in 7 days; analyze in 40 days

Source: EPA, 2004.

[a] Minimum sampling volume may vary depending on specific method.
[b] Acid fixation: Use 1:1 HCl to adjust pH of aqueous samples to less than two. Add approx. 500 μL (2–4 drops) to 40-mL aqueous sample vials; 5-mL to 1-Liter aqueous sample jars. Add acid to vials before collecting sample; use gloves and eye protection when sampling. Other preservatives, such as sulfuric acid or sodium bisulfate, may also be used for this purpose.
[c] Refers to extractable hydrocarbons only.
[d] Generally, C_5 through $C_{10\pm2}$ hydrocarbons detectable through purge-and-trap or headspace analyses; includes most *gasoline range organics* methodologies.
[e] Some states/methodologies require field preservation of soil samples in methanol. In such cases, methanol must be purge-and-trap grade; typically, add 20 mL of methanol to vials prior to sample collection. Use gloves and eye protection when sampling. Shipping of methanol is subject to DOT regulation.
[f] Generally, C_9 through $C_{28\pm7}$ hydrocarbons are detectable through a solvent extraction process; includes most *diesel range organics* methodologies.
[g] SW-846 methodology.
[h] 40 CFR, Part 136.

SAMPLE COLLECTION AND PREPARATION

The first task in any analysis is to separate the analytes from the bulk of the sample. The traditional liquid extraction is the most common means employed, but alternative methods are also available (Table 6.3). A portion of the sample is mixed with an organic solvent into which the analyte is preferentially partitioned. The idea of partitioning is key to the success of this procedure. No organic analyte is completely removed from a sample by a single washing with an organic solvent.

Representative samples are prerequisite for the laboratory evaluation of any type of environmental sample, and many precautions are required in obtaining and handling representative samples (ASTM D270, D1265). The precautions depend on the sampling procedure, the characteristics (low- or high-boiling constituents) of the product being sampled, and the storage tank, container, or tank carrier from which the sample is obtained. In addition, the sample container must be clean, and the type to be used depends not only on the product but also on the data to be produced.

Table 6.3. Extraction Methods Recommended by the Environmental Protection Agency

Extraction Method	EPA SW-846 Method Number	Extraction Matrix	Compounds Extracted	Purpose
Separatory funnel	3510	Water	Semivolatile, nonvolatile	Laboratory
Continuous liquid–liquid	3520	Water	Semivolatile, nonvolatile	Laboratory
Solid-phase extraction	3535	Water	Semivolatile, nonvolatile	Laboratory/ screening
Purge and trap	5030, 5035	Water, soil	Volatile	Laboratory/ field preservation
Headspace	3810, 5021	Water, soil	Volatile	Screening/ laboratory
Shake and vortex	[a]	Soil	Volatile, semivolatile, nonvolatile	Screening/ laboratory
Soxhlet	3540, 3541	Soil	Semivolatile, nonvolatile	Laboratory
Sonication	3550	Soil	Semivolatile, nonvolatile	Laboratory
Supercritical fluid	3560, 3561	Soil	Semivolatile, nonvolatile	Laboratory
Subcritical fluid	3545	Soil	Semivolatile, nonvolatile	Laboratory

Source: Dean, 1998, and references cited therein.

[a] Not an EPA SW-846 approved method.

However, obtaining a representative sample is not just a matter of relying on partition coefficients. Partitioning is controlled thermodynamically, and to achieve equilibrium requires a certain amount of intimate contact time between the sample matrix, the analyte molecules, and the organic solvent. Swirling an aqueous sample with the solvent is not effective for achieving a thermodynamic equilibrium. Mixing the two phases together as thoroughly as possible and then allowing the phases to separate achieves the desired equilibrium. The procedure should be repeated with additional portions of solvent. If equilibrium is not established, the efficiency of the extraction procedure is low and the analytical data will be subject to error and indefensible (Dean, 1998, and references cited therein).

The basic objective of each procedure is to obtain a truly representative sample or, more often, a composite of several samples that can be considered to be a representative sample. In some cases, because of the size of the storage tank and the lack of suitable methods of agitation, several samples are taken from large storage tanks such that the samples represent the properties of the bulk material from different locations in the tank and thus the composite sample will be representative of the entire lot being sampled. This procedure allows for differences in sample that might be due to the stratification of the bulk material due to tank size or temperature at various levels of the storage tank. Solid samples require a different protocol that might involve melting (liquefying) the bulk material (assuming that thermal decomposition is not induced) followed by homogenization. On the other hand, the protocol used for coal sampling (ASTM D346, D2013) might also be applied to sampling petroleum products, such as coke, that are solid and for which accurate analysis is required before sales.

Once the sampling procedure is accomplished, the sample container should be labeled immediately, to indicate the product, time of sampling, location of the sampling point, and any other information necessary for the sample identification. If the samples were taken from different levels of the storage tank, the levels from which the samples were taken and the amounts taken and mixed into the composite should be indicated on the sample documentation.

Sampling records for any procedure must be complete and should include, but is not restricted to, items relating to the origin of the sample, methods of storage, analytical tests performed, the test methods used, and the analyst(s) who performed the test methods (Table 6.1). In summary, there must be a means to identify the sample history as carefully as possible so that each sample is tracked and defined in terms of source and activity. Thus, the accuracy of the data from any subsequent procedures and tests for which the sample is used will be placed beyond a *reasonable doubt*.

Volatile Compounds

A volatile substance is one whose boiling point, or sublimation temperature, is such that it exists to a significant extent in the gaseous phase under ambient conditions. The following are volatile hydrocarbons that are commonly encountered:

Aliphatic	Aromatic
Pentanes	Benzene
Hexanes	Toluene
Heptanes	Ethylbenzene
Octanes	Xylenes
Nonanes	Naphthalene
Decanes	Phenanthrene
	Anthracene
	Acenaphthylene

Several sampling procedures are applicable to volatile compounds but method application often depends on the compound(s) to be sampled (Dean, 2003). Part of the issue of sampling volatile compounds arises because some volatile substances sublime rather than boil, whereas other volatile substances emit significant quantities of vapor well below their boiling point. For sampling volatile hydrocarbons in the field, two procedures are generally recommended: zero headspace and solvent extraction. However, these two procedures do not necessarily give equivalent results.

Zero-headspace procedures involve the collection of a soil sample with immediate transfer to a container into which the sample fits exactly. The only space for gases is that within the soil pores. The volume of sample collected depends on the concentration of volatiles in the soil. It is imperative that the container employed can be interfaced directly with the gas chromatograph. Several commercial versions of zero-headspace sampling devices are available. The sample is transported to the laboratory at 4°C, where it is analyzed directly by purge-and-trap gas chromatography (EPA 5035) or other appropriate techniques, such as vacuum distillation (EPA 5032) or headspace (EPA 5021).

Solvent extraction procedures involve collection of sample by an appropriate device and subsequent immediate placement into a borosilicate glass vessel, which contains a known quantity of ultrapure methanol. The bottle is then transported to the laboratory at 4°C, and the methanol fraction analyzed by purge-and-trap gas chromatography (or a similar procedure).

In general, zero-headspace procedures are employed when the concentrations of volatiles in the soil are relatively low, and solvent extraction methods are used for more polluted soils. Irrespective of which procedure is used, quantitation of volatiles in soil is subject to serious errors if sufficient care is not taken with the sampling operation. Although direct purge-and-trap methods are frequently advocated for the determination of volatiles in samples collected by zero-headspace procedures, there are certain problems associated with this technique. Caution is advised since the procedure really collects only that fraction of the volatile that exists in a free form within the soil pore spaces or is at least in a facile equilibrium with this fraction.

Gas chromatography detectors employed for the determination of volatile organics in soil are generally flame ionization detectors (FIDs), photoionization detectors (PIDs), or mass spectrometry. Flame ionization detectors will respond to all carbon compounds in a sample, whereas photo-ionization detectors are capable of some sensitivity by virtue of the energy of the lamp employed. A 10.2- or 10.0-eV lamp yields more specific responses to unsaturated (including aromatic) hydrocarbons and may also be employed to give a complete BTEX (benzene, toluene, ethylbenzene, and xylene) characterization at sites where this is likely to be an issue. As regards the columns used for the analysis of volatile hydrocarbons, a wide variety can be used. Wide-bore capillary columns of length typically about 105 m are generally employed and they must be capable of resolving 3-methylpentane from methanol as well as ethylbenzene from the xylenes. There may be some variation in the choice of column, however, according to the resolution required by the authority. There is some debate concerning appropriate limits for the gasoline range and this is reflected in disparate legislation among various countries. For instance, the upper range of the gasoline organics may be defined by naphthalene or dodecane.

Typical gas chromatography conditions involve an oven temperature ramped between 40 and 240°C, with a detector maintained at 250°C and an injector at 200°C. There are two methods of calibration for the gas chromatograph. One method consists of analyzing a mixture of individual hydrocarbons that bracket the gasoline range and calculating an average response factor from the response for each component. The other method involves analyzing a standard that contains one or more gasolines.

Condensate Releases

Condensate release might be equated to the release of volatile constituents but are often named as such because of the specific constituents of the condensate, often with some reference to the gas condensate that is produced by certain petroleum and natural gas wells. However, the condensate is often restricted to the benzene, toluene, ethylbenzene, and xylenes (BTEX) family of compounds.

To determine the concentrations of benzene, toluene, ethylbenzene, and xylenes, approved methods (e.g., EPA SW-846 8021B, SW-846 8260) are not only recommended but are insisted upon for regulatory issues. Polynuclear aromatic hydrocarbons (PAHs) may be present in condensate, and evaluation of condensate contamination should include the use of other test methods (EPA SW-846 8270, SW-846 8310) provided that the detection limits are adequate to the task of soil and groundwater protection. Generally, at least one analysis may be required for the most contaminated sample location from each source area. Condensate releases in nonsensitive areas require analysis for naphthalene only. The analysts should ensure that the method has detection limits that are appropriate for risk determinations.

Semivolatile and Nonvolatile Compounds

In almost all cases of hydrocarbon contamination, some attention will have to be paid to the presence of semi- and nonvolatile hydrocarbons. However, the collection, handling of samples, and their ultimate preparation for analysis is entirely different from that used for volatile hydrocarbons. In general, it is not necessary to take such rigorous procedures to prevent loss of analyte following collection, although the procedures should still be verified using appropriate quality control measures (Dean, 2003).

Before analysis of semi- or nonvolatile components can proceed, it is necessary that the hydrocarbon components be brought into solution. In a sample from a contaminated site, semi- and nonvolatile molecules may exist in the soil pores in the free form within the pore spaces, but are far more likely to be adsorbed by organic matter attached to the soil. Indeed, the probability of such adsorption increases with increasing hydrophobicity of the molecules.

A number of procedures are available to help this dissolution and include Soxhlet extraction (EPA 3540C), ultrasonic extraction (EPA 3550B), thermal extraction (EPA 8275A,) and supercritical fluid extraction (EPA 3560, 3561). Although these procedures are well documented, some of their important details are frequently overlooked, with the result that the extraction is unsatisfactory. In the case of ultrasonic extraction, the method (EPA 3550B) stipulates the use of an ultrasonic disrupter of the horn type, with a minimum power of 300 W. Many laboratories, however, wrongly interpret this to mean an ultrasonic bath, used for cleaning glassware. Such baths are of far lower energy and are not capable of separating the hydrocarbons from their association with humic material. As regards the use of supercritical fluid extraction, a methanol modifier is required to achieve complete extraction of polynuclear aromatic hydrocarbons, whereas supercritical carbon dioxide is sufficient to elute normal hydrocarbons (EPA 3560, 3561).

For most analyses, it is necessary to separate the analytes of interest from the matrix (i.e., soil, sediment, and water). Extraction of analytes can be performed using one or more of the following methods: (1) extracting the analytes into a solvent; (2) heating the sample, as may be necessary to remove the solvent and for the analysis of volatile compounds; and (3) purging the sample with an inert gas, as is also used in the analyses of volatile compounds.

Soxhlet, sonication, supercritical fluid, subcritical or accelerated solvent, and purge-and-trap extraction have been introduced into a variety of methods for the extraction of contaminated soil. Headspace is recommended as a screening method. Shaking/vortexing is adequate for the extraction of petroleum hydrocarbons in most environmental samples. For these extraction methods, the ability to extract petroleum hydrocarbons from soil and water samples depends on the solvent and the sample matrix. Surrogates (compounds of known identity and quantity) are frequently added to monitor extraction efficiency. Environmental laboratories also generally perform matrix spikes (addition of target analytes) to determine if the soil or water matrix retains analytes.

Thus, solvents have different extraction efficiencies, and thus extracting the same sample in the same manner by two different solvents may result in different concentrations. The choice of solvents is determined by many factors, including cost, spectral qualities, method regulations, extraction efficiency, toxicity, and availability. Methylene chloride has been the solvent of choice for many semivolatile analyses, due to its high extraction efficiency. Chlorofluorocarbon solvents such as trichlorotrifluoroethane (Freon-113) have been used in the past for oil and grease analyses because of their spectral qualities (they do not absorb in the 2930-cm^{-1} infrared measurement wavelength) and low human toxicity. The use of chlorofluorocarbons is to be questioned because not all of the petroleum constituents are soluble in such solvents. Furthermore, the use of chlorofluorocarbons solvents is being reduced, even phased out of analytical methods, because of their detrimental effects on stratospheric ozone. Tetrachloroethylene and carbon tetrachloride are possible replacements, but caution is advised since these solvents may be sensitive to light and leave residual chlorine in the sample. Methanol is the most common solvent used to preserve and extract volatiles such as benzene, toluene, ethylbenzene, and xylene(s) in soil. But methanol is unsuitable for many constituents of petroleum and petroleum products. In short, no single solvent will satisfy all the criteria necessary for a complete and full extraction or for the solubility of petroleum and its products.

If the release has contaminated a water system, there are several methods that can be employed for sample separation (Patnaik, 2004, Sec. 2):

For Volatiles	**For Semivolatiles**
Purge and trap	Separatory funnel extraction
Headspace	Continuous liquid–liquid extraction
	Solid-phase extraction

Volatile compounds (gasoline, solvents) in water are generally separated from the aqueous matrix by purging with an inert gas and trapping the compounds on a sorbent (EPA 5030, purge-and-trap analysis). The sorbent is later heated to release the volatile compounds, and a carrier gas sweeps the compounds into a gas chromatograph. Headspace analysis is recommended as a screening method (EPA 3810, 5021), although it performs well in particular situations, especially field analysis. In this method, the water sample is placed in a closed vessel with a headspace and heated to drive volatiles into the gas phase, and instrument contamination is minimized because only volatile compounds are introduced into the instrument. Addition of salts or acids may enhance this process.

Samples containing heavy oil, along with the volatile components; can severely contaminate purge-and-trap instrumentation, and caution is advised when interpreting the data. For such samples it may be advisable to use a separatory funnel for the water extraction method for semivolatiles (EPA 3520). In this method, the sample is poured into a funnel-shaped piece of glassware, solvent is added, and the mixture is shaken vigorously. After layer separation, the extract (i.e., the solvent layer) is removed, filtered, dried with a desiccant, and concentrated. Multiple extractions on the same sample may increase overall recovery.

Another commonly used water extraction method for semivolatiles involves continuous liquid–liquid extraction (EPA 3520). In this method, the sample (rather than being shaken with the solvent) is treated with a continuously heated solvent that is nebulized (broken into small droplets) and sprayed on top of the water. Liquid–liquid extraction is excellent for samples containing emulsion-forming solids, but it is more time consuming than separatory funnel extractions. Nevertheless, time consuming or not, the method must produce reliable data that can be used without question for monitoring or regulatory purposes.

Solid-phase extraction (EPA 3535) may also be used for extraction and concentration of semivolatile material. The technique involves passing the water sample through a cartridge or disk containing an adsorbent such as silica or alumina. The adsorbent is often coated with compounds that impart selectivity for particular products or analytes, such as polynuclear aromatic hydrocarbons (PAHs or PNAs). After extraction, the analytes are separated from the solid phase by elution with a small amount of organic solvent. A variant of solid-phase extraction involves dipping a sorbent-coated fiber into the water (solid-phase microextraction). Adsorbed analytes are thermally desorbed directly into a heated chromatographic injection port. Generally, the solid-phase extraction method requires much less solvent and glassware than do the separatory funnel method and liquid–liquid extraction.

For the separation of samples from contaminated soil, there are also several possible methods, depending on whether the contaminant is volatile or semivolatile:

For Volatiles	For Semivolatiles
Purge and trap	Shaking or vortexing (can also be used for volatiles)
Headspace	Soxhlet
	Sonication
	Supercritical fluid
	Subcritical fluid

Volatile compounds (such as BTEX, benzene, toluene, ethylbenzene, xylene, and gasoline) may be extracted from soil using, for example, methanol (EPA 5035, purge-and-trap analysis). In the method, the extraction is usually accomplished by mechanical shaking of the soil with methanol. A portion of the methanol extract is added to a purge vessel and diluted in reagent-grade water. The extract is then purged in a manner similar to that used for a water sample.

Headspace analysis (EPA 3810, 5021) also works well for analyzing volatile petroleum constituents in soil. In the test method, the soil is placed in a headspace vial and heated to drive out the volatiles from the sample into the headspace of the sample container. Salts can be added for more efficient release of the volatile compounds into the headspace. Similar to water headspace analysis, the soil headspace technique is useful when heavy oils and high analyte concentrations are present, which can severely contaminate purge-and-trap instrumentation. Detection limits are generally higher for headspace analysis than for purge-and-trap analysis.

The simplest way to separate semivolatile compounds from soil is to shake or vortex (vigorous mechanical stirring) the soil with a solvent. Adding a desiccant to the soil–solvent mixture can help to break up soil and increase the surface area. The word of caution here is to ensure that the drying agent does not adsorb the solute. Assuming that no such adsorption occurs, the extract can be analyzed directly. Simple shaking is quick and easy, making it an excellent field extraction technique. However, extraction efficiency will vary depending on soil type and whether or not clay minerals (excellent adsorbents for many organic compounds) are present.

Soxhlet extraction (EPA SW-846 3540) is a very efficient extraction process that is commonly used for semivolatile petroleum constituents. In the method, the solvent is heated and refluxed (recirculated) through the soil sample continuously for 16 hours, or overnight. This method generates a relatively large volume of extract that needs to be concentrated. Thus, it is more appropriate for semivolatile constituents than for volatile constituents. Sonication extraction (EPA SW-846 3550) can also be used for semivolatile compounds, and as the name suggests, involves the use of sound waves to enhance analyte transfer from sample to solvent. Sonication is a faster technique than Soxhlet extraction and can require less solvent.

Supercritical fluid extraction (EPA 3540, for total recoverable petroleum hydrocarbons; EPA 3561 for polynuclear aromatic hydrocarbons) is applicable to the extraction of semivolatile constituents. Supercritical fluid extraction involves heating and pressuring a mobile phase to supercritical conditions (where the solvent has the properties of a gas and a liquid). The supercritical fluid is passed through the soil sample, and the analytes are concentrated on a sorbent or trapped cryogenically. The analytes are eluted with a solvent and analyzed using conventional techniques. Carbon dioxide is the most popular mobile phase.

Another method (EPA 3545, accelerated solvent extraction) has been validated using a variety of soil matrixes, ranging from sand to clay. In the method, conventional solvents such as methylene chloride (or a hexane–acetone mixture) are heated [$100°C$, ($212°F$)] and pressurized (2000 psi), then passed through the soil sample (this technique is also suitable for application to petroleum sludge and petroleum sediment). The method has the advantage of requiring smaller solvent volumes than is required by traditional solvent extraction techniques.

In some cases, when petroleum and/or petroleum products are released to the environment, a free phase is formed and sample(s) of the hydrocarbon material can be collected directly for characterization. The ability to analyze free product greatly aids the determination of product type and potential source. The samples may be diluted prior to analysis; EPA SW-846 3580 (waste dilution) gives some guidelines for proper dilution techniques. However, caution is advised since as part of the initial sample collection procedure, water and sediment may be included in the sample inadvertently. Several protocols involved in initial isolation and cleanup of the sample must be recognized. In fact, considerable importance attaches to the presence of *water* or *sediment* in crude oil (ASTM D1796, D4007), for they lead to difficulties in other analyses.

Sediment usually consists of finely divided solids that may be dispersed in the oil or carried in water droplets. The solids may be drilling mud or sand or scale picked up during the transport of the oil, or may consist of chlorides derived from evaporation of brine droplets in the oil. In any event, the sediment can lead to serious plugging of the equipment, corrosion due to chloride decomposition, and a lowering of residual fuel quality.

Water may be found in the crude either in an emulsified form or in large droplets. The quantity is generally limited by pipeline companies and by refiners, and steps are normally taken at the wellhead to reduce the water content as low as possible. However, after a spill, water can be introduced by climatic conditions, and the relevant tests (ASTM D96, D954, D1796; IP, 2004) are regarded as important in crude oil analyses. Prior to analyses, it is often necessary to separate the water from a crude oil sample, and this is usually carried out by one of the procedures described in the preliminary distillation of crude petroleum (IP 24). Overall, there are several methods that can be employed for organic semivolatile sample preparation and cleanup procedures (Table 6.4).

Solids

For homogeneous materials, sampling protocols are relatively simple and straightforward, although caution is always advised lest overconfidence cause errors in the method of sampling as well as the introduction of extraneous material (EPA, 1998). On the other hand, the heterogeneous nature of soil and contaminated soil complicates the sampling procedures. If the soils and the sample are visibly heterogeneous, there is a very strong emphasis on the need to obtain representative samples for testing and analysis.

Thus, the variable composition of contaminated soil, as well as solid sample such as petroleum coke, offers many challenges to environmental analysts, who must ensure that the sample under investigation is representative of the contaminated site or the coke. Furthermore, sample transportation can initiate (due to movement of the sample) processes that result in size and density segregation, in a manner analogous to variations in coal quality from sample to sample (ASTM D346, D2234, D4702, D4915, D4916, D6315, D6518, D6543; ISO 1988).

Therefore, the challenge in sampling solids for environmental analysis is to collect a relatively small portion of the sample that accurately represents the composition of the whole. This requires that sample increments be collected such that no piece, regardless of position (or size) relative to the sampling position and implement, is selectively collected or rejected. Optimization of solids sampling is a function of the many variable constituents of coal and is reflected in the methods by which an unbiased sample can be obtained, as is required by coal sampling (ASTM D197).

Thus, to test any particular environmental solids sample, several criteria must be met: (1) obtain a sample of the solid, (2) ensure that the sample is a true representative of the bulk material, and (last but by no means least) (3) ensure that the sample does not undergo any chemical or physical change after completion

Table 6.4. Organic Semivolatile Sample Preparation and Cleanup Procedures in SW-846

EPA No.	Methodology
3500B	Organic extraction and sample preparation
3510C	Separatory funnel liquid–liquid extraction
3520C	Continuous liquid–liquid extraction
3535	Solid-phase extraction (SPE) (3535A in update IVB)
3540C	Soxhlet extraction
3541	Automated Soxhlet extraction
3542	Extraction of semivolatile analytes collected using modified Method 5 (Method 0010) sampling train
3545	Accelerated solvent extraction (ASE) (3545A in update IVB)
3550B	Ultrasonic extraction
3560	Supercritical fluid extraction of total recoverable petroleum hydrocarbons (TRPH)
3561	Supercritical fluid extraction of polynuclear aromatic hydrocarbons
3562	Supercritical fluid extraction of PCB and organochlorine pesticides (update IVB)
3580A	Waste dilution
3585	Waste dilution for volatile organics
3600C	Cleanup
3610B	Alumina cleanup
3611B	Alumina column cleanup and separation of petroleum wastes
3620B	Florisil cleanup
3630C	Silica gel cleanup
3640A	Gel-permeation cleanup
3650B	Acid–base partition cleanup
3660B	Sulfur cleanup
3665A	Sulfuric acid/permanganate cleanup

of the sampling procedure and during storage prior to analysis. In short, the reliability of a sampling method is the degree of perfection with which the identical composition and properties of the whole coal are obtained in the sample. The reliability of the storage procedure is the degree to which the coal sample remains unchanged, thereby guaranteeing the accuracy and usefulness of the analytical data. At this point, a review of the sampling methods applied to coal, which under favorable conditions can also be applied to environmental solids, is worth inclusion.

The sampling procedures (ASTM D346, D2234, D4702, D4915, D4916, D6315, D6518, D6543) are designed to give a precision such that if gross samples are taken repeatedly from a lot or consignment and prepared according to standard test methods (ASTM D197, D2013) and one ash determination is made on the analysis sample from each gross sample, the majority (usually specified as 95 out of 100) of these determinations will fall within $\pm 10\%$ of the average of all the determinations. When other precision limits are required or when other constituents are used to specify precision, defined special-purpose sampling procedures may need to be employed.

Thus, when a property of the sample (which exists as a large volume of material) is to be measured, there usually will be differences between the analytical data derived from application of the test methods to a *gross lot* or *gross consignment* and the data from the *sample lot*. This difference (the *sampling error*) has a frequency distribution with a mean value and a variance. *Variance* is a statistical term defined as the mean square of errors; the square root of the variance is more generally known as the *standard deviation* or the *standard error of sampling*.

Recognition of the issues involved in obtaining representative samples of coal and minimization of the *sampling error* has resulted in the designation of methods that dictate the correct manner for coal sampling (ASTM D346, D2234, D4702, D4915, D4916, D6315, D6518, D6543; ISO 1988, 2309).

Finally, and this applies to all samples where separation of the solute or released material from the matrix is necessary, it must be recognized that organic compounds in, for example, soil or in a liquid matrix (water or an organic solvent) have to some degree an affinity for the sample matrix. It is this affinity that allows the sample to be retained by the matrix. The affinity may be due to adsorption on the surface of sample particles or solvation in water or other solvent medium.

Therefore, it is necessary in an extraction process to recognize the potential solute–matrix interactions in order to overcome such interactions. A suitable choice of solvent is necessary, making sure that the solvent itself does not reaction with the solid or with the matrix. If the extraction solvent of choice is too powerful (e.g., pyridine or carbon disulfide), switching to a solvent of lesser ability, such as pentane, hexane, or petroleum ether is recommended, remembering that if a high-boiling petroleum sample is being extracted, solvents of lesser ability will leave some of the sample unextracted. In other cases, particularly when the sample is a hydrocarbon petroleum product, a polar solvent (e.g., methanol) can be used to extract polar organic materials such as phenols from the matrix.

On the other hand, organic acids (e.g., carboxylic acids, phenols) and organic bases (e.g., pyridine derivative) may be isolatable by adjustment of the pH to control the direction of partitioning. For example, acidification (pH < 5) of a sample converts the organic bases into salts that move into the aqueous (water) phase, whereas adjusting the pH of the sample to basic (pH > 9 to 11, depending on the nature of the solute) with a suitable base neutralizes the basic analytes and reverses the direction of partitioning to the organic phase. At the same time, organic acids are converted to water-soluble (hydrophilic) salts. Thus, the simple expedient of separate analysis of the acid and basic extractions, rather than combining the extracts into a single sample extract, often serves to reduce matrix interference to a manageable level.

Although widely used, solvent extraction procedures have been demonstrated as sensitive to such variables as the content of humic matter and moisture within samples. Supercritical fluid extraction appears to be a more robust procedure. Thermal extraction procedures are sensitive to the size of the soil sample in some cases since the technique can result in cracking higher-molecular-weight constituents that do not volatilize out of the thermal zone. In the case of solvent extraction procedures, it is necessary to concentrate and also to clean up the samples. With complex mixtures of semivolatile hydrocarbons, it is generally advisable to separate the aliphatic and aromatic fractions.

6.2.2. Extract Concentration

Extract concentration is the one area in the isolation procedure that has the greatest potential for loss of analytes. The operative physical concept during concentration is vapor pressure, not boiling point. However, boiling is uncontrolled insofar as all of the molecules in the sample are attempting to convert from a liquid to a gaseous state. Therefore, equating sample concentration to boiling point leads to erroneous conclusions about the choice of an appropriate method. On the other hand, vapor pressure is a continuous function that relates to the rate of evaporation, and evaporation is what is desired during sample concentration. Thus, the proper technique is to control the rate of solvent evaporation while minimizing analyte loss. A completely successful sample extraction can be performed that is completely negated by an inappropriate method for sample extract concentration.

In many cases, sample extracts are generally filtered, dried with desiccant, and concentrated before analysis. Concentration of the extract may allow for lower sample detection limits. Frequently, sample extracts must be concentrated to obtain detection limits low enough to meet regulatory action limits.

Concentration may be achieved by the following methods:

For Volatiles	For Semivolatiles
Sorbent trapping	Snyder column
Cryogenic trapping	Kuderna Danish concentrator
	Nitrogen evaporation
	Vacuum

The trapping step in a purge-and-trap analysis is essentially a concentration step. Analytes are purged from the matrix into a gas stream and captured on a sorbent trap. The analytes are released by heating the trap. Cryogenic trapping is also used in place of sorbent trapping. In cryogenic trapping, a very cold material (such as liquid nitrogen) surrounds a sample loop and as analytes are purged and swept through the loop, they freeze in the loop. The analytes are released when the trap is heated.

Snyder columns are designed to allow highly volatile solvents to escape while retaining semivolatile analytes of interest. Snyder columns are generally fitted onto the tops of flasks containing extracts, and column design permits solvent to escape as the flask is heated. The analytes of interest condense from a gas to a liquid phase and fall back into the solvent reservoir. The Kuderna–Danish concentrator is a Snyder column with a removable collection tube attached to the bottom. As solvent is evaporated, the extract is collected in the collection tube.

As an alternative to a Snyder column, the sample extracts may be concentrated with nitrogen evaporation by directing a slow stream of gas over the extract surface at room temperature, resulting in minor loss of volatiles. Placing the extract container in warm water helps to speed the process, but then some loss of volatiles can occur. Concentration by evaporating excess solvent with a vacuum is not very common in environmental laboratories. Many semivolatile analytes are lost in the procedure. Additionally, evaporating as a means of concentrating the sample cannot be used if the goal is to detect volatile analytes.

Cleanup steps are an important component of infrared (IR)-based and gravimetric methods because these methods are very sensitive to nonpetroleum hydrocarbon interferences. Cleanup steps are not always a part of the petroleum analytical process, but when they are necessary, the goals of extract cleanup steps typically include one or more of the following: (1) removal of nonpetroleum compounds, (2) isolation of a particular petroleum fraction, and (3) concentration of analytes of interest.

The techniques employed to extract the analytes of interest can frequently extract interfering compounds. Polar compounds such as animal and plant fats, proteins, and small biological molecules may be identified incorrectly as petroleum constituents. Extract cleanup techniques can be used to remove them. In an ideal situation, only interfering compounds are removed. In reality, some polar petroleum constituents can also be removed.

Two techniques are used to clean petroleum extracts. In one technique, interfering compounds are removed by passing the extract through a glass column filled with sorbent. A second technique is to swirl the extract with loose sorbent, then remove the sorbent by filtration. Other methods involve trapping the interfering compounds on a sorbent column such as alumina (EPA SW-846 3611) that is designed to remove interfering compounds and to fractionate petroleum wastes into aliphatic, aromatic, and polar fractions. The fractions can be analyzed separately or combined for measurement of the total petroleum hydrocarbons. Alternatively, silica gel (EPA SW-846 3630) is commonly used for PAHs and phenols. Variations of this technique are used to clean (EPA 418.1) extracts before infrared analysis. In addition, the gel permeation technique has been used for cleanup (EPA SW-846 3640) and works on the principle of size exclusion. Large macromolecules such as lipids, polymers, and proteins are removed from the sample extract. Extracts obtained from soil that have (or have had) high biological activity may be cleaned by this method.

There are two noncolumn cleanup methods, one of which uses acid partition (EPA SW-846 3650) to separate the base/neutral and acid components by adjusting pH. This method is often used before alumina column cleanup to remove acid components. The other method (EPA SW-846 3660) is used for sulfur removal and uses copper, mercury, and tetrabutylammonium sulfite as desulfurization compounds. Sulfur is a common interfering compound for petroleum hydrocarbon analysis, particularly for sediments. Sulfur-containing compounds are very common in crude oil and heavy fuel oil. Elemental sulfur is often present in anaerobically biodegraded fuels. Thus, abnormally high levels of sulfur may be measured as part of the total petroleum hydrocarbon measurement if the cleanup technique is not used.

Even though cleanup procedures are advocated before sample analysis, there can be several limitations to various cleanup steps. The reasons for decreased effectiveness of cleanup procedures include: (1) Sample loading may exceed the capacity of cleanup columns, (2) nonpetroleum compounds may have chemical structures similar to petroleum constituents and may behave like a petroleum constituent, (3) analytes of interest may be removed during the cleanup; and (4) no single cleanup technique removes all the chemical interference.

6.2.3. Sample Cleanup

The use in environmental analysis of any sample cleanup is always accompanied by the possibility of analyte loss. Procedures that depend on polarity interactions between the eluting solvent, the solid adsorbent, and the target analytes to achieve selective isolation are particularly prone to having the desired compounds ending up in the wrong fraction. Sources of these errors include mistakes in the preparation of the eluting solvent, use of an incorrect or deactivated absorbent, the presence of traces of polar solvents in the sample solution, and the structure of the analyte molecule that will determine the behavior of the analytes during separation (Speight, 1999). Attention to detail and procedure is required for successful use of sample cleanup techniques, and the aspects of the procedure that need to be examined include (1) the suitability of the materials to achieve a cleanup that will be sample dependent, (2) the introduction of laboratory contamination, and (3) the success of the procedure on the particular sample.

The introduction of laboratory contamination is a significant but often overlooked concern in sample cleanup. The introduction of extractable materials from plastics used for joining tubes and the introduction of contaminated solvents are only two of the potential sources of laboratory contamination.

6.3. MEASUREMENT

The issues that face environment analysts include the need to provide higher-quality results. In addition, environmental regulations may influence the method of choice. Nevertheless, the method of choice still depends to a large extent on the boiling range (or carbon number) of the sample to be analyzed. For example, there is a large variation in the carbon number range and boiling points (of normal paraffins) for some of the more common petroleum products, and thus a variation in the methods that may be applied to these products (Speight, 2001, 2002).

The predominant methods of measuring the properties of petroleum products are covered by approximately seven test methods used in the determination of bulk quantities of liquid petroleum and its products (ASTM D96, D287, D1085, D1086, D1087, D1250, D1298).

Testing for suspended water and sediment (ASTM D96) is used primarily with fuel oils, where appreciable amounts of water and sediment may cause fouling of facilities for handling the oil and give trouble in burner mechanisms. Three standard methods are available for this determination. The centrifuge method gives the total water and sediment content of the sample by volume; the distillation method gives the water only, volumetrically, and the extraction method gives the solid sediment in percent by weight.

The determination of density of specific gravity (ASTM D287, D1298) in the measurement and calculation of the volume of petroleum products is important since gravity is an index of the weight of a measured volume of the product. Two scales are in use in the petroleum industry, specific gravity and API gravity, the determination being made in each case by means of a hydrometer of constant weight displacing a variable volume of oil. The reading obtained depends on both the gravity and the temperature of the oil.

Gauging petroleum products (ASTM D1085, discontinued in 1996 but still in use) involves the use of procedures for determining the liquid content of tanks, ships and barges, tank cars, and tank trucks. Depth of liquid is determined by gauging through specified hatches or by reading gauge glasses or other devices. There are two basic types of gauges, innage and outage. The procedures used depend on the type of tank to be gauged, its equipment, and the gauging apparatus.

An *innage gauge* measures the depth of liquid in a tank measured from the surface of the liquid to the tank bottom or to a datum plate attached to the shell or bottom. The innage gauge is used directly with the tank calibration table and temperature of the product to calculate the volume of product (ASTM D1250). An *outage gauge* measures the distance between the surface of the product in the tank and the reference point above the surface, which is usually located in the gauging hatch. An outage gauge is used either directly or indirectly with the tank calibration table and the temperature of the product to calculate the volume of product. The amount of any free water and sediment in the bottom of the tank is also gauged so that corrections can be made when calculating the net volume of the crude oil or petroleum product.

The liquid levels of products that have a Reid vapor pressure of 40 lb or more are generally determined by the use of gauge glasses, rotary or slip-tube gauges, tapes and bobs through pressure locks, or other types of gauging equipment. The type of gauging equipment depends on the size and type of the pressure tank.

There are also procedures for determining the temperatures of petroleum and its products when in a liquid state. Temperatures are determined at specified locations in tanks, ships and barges, tank cars, and tank trucks. For a nonpressure tank, a temperature is obtained by lowering a tank thermometer of proper range through the gauging hatch to the liquid level specified. After the entire thermometer assembly has had time to attain the temperature of the product, the thermometer is withdrawn and read quickly. This procedure is also used for low-pressure tanks equipped with gauging hatches or standpipes, and for any pressure tank that has a pressure lock. For tanks equipped with thermometer wells, temperatures are obtained by reading thermometers placed in the wells with their bulbs at the desired tank levels. If more than one temperature is determined, the average temperature of the product is calculated from the temperatures observed. Electrical-resistance thermometers are sometimes used to determine both average and spot temperatures.

In general, the volume received or delivered is calculated from the gauge readings observed. Corrections are made for any *free* water and sediment as determined by the gauge of the water level in the tank. The resulting volume is then corrected to the equivalent volume at 15.6°C (60°F) by use of the average temperature and the appropriate volume correction table (ASTM D1250). When necessary, further correction is made for any suspended water and sediment that may be present in materials such as crude petroleum and heavy fuel oils.

For the measurement of other petroleum products, a wide variety of tests are available. In fact, approximately 350 tests (ASTM, 2000) are used to determine the different properties of petroleum products. Each test has its own limits of accuracy and precision that must be adhered to if the data are to be accepted.

6.4. ACCURACY

The *accuracy* of a test is a measure of how close the test result will be to the true value of the property being measured (ASTM, 2004; Patnaik, 2004). As such, the accuracy can be expressed as the *bias* between the test result and the true value. However, the *absolute accuracy* can be established only if the true value is known.

In the simplest sense, a convenient method to determine a relationship between two measured properties is to plot one against the other. Such an exercise will provide either a line fit of the points or a spread that may or may not be within the limits of experimental error. The data can then be used to determine the approximate accuracy of one or more points employed in the plot. For example, a point that lies outside the limits of experimental error (a *flyer*) will indicate an issue of accuracy with that test and the need for a repeat determination.

However, the graphical approach is not appropriate for finding the absolute accuracy between more than two properties. The well-established statistical technique of regression analysis is more pertinent to determining the accuracy of points derived from one property and any number of other properties. There are many instances in which relationships of this sort enable properties to be predicted from other measured properties with as good precision as they can be measured by a single test. It would be possible to examine in this way the relationships between all the specified properties of a product and to establish certain key properties from which the remainder could be predicted, but that would be a tedious task.

However, the application of statistical analysis to experiment data is not always straightforward and may be fraught with inconsistencies due to the assumptions that are involved in the statistical development an interpretation of the data. Statistical analysis of the data is only a part of the picture; the decision process has to be viewed as a coherent whole. A decision can be made only by taking into account the complex interrelations among the chemical species being consumed and formed and the legal, economic, scientific, and environmental characteristics of the analytical process (Baker, 1966; Dixon and Massey, 1969; Alder and Roessler, 1972; Box et al., 1978; Caulcutt and Boddy, 1983; Jaffe and Spirer, 1987; Meier and Zünd, 2000; Patnaik, 2004, Sec. 3).

An alternative approach to that of picking out the essential tests in a specification using regression analysis is to take a look at the specification as a whole and extract the essential features, termed *principal components analysis*. Principal components analysis involves an examination of a set of data as points in n-dimensional space (corresponding to n original tests) and determines (first) the direction that accounts for the biggest variability in the data (the *first principal component*). The process is repeated until n principal components are evaluated, but not all of these are of practical importance since some may be attributable purely to experimental error. The number of significant principal components shows the number of independent properties being measured by the tests considered.

Following from this, it is necessary to establish the number of independent properties that are necessary to predict product performance in service with the goals of rendering any specification more meaningful and allowing a high degree of predictability of product behavior. On a long-term approach it might be possible to obtain new tests of a fundamental nature to replace, or certainly to supplement, existing tests. In the short term, selecting the best of the existing tests to define product quality is the most beneficial route to predictability.

6.5. PRECISION

The *precision* of a test method is the variability between test results obtained on the same material using a specific test method (ASTM, 2004; Patnaik, 2004). The precision of a test is usually unrelated to its accuracy. The results may be precise, but not necessarily accurate. In fact, the precision of an analytical method is the amount of scatter in the results obtained from multiple analyses of a homogeneous sample. To be meaningful, the precision study must be performed using the exact sample and standard preparation procedures that will be used in the final method. Precision is expressed as *repeatability* and *reproducibility*.

The precision of sampling solids, for example, is a function of the size of increments collected and the number of increments included in a gross sample, improving as both are increased, subject only to the constraint that increment size not be small enough to cause selective rejection of the largest particles present. The manner in which solids sampling is performed as it relates to the precision of the sample thus depends on the number of increments collected from all parts of a lot and the size of the increments. In fact, the number and size of the increments are operating variables that can, within certain limits, be regulated by the sampler.

Intralaboratory or *within-laboratory precision* refers to the precision of a test method when the results are obtained by the same operator in the same laboratory using the same apparatus. In some cases, the precision is applied to data gathered by a different operator in the same laboratory using the same apparatus. Thus, intralaboratory precision has an expanded meaning insofar as it can be applied to laboratory precision.

Repeatability or the *repeatability interval* of a test (r) is the maximum permissible difference due to test error between two results obtained on the same material in the same laboratory.

$$r = 2.77 \times \text{standard deviation of test}$$

The repeatability interval (r) is, statistically, the 95% probability level; that is, the differences between two test results are unlikely to exceed this repeatability interval more than five times in a hundred.

Interlaboratory or *between-laboratory precision* is defined in terms of the variability between test results obtained on the aliquots of the same homogeneous material in different laboratories using the same test method.

The term *reproducibility* or *reproducibility interval* (R) is analogous to the term *repeatability*, but it is the maximum permissible difference between two results obtained on the same material but in different laboratories. Therefore, differences between two or more laboratories should not exceed the reproducibility interval more than five times in a hundred:

$$R = 2.77 \times \text{standard deviation of test}$$

The repeatability and reproducibility values have important implications for quality. As the demand for clear product specifications, and hence control over product consistency, grows, it is meaningless to establish product specifications that are more restrictive than the reproducibility/repeatability values of the specification test methods.

6.6. METHOD VALIDATION

Method validation is the process of proving that an analytical method is acceptable for its intended purpose. Many organizations provide a framework for performing such validations (ASTM, 2004). In general, methods for product specifications and regulatory submission must include studies on specificity, linearity, accuracy, precision, range, detection limit, and quantitation limit.

The process of method development and validation covers all aspects of the analytical procedure and the best way to minimize method problems is to perform validation experiments during development. To perform validation studies, the approach should be viewed with the understanding that validation requirements are continually changing and vary widely, depending on the type of product under test and compliance with any necessary regulatory group.

In the early stages of new product development, it may not be necessary to perform all of the various validation studies. However, the process of validating a method cannot be separated from the actual development of the method conditions, because the developer will not know whether the method conditions are acceptable until validation studies are performed. The development and validation of a new analytical method may therefore be an iterative process. Results of validation studies may indicate that a change in the procedure is necessary, which may then require revalidation. During each validation study, key method parameters are determined and then used for all subsequent validation steps.

The first step in the method development and validation cycle should be to set minimum requirements, which are essentially acceptance specifications for the method. During method development, a complete list of criteria should be agreed on by the end users so that expectations are clear. Once the validation studies are complete, the method developers should be confident in the ability of the method to provide good quantitation in their own laboratories. The remaining studies should provide greater assurance that the method will work well in other laboratories, where different operators, instruments, and reagents are involved and where the method will be used over much longer periods of time.

The remaining precision studies comprise much of what is often referred to as *ruggedness*. *Intermediate precision* is the precision obtained when an assay is performed by multiple analysts using several instruments on different days in one laboratory. Intermediate precision results are used to identify which of the foregoing factors contribute significant variability to the final result.

The last type of precision study is *reproducibility*, which is determined by testing homogeneous samples in multiple laboratories, often as part of interlaboratory crossover studies. The evaluation of reproducibility results often focuses more on measuring bias in results than on determining differences in precision alone. Statistical equivalence is often used as a measure of acceptable interlaboratory results. An alternative, more practical approach is the use of *analytical equivalence*, in which a range of acceptable results is chosen prior to the study and used to judge the acceptability of the results obtained from the various laboratories.

Performing a thorough method validation can be a tedious process, but the reliability of the data generated with the method is linked directly to the application of quality assurance and quality control protocols, which must be followed assiduously (Table 6.5).

Briefly, to assure quality assurance and quality control, samples are analyzed using standard analytical procedures. A continuing program of analytical laboratory quality control verifies data quality and involves participation in interlaboratory crosschecks, and replicate sampling and analysis. When applicable, it is advisable, even insisted upon by the EPA, that analytical labs be certified to complete the analysis requested. However, in many cases, time constraints often do not allow for sufficient method validation. Many researchers have experienced the consequences of invalid methods and realized that the amount of time and resources required to solve problems discovered later exceeds what would have been expended initially if the validation studies had been performed properly.

Table 6.5. Comprehensive List of EPA Quality Control Methods that Apply to Situations within and Outside the Petroleum Refining Industry

EPA No.	Description of Method
814V96005	An Introduction to the ICR[a] Laboratory QC Database System
600987030	Availability Adequacy and Comparability of Testing Procedures for the Analysis of Pollutants Established under Section 304(h) of the Federal Water Pollution Control Act: Report to Congress
503990009	Citizen Volunteers in Environmental Monitoring: Summary Proceedings of the Second National Workshop, New Orleans, Louisiana
506690003	Contaminated Sediments: Relevant Statutes and EPA Program Activities
814P94001	DBP/ICR[b] Analytical Methods Guidance Manual: Public Comment
625R02010	Developing and Implementing an Estuarine Water Quality Monitoring, Assessment, and Outreach Program: The MYSound Project
625C02010	Developing and Implementing an Estuarine Water Quality Monitoring, Assessment, and Outreach Program: The MYSound Project [CDROM]
600486011	Eastern Lake Survey Phase 1: Quality Assurance Report
600486008	Eastern Lake Survey Phase I: Quality Assurance Plan
600389013	Ecological Assessment of Hazardous Waste Sites: A Field and Laboratory Reference
600879006	EPA Manual for Organics Analysis Using Gas Chromatography–Mass Spectrometry
823B00007	Guidance for Assessing Chemical Contaminant Data for Use in Fish Advisories, Volume 1: Fish Sampling and Analysis, Third Edition
823R93002	Guidance for Assessing Chemical Contaminant Data for Use in Fish Advisories, Volume 1: Fish Sampling and Analysis
600479019	Handbook for Analytical Quality Control in Water and Wastewater Laboratories
625689023	Handbook: Quality Assurance/Quality Control (QA/QC) Procedures for Hazardous Waste Incineration
600R95178	ICR Microbial Laboratory Manual
814B96001	ICR Sampling Manual
600483004	Interim Guidelines and Specifications for Preparing Quality Assurance Project Plans
821R01027	Interlaboratory Validation Study Results for *Cryptosporidium* Precision and Recovery for U.S. EPA Method 1622
600R94134	Method 100.2. Determination of Asbestos Structures over 10 Micrometers in Length in Drinking Water
821R02019	Method 1631, Revision E: Mercury in Water by Oxidation, Purge and Trap, and Cold Vapor Atomic Fluorescence Spectrometry
821R96013	Method 1632: Determination of Inorganic Arsenic in Water by Hydride Generation Flame Atomic Absorption
821R95028	Method 1632: Determination of Inorganic Arsenic in Water by Hydride Generation Flame Atomic Absorption [Draft]
821R95033	Method 1640: Determination of Trace Elements in Ambient Waters by On-Line Chelation Pre-concentration and Inductively Coupled Plasma–Mass Spectrometry

Table 6.5. (*continued*)

EPA No.	Description of Method
821B94004b	Method 1664: *n*-Hexane Extractable Material (HEM) and Silica Gel Treated *n*-Hexane Extractable Material (SGT-HEM) by Extraction and Gravimetry (Oil and Grease and Total Petroleum Hydrocarbons)
821R95026	Monitoring Trace Metals at Ambient Water Quality Criteria Levels: Briefing Book
503890006	Ocean Data Evaluation System (ODES) Data Submissions Manual
823B95001	QA/QC Guidance for Sampling and Analysis of Sediments Water and Tissue for Dredged Material Evaluations: Chemical Evaluations
810B92003	Quality Assurance Plan for the National Pesticide Survey of Drinking Water Wells: Analytical Method 2, Chlorinated Pesticides
823R02006	Quality Assurance Project Plan for Analytical Control and Assessment Activities in the National Study of Chemical Residues in Lake Fish Tissue
823R02005	Quality Assurance Project Plan for Sample Collection Activities for a National Study of Chemical Residues in Lake Fish Tissue
810B92001	Quality Assurance Project Plan for the National Pesticide Survey of Drinking Water Wells
810B92010	Quality Assurance Project Plan for the National Pesticide Survey of Drinking Water Wells: Analytical Method 1
810B92002	Quality Assurance Project Plan for the National Pesticide Survey of Drinking Water Wells: Analytical Method 1, Nitrogen/Phosphorous Pesticides, and Analytical Method 3, Chlorinated Acid Herbicides
810B92011	Quality Assurance Project Plan for the National Pesticide Survey of Drinking Water Wells: Analytical Method 3
810B92005	Quality Assurance Project Plan for the National Pesticide Survey of Drinking Water Wells: Analytical Method 5, Methyl Carbamates
810B92006	Quality Assurance Project Plan for the National Pesticide Survey of Drinking Water Wells: Analytical Method 6
810B92012	Quality Assurance Project Plan for the National Pesticide Survey of Drinking Water Wells: Analytical Method 6, Ethylene Thiourea
810B92007	Quality Assurance Project Plan for the National Pesticide Survey of Drinking Water Wells: Analytical Method 7, Fumigants
810B92008	Quality Assurance Project Plan for the National Pesticide Survey of Drinking Water Wells: Analytical Method 9, Nitrate and Nitrite
810B92015	Quality Assurance Project Plan for the National Pesticide Survey of Drinking Water Wells: Survey Statistics, Data Collection, and Processing
810B92014	Quality Assurance Project Plan for the National Survey of Pesticides in Drinking Water Wells: Well Sampling, Data Collection, and Processing
430986004	Quality Assurance/Quality Control (QA/QC) for 301(h) Monitoring Programs: Guidance on Field and Laboratory Methods.
812B92004	Recalculation of Screening Level Concentrations for Nonpolar Organic Contaminants in Marine Sediments [Final Report]

(*continued overleaf*)

Table 6.5. (*continued*)

EPA No.	Description of Method
600A94023	Regulation of Municipal Sewage Sludge under the Clean Water Act Section 503: A Model for Exposure and Risk Assessment for MSW-Compost
821R01028	Results of the Interlaboratory Method Validation Study Results for Determination of *Cryptosporidium* and *Giardia* Using U.S. EPA Method 1623
820R82103	Sampling Protocols for Collecting Surface Water, Bed Sediment, Bivalves, and Fish for Priority Pollutant Analysis: Final Report
823R92006	Sediment Classification Methods Compendium
832C86100	Sludge Compost Marketing and Distribution Regulatory Requirements in the United States
815R00002	Supplement A to the Unregulated Contaminant Monitoring Regulation Analytical Methods and Quality Control Manual
600491016	Test Methods for *Escherichia coli* in Drinking Water: EC Medium with Mug Tube Procedure Nutrient Agar with Mug Membrane Filter Procedure
440183079c	Test Methods: Methods for Non-conventional Pesticides Chemicals Analysis of Industrial and Municipal Wastewater
815R01028	UCMR[c] List 1 and List 2 Chemical Analytical Methods Quality Control Manual (1999)
815R00006	Unregulated Contaminant Monitoring Regulation Analytical Methods and Quality Control Manual
815R99004	Unregulated Contaminant Monitoring Regulation Analytical Methods and Quality Control Manual
815R99004	Unregulated Contaminant Monitoring Regulation Analytical Methods and Quality Control Manual [Draft]
841B04005	Wadeable Streams Assessment: Quality Assurance Project Plan
600487037	Western Lake Survey: Phase 1, Quality Assurance Report

Source: Quevauviller, 2002.
[a] ICR, information collection requests.
[b] DBP, disinfectant by-product.
[c] UCMR, unregulated contaminant monitoring rule.

Putting in time and effort up front will help any environmental analysts to find a way through the method validation maze and eliminate many of the problems common to inadequately validated analytical methods.

One method that is often used for method validation is the addition of a surrogate to the sample matrix, then following the track of the surrogate through the separation procedure. It is appropriate at this point, to comment on such a procedure.

Because of the nature of sampling protocols and the location of the sites from which samples are taken, it should be assumed that no two samples are identical. Factors such as (1) depth from which the sample is taken, (2) distance of the sampling point from the initial location of the spill, and (3) ambient conditions play a role in determining sample character. In fact, each sample is a unique combination of matrix–analyte interactions. Adding surrogate compounds to each sample and then determining their recovery is used as a benchmark for gauging the success of the extraction procedure. The best surrogates are those that are most like the target analytes, such as isotopically labeled versions of each target molecule. Whether or not use of such molecules is justified relates to the data that are required.

As an alternative, similar chemical compounds can be used, but the choice of surrogate compounds can limit or can maximize data interpretation. For most analyses, it is possible to choose surrogates that will always generate excellent recoveries, regardless of the complexity of the sample. Choosing surrogate molecules to produce *acceptable quality control* is not realistic. The use of surrogates is to obtain reliable information about the overall strengths and weaknesses of the analytical method. In addition, the surrogate(s) should reflect the chemical behavior and physical properties of the analytes. Important chemical behavior and properties include (1) acidic and basic properties of the analytes, (2) the range of polarity of the analytes, (3) reactivity in chemical derivatization procedures, and (4) the sensitivity of the analytes to decomposition caused by extremes of chemical or physical environment.

Surrogates chosen to monitor any of these areas should ideally bracket the range of the property. However, it should be pointed out that very few individual methods specify surrogates that provide information on all these areas, let alone bracket the property. The analyst must determine the suitability of the surrogate to reflect the properties and behavior of the analytes as well as the ability of the surrogate to be used for the collection of reliable and defensible data.

6.7. QUALITY CONTROL AND QUALITY ASSURANCE

Quality control (QC) and quality assurance (QA) programs are key components of all analytical protocols in all areas of analysis, including environmental, pharmaceutical, and forensic testing, among others (Patnaik, 2004). These programs mandate that laboratories follow a set of well-defined guidelines to achieve valid analytical results to a high degree of reliability and accuracy within an acceptable range. Although such programs may vary depending on the regulatory authority, certain key features of these programs are more or less the same (see below). However, there is often confusion between the terms *quality assurance* and *quality control*, perhaps because there is also considerable overlap between certain aspects of quality assurance and quality control programs.

6.7.1. Quality Control

Quality controls are single procedures that are performed in conjunction with an analysis to help assess the success of the analysis in a quantitative manner. Examples of quality controls are blanks, calibration, calibration verification, surrogate additions, matrix spikes, laboratory control samples, performance evaluation samples, and determination of detection limits. The success of the quality control is evaluated against an acceptance limit. Actual generation of the acceptance limit is a function of quality assurance; it would not be termed a quality control.

Unlike quality assurance plans, which address primarily regulatory requirements involving comprehensive documentation, quality control programs are science based, the components of which may be defined statistically. The two most important components of quality control are (1) determination of the precision of analysis and (2) determination of the accuracy of measurement.

Whereas *precision* (Section 6.5) measures the reproducibility of data from replicate analyses, the *accuracy* (Section 6.4) of a test estimates how accurate the data are, that is, how close the data would represent probable true values or how accurate the analytical procedure is to giving results that may be close to true values. Precision and accuracy are both measured on one or more samples selected at random for analysis from a given batch of samples. The precision of analysis is usually determined by running duplicate or replicate tests on one of the samples in a given batch of samples. It is expressed statistically as standard deviation, relative standard deviation (RSD), coefficient of variance (CV), standard error of the mean (M), and relative percent difference (RPD).

The standard deviation in measurements, however, can vary with the analyte concentrations. On the other hand, RSD, which is expressed as the ratio of standard deviation to the arithmetic mean of replicate analyses and is given as a percentage, does not have this problem and is a more rational way of expressing precision:

$$\text{RSD} = -\frac{\text{standard deviation}}{\text{arithmetic mean of replicate analysis}} \times 100\%$$

The standard error of the mean, M, is the ratio of the standard deviation to the square root of the number of measurements (n):

$$M = \frac{\text{standard deviation}}{\sqrt{n}}$$

This scale, too, will vary in the same proportion as standard deviation with the size of the analyte in the sample.

In routine testing, many repeat analyses of a sample aliquot may not be possible. Alternatively, therefore, the precision of a test may be determined from duplicate analyses of sample aliquots and expressed as RPD:

$$\text{RPD} = \frac{a_1 - a_2}{(a_1 + a_2)/2} \times 100\%$$

or

$$\text{RPD} = \frac{a_2 - a_1}{(a_1 + a_2)/2} \times 100\%$$

where a_1 and a_2 are the results of duplicate analyses of a sample. Since only two tests are performed on a selected sample, the RPD may not be as accurate a measure of precision as the relative standard deviation, and since the relative standard deviation does not vary with sample size, it should be used whenever possible to estimate the precision of analysis from replicate tests.

The accuracy of an analysis can be determined by several procedures. One common method is to analyze a *known sample*, such as a standard solution or a quality control check standard solution that may be available commercially, or a laboratory-prepared standard solution made from a neat compound, and to compare the test results with the true values (values expected theoretically). Such samples must be subjected to all analytical steps, including sample extraction, digestion, or concentration, similar to regular samples. Alternatively, accuracy may be estimated from the recovery of a known standard solution *spiked* or added into the sample in which a known amount of the same substance that is to be tested is added to an aliquot of the sample, usually as a solution, prior to the analysis. The concentration of the analyte in the spiked solution of the sample is then measured. The percent spike recovery is then calculated. A correction for the bias in the analytical procedure can then be made, based on the percent spike recovery. However, in most routine analysis such bias correction is not required. Percent spike recovery may then be calculated as follows:

$$\text{recovery}(\%) = \frac{\text{measured concentration}}{\text{theoretical concentration}} \times 100\%$$

The percent spike recovery to measure the accuracy of analysis may also be determined by the EPA method often used in environmental analysis:

$$\text{recovery}(\%) = \frac{100(X_s - X_u)}{K}$$

where X_s is the measured value for the spiked sample and X_u is the measured value for the unspiked sample adjusted for the dilution of the spike, and K is the known value of spike in the sample.

6.7.2. Quality Assurance

Quality assurance is an umbrella term that is applied correctly to everything that the laboratory does to assure product reliability. As the product of a laboratory is information, anything that is done to improve the reliability of the information generated falls under quality assurance.

Quality assurance includes all the quality controls, the generation of expectations (acceptance limits) from the quality controls, plus a great number of other activities, such as (1) analyst training and certification; (2) data review and evaluation; (3) preparation of final reports of analysis; (4) information given to clients about tests that are needed to fulfill regulatory requirements; (5) use of the appropriate tests in the laboratory; (6) obtaining and maintaining laboratory certifications/accreditations; (7) conducting internal and external audits; (8) preparing responses to the audit results; (9) receipt, storage, and tracking of samples; and (10) tracking the acquisition of standards and reagents.

Thus, the objective of quality assurance (usually in the form of a *quality assurance plan*) is to obtain reliable and accurate analytical results that may be stated with a high level of confidence (statistically), so that such results are legally defensible. The key features of any plan involve essentially documentation and record keeping. In short, quality assurance program involves documentation of sample collection for testing, the receipt of samples in the laboratory, and their transfer to those who will perform the analyses. The information is recorded on chain-of-custody forms, stating dates and times together with the names and signatures of persons carrying out these tasks. Other pertinent information is also, recorded, such as preservatives added to the sample to prevent degradation of test analytes, the temperature at which the sample is stored, the temperature to which the sample is brought prior to its analysis, the nature of the container (which may affect the stability of the sample), and its holding time prior to testing.

In fact, the performance of quality control is just one small aspect of the quality assurance program. The functions of quality assurance are embodied in the terms *analytically valid* and (in these days of perpetual litigation) *legally defensible*. *Analytically valid* means that the target analyte has been (1) identified correctly and (2) quantified using fully calibrated tests. In addition, the sensitivity of the test method detection limit has been established, and the analysts have demonstrated that they are capable of performing the test. The accuracy and precision of the test on the particular sample must also have been determined, and the possibility of false positive and false negative results has been evaluated through performance of blanks and other test-specific interference procedures.

6.8. METHOD DETECTION LIMIT

The *method detection limit* (MDL) is the smallest quantity or concentration of a substance that a particular instrument can measure (Patnaik, 2004). It is related to the *instrument detection limit* (IDL), which depends on the type of instrument and its sensitivity, and on the physical and chemical properties of the test substance.

The method detection limit is, in reality, a statistical concept that is applicable only in trace analysis of certain types of substances, such as organic pollutants by gas chromatographic methods. The method detection limit measures the minimum detection limit of the method and involves all analytical steps, including sample extraction, concentration, and determination by an analytical instrument. Unlike the instrument detection limit, the method detection limit is not confined only to the detection limit of the instrument.

In the environmental analysis of organic pollutants, the method detection limit is the minimum concentration of a substance that can be measured and reported with 99% confidence that the analyte concentration is greater than zero and is determined from the analysis of a sample in a given matrix containing the analyte. For determination of the method detection limit, several replicate analyses are performed at the concentration level of the instrument detection limit or at a level equivalent to two to five times the background noise level. The standard deviation of the replicate tests is found. The method detection limit is determined by multiplying the standard derivation by the *t-factor*. In environmental analysis, however, periodic determination of the method detection limit (e.g., once per year or with any change in personnel, location, or instrument) is part of the quality control requirement.

REFERENCES

Alder, H. L., and Roessler, E. B. 1972. *Introduction to Probability and Statistics*. W.H. Freeman, San Francisco.

ASTM. 2004. *Annual Book of ASTM Standards*. American Society for Testing and Materials, West Conshohocken, PA.

Baker, C. C. T. 1966. *An Introduction to Mathematics*. Arco Publishing Company, New York.

Box, G., Hunter, W., and Hunter, J. 1978. *Statistics for Experimenters*. Wiley, New York.

Caulcutt, R., and Boddy, R. 1983. *Statistics for Analytical Chemists*. Chapman & Hall, London.

Dean, J. R. 1998. *Extraction Methods for Environmental Analysis*. Wiley, New York.

Dean, J. R. 2003. *Methods for Environmental Trace Analysis*. Wiley, Hoboken, NJ.

Dixon, W. J., and Massey, F. J. 1969. *Introduction to Statistical Analysis*. McGraw-Hill, New York.

EPA. 1998. *Test Methods for Evaluating Solid Waste: Physical/Chemical Methods*. EPA/SW-846. 3rd ed., 1986; Update I, 1992; Update II, 1994; Update III, 1996; Update IV, 1998. Environmental Protection Agency, Washington, DC.

EPA. 2004. Environmental Protection Agency, Washington, DC. Web site: http://www.epa.gov.

IP. 2004. *IP Standard Methods 2004*. Institute of Petroleum, London.

Jaffe, A. J., and Spirer, H. F. 1987. *Misused Statistics: Straight Talk for Twisted Numbers*. Marcel Dekker, New York.

Meier, P. C., and Zünd, R. E. 2000. *Statistical Methods in Analytical Chemistry*. Wiley, New York.

Patnaik, P. (Ed.). 2004. *Dean's Analytical Chemistry Handbook*, 2nd ed. McGraw-Hill, New York.

Quevauviller, P. 2002. *Quality Assurance for Water Analysis*. Wiley, Hoboken, NJ.

Smith, R. K. 1999. *Handbook of Environmental Analysis*, 4th ed. Genium Publishing, Schenectady, NY.

Speight, J. G. 2001. *Handbook of Petroleum Analysis*. Wiley, New York.

Speight, J. G. 2002. *Handbook of Petroleum Product Analysis*. Wiley, Hoboken, NJ.

CHAPTER 7

ANALYTICAL METHODS

As already noted, the chemical composition of petroleum and petroleum products is complex and may change over time following release into the environment. These factors make it essential that the most appropriate analytical methods are selected from a comprehensive list of methods and techniques that are used for the analysis of environmental samples (Dean, 1998; Miller, 2000; Budde, 2001; Sunahara et al., 2002; Nelson, 2003; Smith and Cresser, 2003). But once a method is selected, it may not be the ultimate answer to solving the problem of identification and, hence, behavior (Patnaik, 2004). There are a significant number of petroleum hydrocarbon–affected sites, and evaluation and remediation of these sites may be difficult because of the complexity of the issues (analytical, scientific, and regulatory not to mention economic) regarding water and soil affected.

Once the sample preparation is complete, there are several approaches to the analysis of petroleum constituents in the water and soil: (1) leachability or toxicity of the sample, (2) the amounts of total petroleum hydrocarbons in the sample, (3) petroleum group analysis, and (4) fractional analysis of the sample. These methods measure different petroleum constituents that might be present in petroleum-contaminated environmental media.

The methods that measure the concentration of total petroleum hydrocarbons (Chapter 4) generate a single number that represents the combined concentration of all petroleum hydrocarbons in a sample that are measurable by the particular method. Therefore, determination of the total petroleum hydrocarbons in a sample is method dependent. On the other hand, methods that measure a petroleum group type concentration separate and quantify different categories of hydrocarbons (e.g., saturates, aromatics, and polars/resins) (Chapter 2) (Speight, 1999, 2001). The results of petroleum group type analyses can be useful for product identification because products such as, for example, gasoline, diesel fuel, and fuel oil have characteristic levels of various petroleum groups. The methods that measure identifiable petroleum fractions can be used to indicate and/or quantify the changes that have occurred through weathering of the sample.

Although these methods measure different petroleum hydrocarbon categories, there are several basic steps that are common to the analytical processes for all methods, no matter the method type or the environmental matrix. In general, these steps are (1) collection and preservation—requirements specific to environmental matrix and analytes of interest; (2) extraction so that separations of the analytes

---

Environmental Analysis and Technology for the Refining Industry, by James G. Speight
Copyright © 2005 John Wiley & Sons, Inc.

of interest from the sample matrix can be achieved; (3) concentration—enhances the ability to detect analytes of interest; (4) cleanup, dependent on the need to remove interfering compounds; and (5) measurement, or quantification, of the analytes (Dean, 1998). Each step affects the final result, and a basic understanding of the steps is vital to data interpretation.

7.1. LEACHABILITY AND TOXICITY

As a start and for regulatory and remediation purposes, a standard test is needed to measure the likelihood of toxic substances getting into the environment and causing harm to organisms. The test (required by the U.S. Environmental Protection Agency) is the *toxicity characteristic leaching procedure* (TCLP, EPA SW-846 Method 1311), designed to determine the mobility of both organic and inorganic contaminants present in liquid, solid, and multiphase wastes.

The method was developed to estimate the mobility of specific inorganic and organic contaminates destined for disposal in municipal landfills. The extraction is performed using acetic acid as the extraction fluid. The pH of the acetic acid/sodium acetate buffer solution is maintained at 4.93. This sample/acetic acid mixture is subjected to rotary extraction, designed to accelerate years of material exposure in the shortest possible time. After extraction, the resulting liquid is subjected to analysis utilizing a list of contaminants that includes metals, volatile organic compounds, semivolatile organic compounds, pesticides, and herbicides.

The toxicity characteristic leaching procedure may be subject to misinterpretation if the compounds under investigation are not included in the methods development or the list of contaminants leading to the potential for technically invalid results. However, an alternative procedure, the synthetic precipitation leaching procedure (SPLP, EPA SW-846 Method 1312) may be appropriate. This procedure is applicable for materials where the leaching potential due to normal rainfall is to be determined. Instead of the leachate simulating acetic acid mixture, nitric and sulfuric acids are utilized in an effort to simulate the acid rains resulting from airborne nitric and sulfuric oxides.

7.2. TOTAL PETROLEUM HYDROCARBONS

Total petroleum hydrocarbon (TPH) (Chapter 4) analyses (Tables 7.1 and 7.2) are conducted to determine the total amount of hydrocarbon present in the environment. There are a wide variety of methods for measurement of the total petroleum hydrocarbon in a sample, but analytical inconsistencies must be recognized because of the definition of total petroleum hydrocarbons and the methods employed for analysis (Rhodes et al., 1994). Thus, in practice, the term *total petroleum hydrocarbon* is defined by the analytical method since different methods often give different results because they are designed to extract and measure slightly different subsets of petroleum hydrocarbons.

Table 7.1. Summary of Analytical Methods for Determining the Total Petroleum Hydrocarbons in a Sample

Method Name	Method Type/ Environmental Media	Typical Products Detected	Typical Carbon Ranges Detected	Methods
GC-based TPH methods	Primarily laboratory but also field applications—can be adapted for all media	Primarily gasolines, diesel fuel, and fuel oil No. 2; can be modified for heavier hydrocarbon mixtures (e.g., lubricating oils, heavy fuel oils)	Normally between C_6 and C_{25} or C_{36} (can be modified for higher carbon numbers)	EPA Method 8015B, state-modified 8015 methods
IR-based TPH methods	Laboratory and field screening—most appropriate for water and soil	Primarily diesel and fuel oils	Most hydrocarbons, with exception of volatile and very high hydrocarbons	EPA Method 418.1
Gravimetric TPH methods	Laboratory—most appropriate for wastewaters, sludges, and sediment	Most appropriate for heavier petroleum products (e.g., crude oils, lubricating oils)	Anything that is extractable, with exception of volatiles, which are lost	EPA Method 9071; EPA Method 1664
Immunoassay TPH methods	Field screening—most appropriate for soil and water	Various products (but yields only screening numbers)	Aromatic hydrocarbons (e.g., BTEX, PAHs)	EPA Method 4030

Source: EPA, 2004.

Table 7.2. Summary of Methods for Determining the Total Petroleum Hydrocarbons in a Sample

Method Name	Matrix	Scope of Method
Gravimetric methods		
EPA Method 413.1	Water and wastewater	Petroleum fuels from gasoline through No. 2 fuel oils are completely or partially lost in the solvent removal operation. Recoveries of some crude oils and heavy fuel oils will be low.
EPA Method 9070	Solid waste	Applicable to determination of relatively nonvolatile hydrocarbons. Not applicable to measurement of light hydrocarbons; petroleum fuels, from gasoline through No. 2 fuel oils, are completely or partially lost. Recoveries of some crude oils and heavy fuel oils will be low.
EPA Method 9071A	Sludge	Used to recover low levels of oil and grease from sludge. Used when relatively polar, heavy petroleum fractions are present, or when the levels of nonvolatile greases challenge the solubility limit of the solvent. Not recommended for measurement of low-boiling fractions.
Standard Method 5520B	Water and wastewater	Suitable for biological lipids and mineral hydrocarbons. Not suitable for low-boiling fractions.
Standard Method 5520D	Water and wastewater	Suitable for biological lipids and mineral hydrocarbons. Method D is the method of choice when relatively polar, heavy petroleum fractions are present, or when the levels of nonvolatile greases may challenge the solubility of the solvent.
Standard Method 5520E	Sludge	Suitable for biological lipids and mineral hydrocarbons. Not suitable for low-boiling fractions. Method E is a modification of Method D.
Standard Method 5520F	Water and wastewater	Suitable for biological lipids and mineral hydrocarbons. Not suitable for low-boiling fractions. May be used in conjunction with Method B, D, or E.
Infrared methods		
EPA Method 413.2	Water and wastewater	Applicable to the measurement of light fuels, although loss of about half of any gasoline present during the extraction manipulation can be expected.
EPA Method 418.1	Water and wastewater	Applicable to a wide range of hydrocarbons, although volatile components will be lost. Modifications exist for measurement of TPH in soil.

The analysis for the total petroleum hydrocarbons (TPHs) in a sample as a means of evaluating petroleum-contaminated sites is also an analytical method in common use. The data are used to establish target cleanup levels for soil or water by regulatory agencies in the United States and in many other countries.

The data obtained by the analysis have become key remediation criteria and it is essential that the environmental analyst (and others who may use the data) be knowledgeable about the various analytical methods. It is also important to know that minor method deviations may be found from region to region. For example, in terms of nomenclature, itself a complex and often ill-defined area of petroleum science (Chapter 1) (Speight, 1999), the analytical methods may refer to *total petroleum hydrocarbons* as *mineral oil, hydrocarbon oil, extractable hydrocarbon*, and *oil and grease*.

Thus, as often occurs in petroleum science (Speight, 1999), the definition of total petroleum hydrocarbons depends on the analytical method used because the total petroleum hydrocarbons measurement is the total concentration of the hydrocarbons extracted and measured by a particular method. The same sample analyzed by different methods may produce different values. For this reason, it is important to know exactly how each determination is made since interpretation of the results depends on understanding the capabilities and limitations of the method. If used indiscriminately, measurement of the total petroleum hydrocarbons in a sample can be misleading, leading to an inaccurate assessment of risk.

There are several reasons why the data for total petroleum hydrocarbons do not provide ideal information for investigating sites and establishing target cleanup criteria. For example, use of the term *total petroleum hydrocarbons* suggests that the analytical method measures the combined concentration of all petroleum-derived hydrocarbons, thereby giving an accurate indication of site contamination. But this is not always the case. Furthermore, target cleanup levels based on total petroleum hydrocarbons concentrations implicitly assume (1) that the data are an accurate measurement of petroleum-derived hydrocarbon concentration, and (2) the data also indicate the level of risk associated with the contamination. These assumptions are not correct due to many factors, including the nonspecificity of some of the methods used and the complex nature of petroleum hydrocarbons and their interaction with the environment over time.

One significant difficulty in measuring the concentration of total petroleum hydrocarbons in a sample (for different petroleum products) is the fact that the boiling ranges and carbon number ranges of refined petroleum products often overlap. Refined petroleum products are manufactured primarily through various processes, including distillation processes that separate fractions from crude oil by their respective boiling ranges.

Manufacturing processes may also increase the yield of low-molecular-weight fractions, reduce the concentration of undesirable sulfur and nitrogen components, and incorporate performance-enhancing additives. Additionally, because it is impossible to identify all constituents of a petroleum product, their respective boiling ranges are often used to describe these constituents. And because distillation, as practiced in the industry, is often not capable of producing sharp distinctions in boiling-point cutoffs, there is overlap between distillate fractions. The boiling-point range correlates to the carbon number since the higher the carbon number, the higher the boiling point. However, chemical structure also influences boiling point. For example, branched and aromatic compounds of the same carbon number differ in boiling point from their corresponding n-alkane analogs. For these reasons, boiling point actually defines an approximate carbon range.

Ambiguous terminology associated with the term *total petroleum hydrocarbons* and the methods of analysis also present additional difficulty in interpreting results. Each method has its own designation. For example, there are terms such as *total recoverable petroleum hydrocarbons* (TRPHs), *diesel range organics* (DROs), *gasoline range organics* (GROs), and *total petroleum hydrocarbons–gasoline* (TPH-G). However, to confuse the issue even further, a method name that cites a product (such as gasoline or diesel) only implies a carbon range. For example, *total petroleum hydrocarbons–gasoline* does not necessarily imply that gasoline is present but rather, that hydrocarbons boiling in the gasoline range (whether or not they would actually be found in gasoline) are present in the sample. These abbreviations may imply different carbon ranges to different laboratories, and therefore the methods are not optimized to identify product type. Even with improved, more detailed analytical methods, identification of aged products may prove difficult.

The reason for the availability of a large number of methods for the measurement of total petroleum hydrocarbons centers on the compositional complexity of petroleum and petroleum products, and subsequently, there is no single *suitable* or *adequate* method for measuring all types of petroleum-derived contamination. For example, methods that are appropriate for samples contaminated by gasoline are not often suitable for the measurement of diesel fuel contamination in other samples.

Some methods measure more compounds than other methods because they employ more rigorous extraction techniques or more efficient solvents for the extraction procedure(s). Other methods are subject to interferences from naturally occurring materials such as animal and vegetable oils, peat moss, or humic material, which may result in artificially high reported concentrations of the total petroleum hydrocarbons. Some methods use cleanup steps to minimize the effect of nonpetroleum hydrocarbons, with variable success. Ultimately, many of the methods are limited by the extraction efficiency and the detection limits of the instrumentation used for measurement.

Thus, the choice of a specific method should be based on compatibility with the particular type of hydrocarbon contamination to be measured, and furthermore, the choice may depend on local or regional regulatory requirements for the type of hydrocarbon contamination that is known, or suspected, to be present and the risk that a specific site will change with time as contaminants evaporate, dissolve, biodegrade, and/or become sequestered.

Although the utility of data for total petroleum hydrocarbons for risk assessment is minimal, it is an inexpensive tool that can be used for three purposes: (1) determining if there is a problem; (2) assessing the severity of contamination, and (3) following the progress of a remediation effort. If the data for the total petroleum hydrocarbons indicate that there may be significant contamination of environmental media, other data can be collected so that harm to human health can be assessed quantitatively. These other data can include target analyte concentration data, and petroleum fraction concentration data obtained using the evolving fraction-based analytical methods.

There are many analytical techniques available that measure total petroleum hydrocarbon concentrations in the environment, but no single method is satisfactory for measurement of the entire range of petroleum-derived hydrocarbons. In addition, and because the techniques vary in the manner in which hydrocarbons are extracted and detected, each method may be applicable to the measurement of different subsets of the petroleum-derived hydrocarbons present in a sample. The four most commonly used total petroleum hydrocarbon analytical methods include (1) gas chromatography (GC), (2) infrared spectrometry (IR), (3) gravimetric analysis, and (4) immunoassay (Table 7.1) (Miller, 2000, and references cited therein).

7.2.1. Gas Chromatographic Methods

Gas chromatographic methods are currently the preferred laboratory methods for measurement of total petroleum hydrocarbon measurement because they detect a broad range of hydrocarbons and provide both sensitivity and selectivity. In addition, identification and quantification of individual constituents of the total petroleum hydrocarbon mix is possible.

Methods based on gravimetric analysis (Table 7.2) are also simple and rapid, but they suffer from the same limitations as those of infrared spectrometric methods (Table 7.2). Gravimetric-based methods may be useful for oily sludge and wastewaters, which will present analytical difficulties for other, more sensitive methods. Immunoassay methods for the measurement of total petroleum hydrocarbon are also popular for field testing because they offer a simple, quick technique for in situ quantification of the total petroleum hydrocarbons.

For methods based on gas chromatography (Table 7.3), the total petroleum hydrocarbons fraction is defined as any chemicals extractable by a solvent or purge gas and detectable by gas chromatography–flame ionization detection (GC/FID) within a specified carbon range. The primary advantage of such methods is that they provide information about the type of petroleum in the sample in addition to measuring the amount. Identification of product type(s) is not always straightforward, however, and requires an experienced analyst of petroleum products (Sullivan and Johnson, 1993; Speight, 1999, 2001, 2002). Detection limits are method dependent as well as matrix dependent and can be as low as 0.5 mg/L in water or 10 mg/kg in soil.

Chromatographic columns are commonly used to determine total petroleum hydrocarbon compounds approximately in the order of their boiling points. Compounds are detected by means of a flame ionization detector, which responds to virtually all compounds that can burn. The sum of all responses within a specified range is equated to a hydrocarbon concentration by reference to standards of known concentration.

Table 7.3. EPA Test Methods for Total Petroleum Hydrocarbons

SW-846 Method	Water/ Wastewater Method	Analytes	Primary Equipment	Sample Preparation[a]
4030[b]	n/a	TPHs	Immunoassay	Included in kit
4035[b]	n/a	PAHs	Immunoassay	Included in kit
8015[c]	n/a	Aliphatic and aromatic hydrocarbons; nonhalogenated VOCs	GC/FID	Extraction (SVOCs)[d]; purge-and-trap and headspace (VOCs)[d]; azeotropic distillation (nonhalogenated VOCs)[c,d]
8021[c,e]	502.2/602	Aromatic VOCs (not ethers or alcohols)	GC/PID	Purge-and-trap
8100	n/a	PAHs	GC/FID	Extraction[d]
8260[f]	524.2/624	VOCs	GC/MS	Purge-and-trap; static headspace; azeotropic distillation[d]
8270	525/625	SVOCs	GC/MS	Extraction[d]
8310	610	PAHs	High-performance liquid chromatography (HPLC)	Extraction[d]
8440[g]	418.1[h]	TPH	IR spectrophotometer	Supercritical fluid extraction from soils[d]

[a] These are the standard methods of preparation for the corresponding method. They may vary depending on specific analytical needs.
[b] Screening method for soils.
[c] MTBE can be analyzed for with U.S. EPA SW-846 Method 8015 or 8021; however, 8021 has lower detection limits, is subject to less interference in highly contaminated samples; and tends to be more economical by providing BTEX data in the same analysis. Concerns about coelution with some alkanes requires at least one confirmatory analysis with SW-846 Method 8260 per site.
[d] See Chapter 4 of SW-846 for specific appropriate methods.
[e] 8021 replaces 8010 and 8020.
[f] 8260 replaces 8240.
[g] This method is similar to 418.1, however, perchlorethane (PCE) is used as an IR solvent instead of Freon-113.
[h] 418.1 is used extensively, although it is not on the list of promulgated methods.

Two methods, EPA SW-846 8015 and 8015A, were, in the past, often quoted as the source of gas chromatography–based methods for measurement of the total petroleum hydrocarbons in a sample. However, the original methods were developed for nonhalogenated volatile organic compounds and were designed to measure a short target list of chemical solvents rather than petroleum hydrocarbons. Thus, because there was no universal method for total petroleum hydrocarbons, there were many variations of these methods. Recently, an updated method

(EPA 8015B) provides guidance for the analysis of gasoline and diesel range organic compounds.

The current individual methods differ in procedure, compounds detected, extraction techniques, and extraction solvents used. Some methods may include a cleanup step to remove biogenic (bacterial or vegetation-derived) material, while others do not. The methods have in common a boiling point type of column and a flame ionization detector. Selection of a method depends on the type of hydrocarbon suspected to be in the sample.

For example, if gasoline is suspected to be the sole contaminant, the method will use purge-and-trap sample introduction. If higher-boiling petroleum fractions (diesel, middle distillates, motor oil) are the contaminants, the analysis will use direct injection and hotter oven temperatures. Mixtures or unknown contamination may require both volatile range and extractable range analyses. Alternatively, a single injection can be used to analyze the entire sample, but the extraction method must not use a solvent evaporation step.

Gas chromatography–based methods can be broadly used for (various types) of petroleum contamination but are most appropriate for detecting nonpolar hydrocarbons with carbon numbers between C_6 and C_{25} or C_{36}. Many lubricating oils contain molecules with more than 40 carbon atoms. In fact, crude oil itself contains molecules having more than 100 carbons or more. These high-molecular-weight hydrocarbons are outside the detection range of the more common gas chromatographic methods, but specialized gas chromatographs are capable of analyzing such high-molecular-weight constituents.

Accurate quantification depends on adjusting the chromatograph to reach as high a carbon number as possible, then running a calibration standard with the same carbon range as the sample. There should also be a check for mass discrimination, a tendency for higher-molecular-weight hydrocarbons to be retained in the injection port. If a sample is suspected to be heavy oil or to contain a mixture of light oil and heavy oil, the most appropriate method must be used.

Gravimetric or infrared methods are often preferred for high-molecular-weight samples. These methods can even be used as a check on gas chromatographic data if it is suspected that high-molecular-weight hydrocarbons are present but are not being detected.

Calibration standards vary. Most methods specify a gasoline calibration standard for volatile range total petroleum hydrocarbons and a diesel fuel No. 2 standard for extractable range total petroleum hydrocarbons. Some methods use synthetic mixtures for calibration. Because most methods are written for gasoline or diesel fuel, total petroleum hydrocarbons methods may have to be adjusted to measure contamination by heavier hydrocarbons (e.g., heavy fuel oil, lubricating oil, or crude oil). Such adjustments may entail use of a more aggressive solvent, a wider gas chromatographic window that allows detection of molecules containing up to C_{36} or more, and a different calibration standard that more closely resembles the constituents of the sample under investigation.

Gas chromatographic methods can be modified and fine-tuned so that they are suitable for measurement of specific petroleum products or group types. These modified methods can be particularly useful when there is information on the source of contamination, but method results should be interpreted with the clear understanding that a modified method was used for detection of a specific carbon range.

Interpretation of gas chromatographic data is often complicated, and the analytical method should always be considered when interpreting concentration data. For example, a volatile range analysis may be very useful for quantifying total petroleum hydrocarbons at a gasoline release site, but a volatile range analysis will not detect the presence of lube oil constituents. In addition, a modified method that has been specifically selected for detection of gasoline-range organics at a gasoline-contaminated site may also detect hydrocarbons from other petroleum releases because fuel carbon ranges frequently overlap. Gasoline is found primarily in the volatile range. Diesel fuel falls primarily in an extractable range. Jet fuel overlaps both the volatile and semivolatile ranges. However, the detection of different types of petroleum constituents does not necessarily indicate that there have been multiple releases at a site. Analyses of spilled waste oil will frequently detect the presence of gasoline, and sometimes diesel. This does not necessarily indicate multiple spills since all waste oil contain some fuel. As much as 10% of used motor oil can consist of gasoline.

If the type of contaminant is unknown, a *fingerprint* analysis can help in the identification procedure. A *fingerprint* or *pattern recognition* analysis is a direct injection analysis where the chromatogram is compared to chromatograms of reference materials. Certain fuels can be identified by characteristic, reproducible chromatographic patterns. For example, chromatograms of gasoline and diesel differ considerably, but many hydrocarbon streams may have similar fingerprints. Diesel No. 2 and No. 2 fuel oil both have the same boiling-point range and chromatographic fingerprint. A fingerprint can be used to identify a mixture conclusively when a known sample of that mixture or samples of the mixture's source materials are available as references.

Furthermore, as a fuel evaporates or biodegrades, its pattern can change so radically that identification becomes difficult. Consequently, a gas chromatographic fingerprint is not a conclusive diagnostic tool. The methods used for total petroleum hydrocarbon analysis must stress calibration and quality control, whereas pattern recognition methods stress detail and comparability.

The gas chromatographic methods usually cannot quantitatively detect compounds below C_6 because these compounds are highly volatile and interference can occur from the solvent peak. As much as 25% of fresh gasoline can be below C_6 but the problem is reduced for weathered gasoline and/or diesel range contamination because most of the very volatile hydrocarbons ($<C_6$) may no longer be present in the sample. Gas chromatographic methods may also be inefficient for quantification of polar constituents (nitrogen-, oxygen-, and sulfur-containing molecules). Some of the polar constituents are too reactive to pass through a gas chromatograph and thus will not reach the detector for measurement.

Oxygenated gasoline is sometimes analyzed by GC-based methods, but it should be noted that the efficiency of purge methods is lower for oxygenates such as ethers and alcohols because detector response to oxygenates is lower relative to hydrocarbons. Therefore, the data will be biased slightly low for ether-containing fuels compared to equivalent amounts of traditional gasoline. Methanol and ethanol elute before hexane, and consequently are not quantified and may not even be detected due to coelution with the solvent.

On the other hand, gas chromatographic methods may overestimate the concentration of total petroleum hydrocarbons in a sample due to the detection of nonpetroleum compounds. In addition, cleanup steps do not separate petroleum hydrocarbons perfectly from biogenic material such as plant oils and waxes, which are sometimes extracted from vegetation-rich soil. Silica gel cleanup may help to remove this interference but may also remove some polar hydrocarbons.

Because petroleum is made up of so many isomers, many compounds, especially those with more than eight carbon atoms, coelute with isomers of nearly the same boiling point. These unresolved compounds are referred to as the *unresolved complex mixture*. They are legitimately part of the petroleum signal, and unless otherwise specified, should be quantified. Quantifying such a mixture requires a baseline-to-baseline integration mode rather than a peak-to-peak integration mode. The baseline-to-baseline integration quantifies all of the petroleum constituents in the sample, but in peak-to-peak integration, only the individual resolved hydrocarbons (not including the unresolved complex mixture) are quantified.

7.2.2. Infrared Spectroscopy Methods

Infrared methods measure the absorbance of the C–H bond and most methods typically measure the absorbance at a single frequency (usually, 2930 cm^{-1}) that corresponds to the stretching of aliphatic methylene (CH$_2$) groups. Some methods use multiple frequencies, including 2960 cm^{-1} (CH$_3$ groups) and 2900 to 3000 cm^{-1} (aromatic C–H bonds).

Therefore, for infrared spectroscopic methods, the total petroleum hydrocarbons comprise any chemicals extracted by a solvent that are not removed by silica gel and can be detected by infrared spectroscopy at a specified wavelength. The primary advantage of the infrared-based methods is that they are simple and rapid. Detection limits (e.g., for EPA 418.1) are approximately 1 mg/L in water and 10 mg/kg in soil. However, the infrared method(s) often suffer from poor accuracy and precision, especially for heterogeneous soil samples. Also, the infrared methods give no information on the type of fuel present in the sample, and there is little, often no information about the presence or absence of toxic molecules, and no specific information about potential risk associated with the contamination.

Samples are extracted with a suitable solvent (i.e., a solvent with no C–H bonds), and biogenic polar materials are removed with silica gel. Some polar petroleum constituents may be removed as part of the silica gel cleanup. The absorbance of the silica gel eluate is measured at the specified frequency and compared to the absorbance of a standard or standards of known petroleum hydrocarbon concentration. The absorbance is a measurement of the sum of all the compounds contributing to the result. However, infrared methods cannot provide information on the type of hydrocarbon contamination.

For all IR-based TPH methods, the C–H absorbance is quantified by comparing it to the absorbance of standards of known concentration. An assumption is made that the standard has an aliphatic-to-aromatic ratio and an infrared response similar to that of the sample. Consequently, it is important to use a calibration standard as similar to the type of contamination as possible (EPA 418.1).

The infrared method that has been used most frequently (EPA 418.1) is appropriate only for water samples. A separatory funnel liquid–liquid extraction technique is used to extract hydrocarbons from the water. A method (EPA 5520D) using a Soxhlet extraction technique is suitable for sludge. This extraction is frequently used to adapt the method (EPA 418.1) to soil samples. An infrared-based supercritical fluid extraction method for diesel range contamination (EPA 3560) is available.

Similar to gas chromatographic methods, the data from infrared methods must be interpreted after considering certain limitations and interferences that can affect data quality. For example, the C–H absorbance is not always measured in exactly the same way. Within the set of methods that specify a single infrared measurement, some methods call for the measurement at precisely 2930 cm^{-1}, whereas others (including EPA 418.1) require measurement at the absorbance maximum nearest 2930 cm^{-1}. This variation can make a significant difference in the magnitude of the result and can lead to confusion when comparing duplicate sample results. If only C–H absorbance is measured, infrared methods will potentially underestimate the concentration of total petroleum hydrocarbons in samples that contain petroleum constituents such as benzene and naphthalene that do not contain alkyl C–H groups.

Because an infrared result is calculated as if the aromatics in the sample were present in the same ratio as in the calibration standard, accuracy depends on use of a calibration standard as similar to the type of contamination as possible. Use of a dissimilar standard will tend to create a positive bias in highly aliphatic samples and a negative bias in highly aromatic samples.

In summary, infrared methods are prone to interferences (positive bias) from nonpetroleum sources since many organic compounds have some type of alkyl group associated with them, whether petroleum-derived or not.

7.2.3. Gravimetric Methods

Gravimetric methods measure all chemicals that are extractable by a solvent, not removed during solvent evaporation, and capable of being weighed. Some gravimetric methods include a cleanup step to remove biogenic material. The advantage of gravimetric methods is that they are simple and rapid. Detection limits are approximately 5 to 10 mg/L in water and 50 mg/kg in soil.

However, gravimetric methods are not suitable for measurement of low-boiling hydrocarbons that volatilize at temperatures below 70 to 85°C. They are recommended for use with (1) oily sludge, (2) for samples containing heavy-molecular-weight hydrocarbons, or (3) for aqueous samples when hexane is preferred as the solvent.

Gravimetric methods give no information on the type of fuel present, no information about the presence or absence of toxic compounds, and no specific information about potential risk associated with the contamination.

In the method(s), petroleum constituents are extracted into a suitable solvent. Biogenic polar materials typically may be partially or completely removed with silica gel. The solvent is evaporated and the residue is weighed. This quantity is reported as a percentage of the total soil sample dry weight. These methods are better suited for heavy oil because they include an evaporation step.

There are a variety of gravimetric oil and grease methods suitable for testing water and soil samples (e.g., EPA SW-846 9070, 413.1, 9071). Technically, the result is an oil and grease result because no cleanup step is used. One method (EPA 9071) is used to recover low levels of oil and grease by chemically drying a wet sludge sample and then extracting it using Soxhlet apparatus. Results are reported on a dry-weight basis. The method is also used when relatively polar high-molecular-weight petroleum fractions are present, or when the levels of nonvolatile grease challenge the solubility limit of the solvent. Specifically, the method (EPA SW-846 9071) is suitable for biological lipids, mineral hydrocarbons, and some industrial wastewater.

Gravimetric methods for oil and grease (e.g., EPA SW-846 9071) measure anything that dissolves in the solvent and remains after solvent evaporation. These substances include hydrocarbons, vegetable oils, animal fats, waxes, soaps, greases, and related biogenic material. Gravimetric methods for total petroleum hydrocarbons (EPA 1664) measure anything that dissolves in the solvent and remains after silica gel treatment and solvent evaporation.

This method (EPA 1664) is a liquid–liquid extraction gravimetric procedure that employs *n*-hexane as the extraction solvent, in place of 1,1,2-trichloroethane (CFC-113) and/or 1,2,2-trifluoroethane (Freon-113), for determination of the conventional pollutant oil and grease. Because the nature and amount of material determined are defined by the solvent and by the details of the method used for extraction, oil and grease *method-defined* analytes are used. The method may be modified to reduce interferences and take advantage of advances in instrumentation provided that all method equivalency and performance criteria are met. However, *n*-hexane is a poor solvent for high-molecular-weight petroleum constituents (Speight, 1999, 2001). Thus, the method will produce erroneous data for samples contaminated with heavy oils.

All gravimetric methods measure any suspended solids that are not filtered from solution, including bacterial degradation products and clay fines. Method 9071 specifies using cotton or glass wool as a filter. Because extracts are heated to remove solvent, these methods are not suitable for measurement of low-boiling low-molecular-weight hydrocarbons (i.e., hydrocarbons having fewer than 15 carbon atoms) that volatilize at temperatures below 70 to 85°C (158 to 185°F). Liquid fuels, from gasoline through No. 2 fuel oil, lose volatile constituents during solvent removal. In addition, soil results that are reported on a dry-weight basis suffer from potential losses of lower-boiling hydrocarbon constituents during moisture determination, where the matrix is dried at approximately 103 to 105°C (217 to 221°F) for several hours in an oven.

7.2.4. Immunoassay Methods

Immunoassay methods correlate total petroleum hydrocarbons with the response of antibodies to specific petroleum constituents. Many methods measure only aromatics that have an affinity for the antibody, benzene–toluene–ethylbenzene–xylene, and PAH analysis (EPA 4030, Petroleum Hydrocarbons by Immunoassay).

The principle behind the test method(s) is that antibodies are made of proteins that recognize and bind with foreign substances (antigens) that invade host animals. Synthetic antibodies have been developed to complex with petroleum constituents. The antibodies are immobilized on the walls of a special cell or filter membrane. Water samples are added directly to the cell, while soils must be extracted before analysis. A known amount of labeled analyte (typically, an enzyme with an affinity for the antibody) is added after the sample. The sample analytes compete with the enzyme-labeled analytes for sites on the antibodies. After equilibrium is established, the cell is washed to remove any unreacted sample or labeled enzyme. Color development reagents that react with the labeled enzyme are added. A solution that stops color development is added at a specified time, and the optical density (color intensity) is measured. Because the coloring agent reacts with the labeled enzyme, samples with high optical density contain low concentrations of analytes. Concentration is inversely proportional to optical density.

The antibodies used in immunoassay kits are generally designed to bond with selected compounds. A correction factor supplied by the manufacturer must be used to calculate the concentration of total petroleum hydrocarbons. The correction factor can vary depending on product type because it attempts to correlate total petroleum hydrocarbons with the surrogates measured.

Immunoassay tests do not identify specific fuel types and are best used as screening tools. The tests are dependent on soil type and homogeneity. In particular, for clay and other cohesive soils, the tests are limited by a low capacity to extract hydrocarbons from the sample.

7.3. PETROLEUM GROUP ANALYSIS

Petroleum group analyses are conducted to determine amounts of the petroleum compound classes (e.g., saturates, aromatics, and polars/resins) present in petroleum-contaminated samples. This type of measurement is sometimes used to identify fuel type or to track plumes. It may be particularly useful for higher-boiling products such as asphalt. Group-type test methods include multidimensional gas chromatography (not often used for environmental samples), high-performance liquid chromatography (HPLC), and thin-layer chromatography (TLC) (Miller, 2000; Patnaik, 2004).

Test methods that analyze individual compounds (e.g., benzene–toluene–ethylbenzene–xylene mixtures and PAHs) are generally applied to detect the presence of an additive or to provide concentration data needed to estimate environmental and health risks that are associated with individual compounds. Common constituent measurement techniques include gas chromatography with second-column confirmation, gas chromatography with multiple selective detectors, and gas chromatography with mass spectrometry detection (GC/MS) (EPA 8240).

Many common environmental methods measure individual petroleum constituents or *target compound* rather than the entire signal from the total petroleum hydrocarbons. Each method measures a suite of compounds selected because of their toxicity and common use in industry.

For organic compounds, there are three series of target compound methods that must be used for regulatory purposes:

1. *EPA 500 series*: Organic Compounds in Drinking Water, regulated under the Safe Drinking Water Act.
2. *EPA 600 series*: Methods for Organic Chemical Analysis of Municipal and Industrial Wastewater, regulated under the Clean Water Act.
3. *SW-846 series*: Test Methods for Evaluating Solid Waste: Physical/Chemical Methods, promulgated by the EPA's Office of Solid Waste and Emergency Response.

The 500 and 600 series methods provide parameters and conditions for the analysis of drinking water and wastewater, respectively. One method (EPA SW-846) is focused on the analysis of nearly all matrixes, including industrial waste, soil, sludge, sediment, and water-miscible and non-water-miscible wastes. It also provides for the analysis of groundwater and wastewater but is not used to evaluate compliance of public drinking water systems.

Selection of one method over another is often dictated by the nature of the sample and the particular compliance or cleanup program for which the sample is being analyzed. It is essential to recognize that capabilities and requirements vary between methods when requesting any analytical method or suite of methods. Most compound-specific methods use a gas chromatographic–selective detector, high-performance liquid chromatography, or gas chromatography–mass spectrometry.

More correctly, group analytical methods are designed to separate hydrocarbons into categories, such as saturates, aromatics, resins and asphaltenes (SARA) or paraffins, isoparaffins, naphthenes, aromatics, and olefins (PIANO). These chromatographic, gas chromatographic, and high-performance liquid chromatographic methods (HPLC) (Table 7.3) were developed for monitoring refinery processes or evaluating organic synthesis products. Column chromatographic methods that separate saturates from aromatics are often used as preparative steps for further analysis by gas chromatography–mass spectrometry. Thin-layer chromatography is sometimes used as a screening technique for petroleum product identification.

7.3.1. Thin-Layer Chromatography

In the environmental field, thin-layer chromatography (TLC) is best used for screening analyses and characterization of semivolatile and nonvolatile petroleum products. The precision and accuracy of the technique is inferior to those of other methods (EPA 8015, EPA 418.1), but when speed and simplicity are desired, thin-layer chromatography may be a suitable alternative. For characterizations of petroleum products such as asphalt, the method has the advantage of separating compounds that are too high-boiling to pass through a gas chromatograph. Although thin-layer chromatography does not have the resolving power of a gas chromatograph, it is able to separate different classes of compounds. Thin-layer chromatography analysis is fairly simple, and since the method does not give highly accurate or precise results, there is no need to perform the highest-quality extractions.

In the method, soil samples are extracted by shaking or vortexing with the solvent. Water samples are extracted by shaking in a separatory funnel. If there is a potential for the presence of compounds that interfere with the method and make the data suspect, silica gel can be added to clean the extract. Sample extract aliquots are placed close to the bottom of a glass plate coated with a stationary phase. The most widely used stationary phases are made of an organic hydrocarbon moiety bonded to a silica backbone.

For the analysis of petroleum hydrocarbons, a moderately polar material stationary phase works well. The plate is placed in a sealed chamber with a solvent (mobile phase). The solvent travels up the plate, carrying compounds present in the sample. The distance a compound travels is a function of the affinity of the compound to the stationary phase relative to the mobile phase. Compounds with chemical structure and polarity similar to those of the solvent travel well in the mobile phase. For example, the saturated hydrocarbons seen in diesel fuel travel readily up a plate in a hexane mobile phase. Polar compounds such as ketones or alcohols travel a smaller distance in hexane than do saturated hydrocarbons.

After a plate has been exposed to the mobile-phase solvent for the required time, the compounds present can be viewed by several methods. Polynuclear aromatic hydrocarbons, other compounds with conjugated systems, and compounds containing heteroatoms (nitrogen, oxygen, or sulfur) can be viewed with long- and short-wave ultraviolet light. The unaided eye can see other material, or the plates can be developed in iodine. Iodine has an affinity for most petroleum compounds, including the saturated hydrocarbons, and stains the compounds a reddish-brown color.

The method is considered to be a qualitative and useful tool for rapid sample screening. Limitations of the method center on its moderate reproducibility, detection limits, and resolving capabilities. Variability between operators can be as high as 30%. Detection limits (without any concentration of the sample extract) are near 50 ppm (mg/kg) for most petroleum products in soils. When the aromatic content of a sample is high, as with bunker C fuel oil, the detection limit can be near 100 ppm. It is often not possible to distinguish between similar products, such as diesel and jet fuel. As with all chemical analyses, quality assurance tests should be run to verify the accuracy and precision of the method.

7.3.2. Immunoassay

A number of different testing kits based on immunoassay technology are available for rapid field determination of certain groups of compounds, such as benzene–toluene–ethylbenzene–xylene (EPA 4030) or polynuclear aromatic hydrocarbons (EPA 4035, Polycyclic Aromatic Hydrocarbons by Immunoassay). The immunoassay screening kits are self-contained portable field kits that include components for sample preparation, instrumentation to read assay results, and immunoassay reagents.

Unless the immunoassay kit is benzene sensitive, the kit may display strong biases, such as the low affinity for benzene relative to toluene, ethylbenzene, xylenes, and other aromatic compounds. This will cause an underestimation of the actual benzene levels in a sample, and since benzene is often the dominant compound in leachates due to its high solubility, a low sensitivity for benzene is undesirable.

The quality of the analysis of polynuclear aromatic hydrocarbons is often dependent on the extraction efficiency. Clay and other cohesive soils lower the ability to extract polynuclear aromatic hydrocarbons. Another potential problem with polynuclear aromatic hydrocarbon analysis is that the test kits may have different responses for different compounds.

7.3.3. Gas Chromatography

Gas chromatography uses the principle of a stationary phase and a mobile phase. Much attention has been paid to the various stationary phases, and books have been written on the subject as it pertains to petroleum chemistry. Briefly, *gas–liquid chromatography* (GLC) is a method for separating the volatile components of various mixtures (Altgelt and Gouw, 1975; Fowlis, 1995; Grob, 1995). It is, in fact, a highly efficient fractionating technique, and it is ideally suited to the quantitative analysis of mixtures when the possible components are known and the interest lies only in determining the amounts of each present. In this type of application gas chromatography has taken over much of the work previously done by the other techniques; it is now the preferred technique for the analysis of hydrocarbon gases, and gas chromatographic in-line monitors are finding increasing application in refinery plant control. Gas–liquid chromatography is also used extensively for individual component identification, as well as percentage composition, in the gasoline boiling range.

The mobile phase is the carrier gas, and the gas selected has a bearing on the resolution. Nitrogen has very poor resolution ability; helium or hydrogen are better choices, with hydrogen being the best carrier gas for resolution. However, hydrogen is reactive and may not be compatible with all sets of target analytes. There is an optimum flow rate for each carrier gas to achieve maximum resolution. As the temperature of the oven increases, the flow rate of the gas changes, due to the thermal expansion of the gas. Most modern gas chromatographs are equipped with constant-flow devices that change the gas valve settings as the temperature in the oven changes, so changing flow rates are no longer a concern. Once the flow is optimized at one temperature, it is optimized for all temperatures.

For environmental analysis (Bruner, 1993), particularly the volatile samples such as those found in total petroleum hydrocarbons (Table 7.4), the gas chromatograph is generally interfaced with a purge-and-trap system as described in Chapter 7. The photoionization detector works by bombarding compounds with ultraviolet (UV) light, generating a current of ions. Compounds with double carbon bonds, conjugated systems (multiple carbon double bonds arranged in a specific manner), and aromatic rings are easily ionized with the UV light generated by the photoionization detector lamp, while most saturated compounds require higher-energy radiation.

One method (EPA 8020) that is suitable for volatile aromatic compounds is often referred to as benzene–toluene–ethylbenzene–xylene analysis, although the method includes other volatile aromatics. The method is similar to most volatile organic gas chromatographic methods. Sample preparation and introduction is typically by purge-and-trap analysis (EPA 5030). Some oxygenates, such as methyl-t-butyl ether (MTBE), are also detected by a photoionization detector, as well as olefins, branched alkanes, and cycloalkanes.

Table 7.4. Summary of Gas Chromatographic Methods for Determining the Total Petroleum Hydrocarbons in a Sample

Method Name	Matrix	Scope of Method
Direct Injection methods		
EPA Method 8015B	Solid wastes	Used to determine the concentration of petroleum hydrocarbons, including gasoline range organics (GROs). Analysts should use the fuel contaminating the site for quantitation.
ASTM Method D3328-90	Water	Petroleum oils such as distillate fuel, lubricating oil, and crude oil recovered from water or beaches. Identification of a recovered oil is determined by comparison with known oils, selected because of their possible relationship to the recovered oil.
Purge and trap and headspace methods		
EPA Method 8015B	Solid wastes	Used to determine the concentration of petroleum hydrocarbons, including diesel range organics (DROs) and jet fuel. Analysts should use the fuel contaminating the site for quantitation.

Certain false positives are common (EPA 8020). For example, trimethylbenzenes and gasoline constituents are frequently identified as chlorobenzenes (EPA 602, EPA 8020) because these compounds elute with nearly the same retention times from nonpolar columns. Cyclohexane is often mistaken for benzene (EPA 8015/8020) because both compounds are detected by a 10.2-eV photoionization detector and have nearly the same elution time from a nonpolar column (EPA 8015). The two compounds have very different retention times on a more polar column (EPA 8020), but a more polar column skews the carbon ranges (EPA 8015). False positives for oxygenates in gasoline are common, especially in highly contaminated samples.

For semivolatile constituents of petroleum, the gas chromatograph is generally equipped with either a packed or a capillary column. Either neat or diluted organic liquids can be analyzed via direct injection, and compounds are separated during movement down the column. The flame ionization detector uses a hydrogen-fueled flame to ionize compounds that reach the detector. For PAHs a method is available (EPA 8100) in which injection of sample extracts directly onto the column is the preferred method for sample introduction for this packed-column method.

A gas chromatography–flame ionization detector system can be used for the separation and detection of nonpolar organic compounds. Semivolatile constituents are among the analytes that can readily be resolved and detected using the system. If a packed column is used, four pairs of compounds may not be resolved adequately and are reported as a quantitative sum: anthracene and phenanthrene, chrysene and benzo[*a*]anthracene, benzo[*b*]fluoranthene and benzo[*k*]fluoranthene, and dibenzo[*a,h*]anthracene and indeno[1,2,3-*cd*]pyrene. This issue can be resolved through the use of a capillary column in place of a packed column.

7.3.4. High-Performance Liquid Chromatography

A high-performance liquid chromatography system can be used to measure concentrations of target semi- and nonvolatile petroleum constituents. The system only requires that the sample be dissolved in a solvent compatible with those used in the separation. The detector most often used in petroleum environmental analysis is the fluorescence detector. These detectors are particularly sensitive to aromatic molecules, especially PAHs. An ultraviolet detector may be used to measure compounds that do not fluoresce.

In the method, PAHs are extracted from the sample matrix with a suitable solvent, which is then injected into the chromatographic system. Usually, the extract must be filtered because fine particulate matter can collect on the inlet frit of the column, resulting in high back-pressures and eventual plugging of the column. For most hydrocarbon analyses, reverse-phase high-performance liquid chromatography (i.e., using a nonpolar column packing with a more polar mobile phase) is used. The most common bonded phase is the octadecyl (C_{18}) phase. The mobile phase commonly comprises aqueous mixtures of either acetonitrile or methanol.

After the chromatographic separation, the analytes flow through the cell of the detector. A fluorescence detector shines light of a particular wavelength (the excitation wavelength) into the cell. Fluorescent compounds absorb light and reemit light of other, higher wavelengths (emission wavelengths). The emission wavelengths of a molecule are determined primarily by its structure. For polynuclear aromatic hydrocarbons, the emission wavelengths are determined primarily by the arrangement of the rings and vary greatly between isomers.

Some PAHs (e.g., phenanthrene, pyrene, and benzo[g,h,i]perylene) are commonly seen in products boiling in the middle to heavy distillate range. In a method for their detection and analysis (EPA 8310), an octadecyl column and an aqueous acetonitrile mobile phase are used. Analytes are excited at 280 nm and detected at emission wavelengths of >389 nm. Naphthalene, acenaphthene, and fluorene must be detected by a less sensitive UV detector because they emit light at wavelengths below 389 nm. Acenaphthylene is also detected by UV detector.

Methods using fluorescence detection will measure any compounds that elute in the appropriate retention time range and which fluoresce at the targeted emission wavelength(s). In the case of one method (EPA 8310), the excitation wavelength excites most aromatic compounds. These include the target compounds and also many derivatized aromatics, such as alkylaromatics, phenols, anilines, and heterocyclic aromatic compounds containing the pyrrole (indole, carbazole, etc.), pyridine (quinoline, acridine, etc.), furan (benzofuran, naphthofuran, etc.), and thiophene (benzothiophene, naphthothiophene, etc.) structures. In petroleum samples, alkyl polynuclear aromatic hydrocarbons are strong interfering compounds. For example, there are five methylphenanthrenes and over 20 dimethylphenanthrenes. The alkyl substitution does not significantly affect either the wavelengths or intensity of the phenanthrene fluorescence. For a very long time after the retention time of phenanthrene, the alkylphenanthrenes will interfere, affecting the measurements of all later-eluting target PAHs.

Interfering compounds will vary considerably from source to source, and samples may require a variety of cleanup steps to reach required method detection limits. The emission wavelengths used (EPA 8310) are not optimal for sensitivity of the small ring compounds. With modern electronically controlled monochromator, wavelength programs can be used which tune excitation and emission wavelengths to maximize sensitivity and/or selectivity for a specific analyte in its retention time window.

7.3.5. Gas Chromatography–Mass Spectrometry

A gas chromatography–mass spectrometry system is used to measure concentrations of target volatile and semivolatile petroleum constituents. It is not typically used to measure the amount of total petroleum hydrocarbons. The advantage the technique is the high selectivity, or ability to confirm compound identity through retention time and unique spectral pattern.

The current method (EPA SW-846 8260) for the analysis of volatile compounds reveals that most of the compounds listed in these methods are not typically found in petroleum products. However, a method that uses selected ion monitoring (SIM) involves system setup to measure only selected target masses rather than scanning the full mass range. This technique yields lower detection limits for specific compounds. At the same time, it gives the more complete information available from the total ion chromatogram and the full-mass-range spectrum of each compound. The technique is sometimes used to quantify compounds present at very low concentrations in a complex hydrocarbon matrix. It can be used if the target compound's spectrum has a prominent fragment ion at a mass that distinguishes it from the rest of the hydrocarbon compounds.

The most common method for GC/MS analysis of semivolatile compounds (EPA SW-846 8270) includes 16 polycyclic aromatic compounds, some of which commonly occur in middle distillate to heavy petroleum products. The method also quantifies phenols and cresols, compounds that are not hydrocarbons but may occur in petroleum products. Phenols and cresols are more likely found in crude oils and weathered petroleum products.

To reduce the possibility of false positives, the intensities of one to three selected ions are compared to the intensity of a unique target ion of the same spectrum. The sample ratios are compared to the ratios of a standard. If the sample ratios fall within a certain range of the standard, and the retention time matches the standard within specifications, the analyte is considered present. Quantification is performed by integrating the response of the target ion only.

Mass spectrometers are among the most selective detectors, but they are still susceptible to interferences. Isomers have identical spectra, whereas many other compounds have similar mass spectra. Heavy petroleum products can contain thousands of major components that are not resolved by the gas chromatograph. As a result, multiple compounds enter the mass spectrometer simultaneously. Different compounds may share many of the same ions, confusing the identification process. The probability of misidentification is high in complex mixtures such as petroleum products.

7.4. PETROLEUM FRACTIONS

Rather than quantifying a complex total petroleum hydrocarbon mixture as a single number, petroleum hydrocarbon fraction methods break the mixture into discrete hydrocarbon fractions, thus providing data that can be used in a risk assessment and in characterizing product type and compositional changes such as may occur during weathering (oxidation). The fractionation methods can be used to measure both volatile and extractable hydrocarbons.

In contrast to traditional methods for total petroleum hydrocarbons that report a single concentration number for complex mixtures, the fractionation methods report separate concentrations for discrete aliphatic and aromatic fractions. The petroleum fraction methods available are GC-based and are thus sensitive to a broad range of hydrocarbons. Identification and quantification of aliphatic and aromatic fractions allows one to identify petroleum products and evaluate the extent of product weathering. These fraction data also can be used in risk assessment.

One particular method is designed to characterize C_6 to $C_{28}+$ petroleum hydrocarbons in soil as a series of aliphatic and aromatic carbon range fractions. The extraction methodology differs from other petroleum hydrocarbon methods because it uses n-pentane, not methylene chloride, as the extraction solvent. If methylene chloride is used as the extraction solvent, aliphatic and aromatic compounds cannot be separated.

n-Pentane extracts petroleum hydrocarbons in this range efficiently. The entire extract is separated into aliphatic and aromatic petroleum-derived fractions (EPA SW-846 3611, SW-846 3630). The aliphatic and aromatic fractions are analyzed separately by gas chromatography and quantified by summing the signals within a series of specified carbon ranges that represent the fate and transport fractions. The gas chromatograph is equipped with a boiling-point column (nonpolar capillary column). Gas chromatographic parameters allow the measurement of a hydrocarbon range of n-hexane (C_6) to n-octacosane ($C_{28}+$), a boiling-point range of approximately 65 to 450°C.

REFERENCES

Altgelt, K. H., and Gouw, T. H. 1975. In *Advances in Chromatography*, J. C. Giddings, E. Grushka, R. A. Keller, and J. Cazes (Eds.). Marcel Dekker, New York.

Bruner, F. 1993. *Gas Chromatographic Environmental Analysis: Principles, Techniques, and Instrumentation*. Wiley, New York.

Budde, W. L. 2001. *The Manual of Manuals*. Office of Research and Development, Environmental Protection Agency, Washington, DC.

Dean, J. R. 1998. *Extraction Methods for Environmental Analysis*. Wiley, New York.

EPA. 2004. Environmental Protection Agency, Washington, DC. Web site: http://www.epa.gov.

Fowlis, I. A. 1995. *Gas Chromatography*, 2nd ed. Wiley, New York.

Grob, R. L. 1995. *Modern Practice of Gas Chromatography*, 3rd ed. Wiley, New York.

Miller, M. (Ed.). 2000. *Encyclopedia of Analytical Chemistry*. Wiley, New York.

Nelson, P. 2003. *Index to EPS Test Methods*. U.S. EPA New England Region, Boston.

Patnaik, P. (Ed.). 2004. *Dean's Analytical Chemistry Handbook*, 2nd ed. McGraw-Hill, New York.

Rhodes, I. A., Hinojas, E. M., Barker, D. A., and Poole, R. A. 1994. In *Proceedings, 7th Annual Conference: EPA Analysis of Pollutants in the Environment*, Norfolk, VA. Environmental Protection Agency, Washington, DC.

Smith, K. A., and Cresser, M. 2003. *Soil and Environmental Analysis: Modern Instrumental Techniques*. Marcel Dekker, New York.

Speight, J. G. 1999. *The Chemistry and Technology of Petroleum*, 3rd ed. Marcel Dekker, New York.

Speight, J. G. 2001. *Handbook of Petroleum Analysis*. Wiley, New York.

Speight, J. G. 2002. *Handbook of Petroleum Product Analysis*. Wiley, Hoboken, NJ.

Sunahara, G. I., Renoux, A. Y., Thellen, C., Gaudet, C. L., and Pilon, A. (Eds.). 2002. *Environmental Analysis of Contaminated Sites*. Wiley, Hoboken, NJ.

CHAPTER 8

TOTAL PETROLEUM HYDROCARBONS

Petroleum and petroleum products are complex mixtures of hundreds of hydrocarbon compounds, ranging from volatile organic compounds to high-molecular-weight nonvolatile organic compounds (Chapters 2 and 3). Therefore, it is not surprising that the composition of petroleum products varies considerably depending on (1) the source of the petroleum (derived from underground reservoirs, which vary greatly in their chemical composition) and (2) the refining practices used to produce the end product. In short, as long as the products meet the relevant specifications, product composition can vary considerably. This, then, moves on to the issue of the meaning of the term *total petroleum hydrocarbons* and the interpretation of the data.

During the refining process, crude oil is separated into fractions having similar boiling ranges. These fractions are modified by cracking, condensation, polymerization, and alkylation processes and are formulated into commercial products such as naphtha, gasoline, jet fuel, and fuel oils. The composition of any one of these products can vary based on the refinery involved, time of year, variation in additives or modifiers, and other factors. The chemical composition of the product can be further affected by weathering and/or biological modification upon release to the environment.

As already noted (Chapter 6), *total petroleum hydrocarbons* (TPHs) is a term applied to the measurable amount of petroleum-based hydrocarbon in environmental media (Irwin et al., 1997). Furthermore, compounds that are included in the total petroleum hydrocarbons from crude oil are composed of petroleum hydrocarbons in the range C_1 to beyond C_{35}. For example, refined products from crude oil include (1) gasoline, which characteristically has total petroleum hydrocarbons composed of hydrocarbons between the size range of normal hexane (n-C_6) and normal dodecane (n-C_{12}), and (2) diesel fuel, which characteristically is composed of hydrocarbons found between the size range of normal undecane (n-C_{11}) and normal pentacosane (n-C_{25}). However, petroleum and the higher-boiling petroleum products (such as the nondistillable residual fuel oil and asphalt) are not limited to the carbon ranges shown and will contain many higher-carbon-number and higher-molecular-weight constituents (Speight, 1999).

Determination of the total petroleum hydrocarbons in a sample is made by using several laboratory tests that are relatively inexpensive, relatively quick,

---

Environmental Analysis and Technology for the Refining Industry, by James G. Speight
Copyright © 2005 John Wiley & Sons, Inc.

sometimes ineffective, but not usually quantitative. The results are, thus, dependent on analysis of the medium in which the hydrocarbons are found. Since it is a measured, gross quantity without identification of its constituents, the *total petroleum hydrocarbons* data represent a mixture. Thus, the total petroleum hydrocarbons data are not a direct indicator of risk to humans or to the environment. The total petroleum hydrocarbons data value can be a result of one of several analytical methods, some of which have been used for decades and others developed in the past several years.

The screening measures for total petroleum hydrocarbons screening are done differently in various states, regions, and individual laboratories. In fact, the data that are reported as total petroleum hydrocarbons are so variable that much caution should be exercised when attempting to compare or interpret the data. As analytical methods evolve in response to environmental needs, the definition of total petroleum hydrocarbons may become more closely related to reality rather than to the respective analytical method.

Several hundred individual hydrocarbon chemicals defined as petroleum-based have been identified. Furthermore, each individual crude oil and each individual petroleum product has a specific mixture of the various constituents because of the variation in petroleum composition (Chapter 2), and this variation is reflected in the composition of the finished petroleum product. At this point it is worthy of note that the term *petroleum hydrocarbons* (PHC) is widely used to refer to the hydrogen- and carbon-containing compounds originating from crude oil, but *petroleum hydrocarbons* should be distinguished from *total petroleum hydrocarbons* because the term *total petroleum hydrocarbons* is specifically associated with environmental sampling and analytical results (Weisman, 1998; CFR, 2004).

Therefore, methods for the analysis of total petroleum hydrocarbon are frequently used to find areas of gross contamination but are often inadequate even for this task. Indeed, any one of several variables, such as differences in moisture content, can lead to analytical inconsistencies, and therefore the data do not consistently give reliable insights as to which part of the site is most contaminated.

Thus, if the concentration of total petroleum hydrocarbons concentrations is high, it usually signifies that significant amounts of petroleum hydrocarbons are there. However, if the concentration of the total petroleum hydrocarbons values is low or undetectable, it is not certain that a significant petroleum hydrocarbon contamination problem is not present. in fact, few other environmental monitoring parameters have been so widely and consistently misapplied and misinterpreted.

The purpose of this chapter is to describe well-established analytical methods that are available for detecting and/or measuring and/or monitoring total petroleum hydrocarbons and their metabolites, as well as other biomarkers of the exposure and effect of total petroleum hydrocarbons. The intent is not to provide an exhaustive list of analytical methods. Rather, the intention is to identify well-established methods that are used as the standard methods approved by federal agencies and organizations such as the Environmental Protection Agency and the National Institute for Occupational Safety and Health (NIOSH) or methods prescribed by state governments for water and soil analysis. Other methods presented are those that are approved by groups such as American Society for Testing and Materials (ASTM, 2004).

8.1. PETROLEUM CONSTITUENTS

Petroleum can be broadly divided into paraffinic, asphaltic, and mixed crude oils (Speight, 1999, 2001, and references cited therein). Paraffinic crude oil is composed of aliphatic hydrocarbons (paraffins), paraffin wax (longer-chain aliphatic compounds), and high-grade distillate. Naphtha is the lightest of the paraffin fraction, followed by kerosene fractions (Chapter 2). Asphaltic crude oil contains higher concentrations of cycloaliphatic compounds and high-viscosity gas oils and residua. Petroleum solvents are the product of crude oil distillation and are generally classified by boiling-point ranges. Lubricating oil, grease, and wax are high-boiling-point fractions of crude oils. The highest-boiling fractions (that are, in fact, nonvolatile under the conditions of refinery distillation) are the semisolid to solid residua.

Petroleum products themselves are the source many of the components but do not define *total petroleum hydrocarbons*. Knowing the composition of petroleum products does assist in defining the potential hydrocarbons that become environmental contaminants, but any ultimate exposure is determined also by how the product changes with use, by the nature of the release, and by the environmental fate of the hydrocarbons released. When petroleum products are released into the environment, changes occur that significantly affect their potential effects. Physical, chemical, and biological processes change the location and concentration of hydrocarbons at any particular site.

Petroleum hydrocarbons are commonly found in environmental contaminants, although they are not usually classified as hazardous waste. However, soil and groundwater contamination by petroleum hydrocarbon has spurred various analytical and site remediation developments (e.g., risk-based corrective actions).

The assessment of health effects due to exposure to the total petroleum hydrocarbons requires much more detailed information than what is provided by a single total petroleum hydrocarbon value. More detailed physical and chemical properties and analytical information on the total petroleum hydrocarbons fraction and its components are required. Indeed, a critical aspect of assessing the toxic effects of the total petroleum hydrocarbons is the measurement of the compounds, and the first task is to appreciate the origin of the various fractions (compounds) of the total petroleum hydrocarbons. Transport fractions are determined by several chemical and physical properties (i.e., solubility, vapor pressure, and propensity to bind with soil and organic particles). These properties are the basis of measures of leachability and volatility of individual hydrocarbons and transport fractions (Chapters 8, 9, and 10).

Leaking underground storage tanks (LUSTs) are the most frequent causes of regulations for issues related to total petroleum hydrocarbons. Soil contamination has been a growing concern, because it can be a source of groundwater (drinking water) contamination. Contaminated soils can reduce the usability of land for development, and weathered petroleum residuals may stay bound to soils for years. Positive test results for the presence of total petroleum hydrocarbons may require action to remove or reduce the total petroleum hydrocarbons problem.

Specific contaminants that are components of total petroleum hydrocarbons, such as BTEX (benzene, toluene, ethylbenzene, and xylene), *n*-hexane, jet fuels, fuel oils, and mineral-based crankcase oil have been studied and a number of toxicological profiles have been developed on individual constituents and petroleum products. However, the character of the total petroleum hydrocarbons has not been studied extensively and no profiles have been developed. Although several toxicological profiles have been developed for petroleum products and for specific chemicals found in petroleum, the total petroleum hydrocarbon test results have been too nonspecific to be of real value in the assessment of its potential health effects.

The result of these processes is an alteration in the composition of the hydrocarbon discharged into the soil. Clearly, those hydrocarbons that are most strongly sorbed onto soil organic matter will be most resistant to loss or alteration by the other processes. Conversely, the more volatile and soluble hydrocarbons will be the most susceptible to change by volatilization, reaction, leaching, or biodegradation. The ultimate result will be "weathering" of the hydrocarbon mixture discharged into the soil, with an accompanying change in its composition and preferential transport of certain fractions to other environmental compartments.

8.2. ANALYTICAL METHODS

Since the term *total petroleum hydrocarbons* (total petroleum hydrocarbons) includes any petroleum constituent that falls within the measurable amount of petroleum-based hydrocarbons in the environment; the information obtained for total petroleum hydrocarbons depends on the analytical method used. Therefore, the difficulty associated with measurement of the total petroleum hydrocarbons is that the scope of the methods varies greatly (Table 8.1). Some methods are nonspecific, whereas others provide results for hydrocarbons in a boiling-point range. Interpretation of analytical results requires an understanding of how the determination was made (Miller, 2000, and references cited therein; Dean, 2003).

The very volatile gases (compounds with four or fewer carbons), crude oil, and the solid asphaltic materials are not included in this discussion of analytical methods but are included elsewhere (Chapters 7 and 9).

8.2.1. Environmental Samples

Most of the analytical methods discussed here for total petroleum hydrocarbons have been developed within the framework of federal and state regulatory initiatives. The initial implementation of the Federal Water Pollution Control Act (FWPCA) focused on controlling conventional pollutants such as oil and

Table 8.1. Summary of the EPA and ASTM Test Methods for Determination of Total Petroleum Hydrocarbons

Sample Matrix	Preparation Method	Analytical Method	Reference
Ambient air	Collection on Tenax GC cartridge; thermal desorption	Capillary GC/MS	EPA Method TO-1
Water, wastes (oil and grease)	Solvent extraction	Gravimetric	EPA Method 413.1
Water, wastes (oil and grease)	Solvent extraction	IR	EPA Method 413.2
Water and wastes (TRPH)	Solvent extraction; silica gel column separation	IR	EPA Method 418.1
Water, aqueous wastes (oil and grease)	Solvent extraction	Gravimetric	EPA Method 9070
Sludge and sediment (oil and grease)	Sample is dried; Soxhlet extraction	Gravimetric	EPA Method 9071A
Soils, sediments, fly ash (TRPH)	Supercritical fluid extraction	Method 8015B	EPA Method 3560
Soil (TPH)	Extraction; filtration	Immunoassay	EPA Method 4030
Ground or surface water, soil (DRO, GRO)	DRO: solvent extraction; GRO: purge and trap or vacuum distillation or headspace sampling	Capillary GC/FID	EPA Method 8015B
Water (petroleum oils)	Solvent extraction; evaporation	GC/FID	ASTM Method D3328
	Solvent extraction; evaporation	IR	ASTM Method D3414

Source: ASTM, 1995; EPA, 2004.

grease. Methods have been developed for monitoring wastewaters (EPA 413.1, 413.2, 418.1).

The method of analysis often used for the total petroleum hydrocarbons (EPA 418.1) method provides a *one-number* value of the total petroleum hydrocarbons in an environmental medium. It does not, by any stretch of the imagination, provide information on the composition (i.e., individual constituents of the hydrocarbon mixture).

Freon-extractable material is reported as total organic material from which polar components may be removed by treatment with silica gel, and the material remaining, as determined by infrared (IR) spectrometry, is defined as *total recoverable petroleum hydrocarbons* (TRPHs, or total petroleum hydrocarbons–IR). A number of modifications of these methods exist, but one particular method (EPA 418.1; see also EPA 8000 and 8100) has been one of the most widely used for the determination of total petroleum hydrocarbons in soils. Many states use or permit the use of this method (EPA 418.1) for identification of petroleum products and during remediation of sites. This method is subject to limitations, such as interlaboratory variations and inherent inaccuracies. In addition, methods that use Freon-113 as the extraction solvent are being phased out and the method is being replaced by a more recent method (EPA 1664) in which n-hexane is used as the solvent and the n-hexane extractable material (HEM) is treated with silica gel to yield the total petroleum hydrocarbons.

The amount of the total petroleum hydrocarbons measured by this method depends on the ability of the solvent used to extract the hydrocarbon from the environmental media and the absorption of IR light by the hydrocarbons in the solvent extract. In addition, the method (EPA 418.1) is not specific to hydrocarbons and does not always indicate petroleum contamination (e.g., humic acid, a nonpetroleum hydrocarbon, may be detected by this method).

An important feature of the analytical methods for total petroleum hydrocarbons is the use of an *equivalent carbon number index* (EC index, ECN index, or ECNI), which represents equivalent boiling points for hydrocarbons and is the physical characteristic that is the basis for separating petroleum (and other) components in chemical analysis. Petroleum fractions as discussed in this profile are defined by the ECN index.

The conventional methods of analysis for total petroleum hydrocarbons (Chapter 7) have been used widely to investigate sites that may be contaminated with petroleum hydrocarbon products (see also EPA 418.1) for the determination of petroleum hydrocarbons. The important advantage of this method is that excellent sample reproducibility can be obtained, but the disadvantages are that (1) petroleum hydrocarbon composition varies among sources and over time, so results are not always comparable; (2) the more volatile compounds in gasoline and light fuel oil may be lost in the solvent concentration step; (3) there are inherent inaccuracies in the method; and (4) the method provides virtually no information on the types of hydrocarbons present. This is not necessarily the fault of the method but illustrates the nature of the problems insofar as petroleum hydrocarbons in a leak or spill will change over time due to (1) volatility losses, (2) weathering (oxidation), and (3) microbial activity.

Thus, the methods for measurement of total petroleum hydrocarbons (Table 8.1) (see also Chapter 7) provide adequate screening information but do not provide sufficient information on the extent of the contamination and product type.

Methods that use gas chromatography methods do provide some information about the product type. Most of the methods involve a sample preparation procedure followed by analysis using GC techniques. The GC determination is based on selected components or the sum of all components detected within a given range. Frequently, the approach is to use two methods, one for the volatile range and another for the semivolatile range. Volatile constituents in water or solid samples are determined by purge-and-trap gas chromatography–flame ionization.

The analysis is often called the *gasoline range organics* (GRO) *method*. The semivolatile range is determined by analysis of an extract by gas chromatography–flame ionization and is referred to as the *diesel range organics* (DRO) *method*.

With regard to releases from underground storage tanks (USTs), the most common method (EPA 418.1) is still used, but gas procedures have been developed to provide more specific information on the hydrocarbon content of water and soil, as per local and/or regional legislation. These methods, coupled with specific extraction techniques, can provide information on product type by comparison of the chromatogram with standards. Quantitative estimates may be made for a boiling range or for a range of carbon numbers by summing peaks within a specific window. However, these methods do have limitations, such as erroneous data caused by interferences, low recovery due to the standard selected, and petroleum product changes caused by volatility, weathering, and microbial activity.

Another method (EPA 3611) that focuses on the to separation of groups or fractions with similar mobility in soils is based on the use of alumina and silica gel (EPA 3630) that are used to fractionate the hydrocarbon into aliphatic and aromatic fractions. A gas chromatograph equipped with a boiling-point column (nonpolar capillary column) is used to analyze whole soil samples as well as the aliphatic and aromatic fractions to resolve and quantify the fate-and-transport fractions. The method is versatile and performance based and therefore can be modified to accommodate data quality objectives.

Another analytical method (EPA 8015, Modified) commonly used for determining the total petroleum hydrocarbons reports the concentration of purgeable and extractable hydrocarbons; these are sometimes referred to as gasoline and diesel range organics because the boiling-point ranges of the hydrocarbon in each roughly correspond to those of gasoline and diesel fuel, respectively. Purgeable hydrocarbons are measured by purge-and-trap gas chromatography (GC) analysis using a flame ionization detector (FID), and the extractable hydrocarbons are extracted and concentrated prior to analysis. The results are most frequently reported as single numbers for purgeable and extractable hydrocarbons.

Higher-boiling hydrocarbons (C_{12} to C_{26}) are analyzed using an extraction procedure followed by column separation using silica gel (EPA 3630 Modified) of the aromatic and aliphatic groupings or fractions. The two fractions are then analyzed using gas chromatography–flame ionization. Polynuclear aromatic markers and n-alkane markers are used to divide the higher-boiling aromatic and aliphatic fractions by carbon number, respectively.

It is possible that the photoionization detector (1) may not be completely selective for aromatics and can lead to an overestimate of the more mobile and toxic aromatic content; and (2) the results from the two analyses, purgeable and extractable hydrocarbons, can overlap in carbon number and cannot simply be added together to get a total concentration of the total petroleum hydrocarbons.

Air

Because of the relative complexity of the analytical methods for total petroleum hydrocarbons, there is a need for devising methods for the determination of total petroleum hydrocarbons. But the major problem lies in the range of compounds covered by the term *hydrocarbons*. Again, the most notable variation is in the relative volatility and other properties of the hydrocarbons under investigation. Although instrumental detection methods are available (Sadler and Connell, 2003), another approach involves collection of the contaminated soil and sealing it in a container, where the soil gas can accumulate. This gas is then analyzed by one of several reliable instrumental procedures.

Some methods for determining hydrocarbons in air matrices usually depend on adsorption of the constituents of the total petroleum hydrocarbon fraction onto a solid sorbent, subsequent desorption, and determination by gas chromatographic methods. Hydrocarbons within a specific boiling range (*n*-pentane to *n*-octane) in occupational air are collected on a sorbent tube, desorbed with solvent, and determined using gas chromatography–flame ionization. Although method precision and accuracy are usually high, performance may be reduced at high humidity.

On the other hand, the complex mixture of petroleum hydrocarbons potentially present in an air sample can be minimized by separation of the sample into aliphatic and aromatic fractions, and then these two major fractions are separated into smaller fractions based on carbon number. Individual compounds (e.g., benzene, toluene, ethylbenzene, xylenes, MTBE, naphthalene) are also identified using this method. The range of compounds that can be identified includes C_4 (1,3-butadiene) through C_{12} (*n*-dodecane).

As a partial compromise between the use of on-site instrumental analysis and laboratory analysis, a passive sampler can be immersed into the soil (at a specified depth or at several depths) to collect the evolved gases that are adsorbed onto a solid-phase support. The sampler is then removed to the laboratory, where the gases are transferred by Curie point desorption, directly into the ion source of an interfaced quadrupole mass spectrometer. This procedure has its origin in the petroleum exploration industry, and the samplers can be used at a considerable range of depths (Einhorn et al., 1992).

A number of procedures, based on microanalysis of samples for known physical properties (Chapter 8, 9, and 10), have also been employed. For example, field screening, which uses infrared spectroscopy, employing a portable version of the laboratory procedure has been used (Kasper et al., 1991). Field turbidometric methods favor the determination of high-boiling hydrocarbons and are of some use in delineating such pollution within the soil (Kahrs et al., 1999). The fluorescence spectra exhibited by the aromatic components provide the basis for laser-induced fluorescence spectroscopy (Apitz et al., 1992; Löhmannsröben et al., 1999). They allow detection of polycyclic aromatic compounds and thus are able to take account of a fraction not measured by other field screening techniques.

Soils and Sediments

Hydrocarbon species can enter the soil environment from a number of sources. The origin of the contaminants has a significant bearing on the species present and hence the analytical methodology to be used (Driscoll et al., 1992). Unlike other chemicals (notably pesticides), hydrocarbons were generally not applied to soils for a purpose, and thus hydrocarbon contamination results almost entirely from misadventure. The source that is probably most familiar to persons involved in the study of contaminated sites is leakage from underground storage tanks. This is particularly important at the site of former service stations, and the hydrocarbons involved are generally in the gasoline or diesel range. Other major sources include spillage during refueling and lubrication, the hydrocarbons being within the diesel and heavy oil range. Places in which transfer and handling of crude oils take place (such as tanker terminals and oil refineries) are also potential places of contamination, the oil being largely of the heavier hydrocarbon type.

Since the group of chemicals generally referred to as *total petroleum hydrocarbons* have widely differing properties, they are likely to present a significant analytical challenges. Additionally, the hydrocarbons will be associated with the soil in different ways and hence the strength of the hydrocarbon interaction (usually, sorption) will vary according to the nature of the hydrocarbon as well as with the nature of any other organic matter present in the soil.

Thus, the relevant chemistry of hydrocarbons likely to be encountered at contaminated sites is reviewed briefly and the importance of hydrocarbon speciation is noted in terms of a toxicological basis for risk assessment. Hydrocarbon interaction with soil contaminants is important in terms of both their toxicology and their accessibility by analytical methods. There is no simple procedure that will give an overall picture of hydrocarbons present at contaminated sites. This is largely because the molecules are present in two separate categories: volatile and semi- or nonvolatile. These two categories require significantly different sample collection, handling, and management techniques (Siegrist and Jenssen, 1990). Volatile hydrocarbons may be collected by zero-headspace procedures or by immediate immersion of the soil into methanol.

The analysis involves gas chromatographic methods such as purge and trap, vacuum distillation, and headspace (Askari et al., 1996). On the other hand, samples for the determination of semi- and nonvolatile hydrocarbons need not be collected in such a rigorous manner. On arrival at the laboratory, they require extraction by techniques such as solvent or supercritical fluid. Some cleanup of extracts is also necessary in most cases and the analytical finish is again by gas chromatography. Detectors used range from flame ionization to Fourier transform infrared and mass spectrometric, the latter types being necessary to achieve speciation of the component hydrocarbons.

The determination of hydrocarbon contaminants in soil is one of the most frequently performed analyses in the study of contaminated sites and is also one of the least standardized. Given the wide variety of hydrocarbon contaminants that can potentially enter and exist in the soil environment, a need exists for methods that quantify these chemicals satisfactorily. Formerly, the idea of total hydrocarbon determination in soil was seen as providing a satisfactory tool for assessing contaminated sites, but the nature of the method and the site specificity dictate a risk-based approach in data assessment. Quantitation of particular hydrocarbon species may be required.

Currently, many regulatory agencies recommend the common methods (EPA 418.1, EPA 801.5 Modified) or similar methods for analysis during remediation of contaminated sites. In reality, there is no standard for the measurement of total petroleum hydrocarbons since each method may need to be chosen or adapted on the basis of site specificity.

There is a trend toward the use of GC techniques in the analysis of soils and sediments. One aspect of these methods is that *volatiles* and *semivolatiles* are determined separately. The volatile or gasoline range organic constituents are recovered using purge-and-trap or other stripping techniques. Semivolatiles are separated from the solid matrix by solvent extraction. Other extraction techniques have been developed to reduce the hazards and the cost of solvent use and to automate the process, and techniques include supercritical fluid extraction (SFE), microwave extraction, Soxhlet extraction, sonication extraction, and solid-phase extraction (SPE) (EPA 3540C.). Capillary column techniques have largely replaced the use of packed columns for analysis, as they provide resolution of a greater number of hydrocarbon compounds.

Because of the overall complexity of the problem and of the spectrum of hydrocarbons likely to be encountered, it is impossible to view the total petroleum hydrocarbons as a single entity. There have been many approaches to the problem, but the simplest and the one used most frequently is the one based on the vapor pressure ranges of the relevant organic constituents. This also relates to the sampling methodology employed, and the approach consists of subdividing the hydrocarbons into the most volatile fraction [referred to as gasoline range organics (GROs)] and the less volatile fraction. In the case of monitoring of storage tanks, a subfraction [known as diesel range organics (DROs)] is often distinguished among the semivolatile fraction.

As regards a contaminated soil, this type of analysis may not be possible because the various hydrocarbons cannot be extracted from the sample with equal efficiency. Volatile organic compounds require special procedures to achieve satisfactory recovery from the soil matrix. It thus becomes important to distinguish between those compounds that are considered to be volatile and those that rank as semi- or nonvolatile compounds.

Water and Wastewater

The overall method includes sample collection and storage, extraction, and analysis steps. Sampling strategy is an important step in the overall process. Care must be taken to assure that the samples collected are representative of the environmental medium and that they are collected without contamination. There is an extensive list of test methods for water analysis (Tables 8.2, 8.3, and 8.4), which includes numerous modifications of the original methods, but most involve alternative extraction methods developed to improve overall method performance for the analysis. Solvent extraction methods with hexane are also in use.

Release in a Nonsensitive Area

For petroleum and petroleum product releases in a nonsensitive area (if there is such an area), the analytical methods preferred to determine the concentration of total petroleum hydrocarbons in environmental media is the standard EPA test method (EPA 418.1). For initial delineation of the area, test field kits may be used in nonsensitive areas, provided that the results are comparable to laboratory data. Final confirmatory sampling and analyses should be carried out using laboratory analyses.

Release in a Sensitive Area

For petroleum and petroleum product releases in a sensitive area, the preferred analytical method to determine concentrations of total petroleum hydrocarbons in environmental media is the standard EPA test method (EPA 418.1). To determine concentrations of benzene, toluene, ethylbenzene, and xylenes in environmental media, other methods (EPA SW-846, SW-846 8021B, SW-846 8260) are preferred, provided that the detection limits are adequate for soil and for groundwater protection.

To determine concentrations of PAHs in environmental media, approved methods (EPA SW-846 8270, SW-846 8310) are necessary provided that the detection limits are adequate for soil and groundwater protection. Generally, regulatory agencies will require at least one PAH analysis from the most contaminated sample from each source area, and the analysts must ensure that lab detection limits are appropriate for risk determination.

8.2.2. Biological Samples

Few analytical methods are available for the determination of total petroleum hydrocarbons in biological samples, but analytical methods for several important hydrocarbon components of total petroleum hydrocarbons may be modified. Most involve solvent extraction and saponification of lipids, followed by separation into aliphatic and aromatic fractions on adsorption columns. Hydrocarbon groups or target compounds are determined by gas chromatography–flame ionization or

Table 8.2. EPA Test Methods for Organic Compounds in Water

EPA No.	Method
813R93002	A Review of Methods for Assessing Aquifer Sensitivity and Ground Water Vulnerability to Pesticide Contamination
600483039	Addendum to Handbook for Sampling and Sample Preservation, EPA-600/4-82-029
660274021	Analysis of Coprostanol: An Indicator of Fecal Contamination
815R03003	Analytical Feasibility Support Document for the Six-Year Review of Existing National Primary Drinking Water Regulations (Reassessment of Feasibility for Chemical Contaminants)
821R00003	Analytical Method Guidance for EPA Method 1664: A Implementation and Use (40 CFR Part 136)
600987030	Availability Adequacy and Comparability of Testing Procedures for the Analysis of Pollutants Established under Section 304(h) of the Federal Water Pollution Control Act: Report to Congress
600284101	Characterization of Soil Disposal System Leachates
503289001	Compendium of Methods for Marine and Estuarine Environmental Studies
440582019	Compilation of Water Quality Standards for Marine Waters
910992029	Consensus Method for Determining Groundwaters under the Direct Influence of Surface Water Using Microscopic Particulate Analysis (MPA)
440185079	Development Document for Effluent Limitations Guidelines and Standards for the Pesticide Point Source Category [Final]
570990013	Drinking Water from Household Wells
600488025	Eastern Lake Survey: Phase II, National Stream Survey; Phase I, Processing Laboratory Operations Report
600777113	Environmental Pathways of Selected Chemicals in Freshwater Systems: Part I; Background and Experimental Procedures
821R98008	Evaluating Field Techniques for Collecting Effluent Samples for Trace Metals Analysis
823B98004	Evaluation of Dredged Material Proposed for Discharge in Waters of the U.S., Testing Manual (The Inland Testing Manual)
821F03019	Fact Sheet: Guidelines Establishing Test Procedures for the Analysis of Pollutants Under the Clean Water Act; National Primary Drinking Water Regulations; and National Secondary Drinking Water Regulations; Analysis, Sampling, and Monitoring Procedures; Proposed Rule
815F03003	Fact Sheet: U.S. EPA Drinking Water Methods for Chemical Parameters
600978038	First American–Soviet Symposium on Chemical Pollution of the Marine Environment
833B94001	Guidance for the Determination of Appropriate Methods for the Detection of Section 313 Water Priority Chemicals
821F03009	Guidelines Establishing Test Procedures for the Analysis of Pollutants: Analytical Methods for Biological Pollutants in Ambient Water; Final Rule [Fact Sheet]

ANALYTICAL METHODS 219

Table 8.2. (*continued*)

EPA No.	Method
600479019	Handbook for Analytical Quality Control in Water and Wastewater Laboratories
625673002	Handbook for Monitoring Industrial Wastewater
821R01027	Interlaboratory Validation Study Results for *Cryptosporidium* Precision and Recovery for U.S. EPA Method 1622
812B92004	Lead and Copper Monitoring Guidance for Water Systems Serving 101 to 500 Persons
812B92005	Lead and Copper Monitoring Guidance for Water Systems Serving 501 to 3,300 Persons
816F00009	Lead and Copper Rule Minor Revisions Fact Sheet for Public Water Systems That Serve More Than 50,000 Persons
816F00010	Lead and Copper Rule Minor Revisions: Fact Sheet for Large System Owners and Operators [Draft]
816F00007	Lead and Copper Rule Minor Revisions: Fact Sheet for Public Water Systems That Serve 3300 or Fewer Persons
816F00008	Lead and Copper Rule Minor Revisions: Fact Sheet for Public Water Systems That Serve 3301 to 50,000 Persons
816F99011	Lead and Copper Rule Minor Revisions: Fact Sheet for State Primacy Agencies
816F00010	Lead and Copper Rule Minor Revisions: Fact Sheet for Tribal Water System Owners and Operators
570989001	Lead in School Drinking Water
600375009	Limnological Investigation of the Muskegon County Michigan Wastewater Storage Lagoons: Phase 1
815B97001	Manual for the Certification of Laboratories Analyzing Drinking Water Criteria and Procedures Quality Assurance [Fourth Edition]
600878008	Manual for the Interim Certification of Laboratories Involved in Analyzing Public Drinking Water Supplies: Criteria and Procedures
815D03008	Membrane Filtration Guidance Manual
600R94134	Method 100.2. Determination of Asbestos Structures over 10 Micrometers in Length in Drinking Water
821R02020	Method 1103.1: *Escherichia coli* (*E. coli*) in Water by Membrane Filtration Using Membrane-Thermotolerant *Escherichia coli* Agar (mTEC)
821R02021	Method 1106.1: Enterococci in Water by Membrane Filtration Using Membrane-Enterococcus-Esculin Iron Agar (mE-EIA)
821R02024	Method 1604: Total Coliforms and *Escherichia coli* in Water by Membrane Filtration Using a Simultaneous Detection Technique (MI Medium)
821R01026	Method 1622: *Cryptosporidium* in Water by Filtration/IMS/FA
821R97023	Method 1622: *Cryptosporidium* in Water by Filtration/IMS/FA [Draft]
821R01025	Method 1623: *Cryptosporidium* and *Giardia* in Water by Filtration/IMS/FA
821R96013	Method 1632: Determination of Inorganic Arsenic in Water by Hydride Generation Flame Atomic Absorption

(*continued overleaf*)

Table 8.2. (*continued*)

EPA No.	Method
821R95031	Method 1638: Determination of Trace Elements in Ambient Waters by Inductively Coupled Plasma–Mass Spectrometry
821R96005	Method 1638: Determination of Trace Elements in Ambient Waters by Inductively Coupled Plasma–Mass Spectrometry
821R96006	Method 1639: Determination of Trace Elements in Ambient Waters by Stabilized Temperature Graphite Furnace Atomic Absorption
821R96007	Method 1640: Determination of Trace Elements in Ambient Waters by On-Line Chelation Preconcentration and Inductively Coupled Plasma–Mass Spectrometry
821R96011	Method 1669: Sampling Ambient Water for Trace Metals at EPA Water Quality Criteria Levels
821R96008	Method 1669: Sampling Ambient Water for Trace Metals at EPA Water Quality Criteria Levels [Draft]
815R03007	Method 326.0: Determination of Inorganic Oxyhalide Disinfection By-Products in Drinking Water Using Ion Chromatography Incorporating the Addition of a Suppressor Acidified Post Column Reagent for Trace Bromate Analysis, Revision 1.0
821R99013	Method OIA-1677: Available Cyanide by Flow Injection, Ligand Exchange, and Amperometry
821C97001	Methods and Guidance for the Analysis of Water (Includes EPA Series 500, 600, 1600 Methods) Single User (on CDROM)
600479020	Methods for Chemical Analysis of Water and Wastes
821B96005	Methods for Organic Chemical Analysis of Municipal and Industrial Wastewater
821R92008	Methods for the Determination of Diesel Mineral and Crude Oils in Offshore Oil and Gas Industry and Discharges
821R93010a	Methods for the Determination of Non-conventional Pesticides in Municipal and Industrial Wastewater, Volume I
600488039	Methods for the Determination of Organic Compounds in Drinking Water
600878017	Microbiological Methods for Monitoring the Environment: Water and Wastes
910R96001	Microscopic Particulate Analysis (MPA) for Filtration Plant Optimization
812B92007	Monitoring Requirements for Lead and Copper Rules: Water Systems Serving 10,001 to 50,000 Persons
812B92008	Monitoring Requirements for Lead and Copper Rules: Water Systems Serving 50,001 to 100,000 Persons
812B92009	Monitoring Requirements for Lead and Copper Rules: Water Systems Serving > 100,000 Persons
821R95026	Monitoring Trace Metals at Ambient Water Quality Criteria Levels: Briefing Book
570990NPS7	National Pesticide Survey: Glossary
570990NPS6	National Pesticide Survey: Project Summary
570990NPS5	National Pesticide Survey: Summary Results of EPA's National Survey of Pesticides in Drinking Water Wells

Table 8.2. (*continued*)

EPA No.	Method
570990NPS1	National Pesticide Survey: Survey Design
570984003	National Statistical Assessment of Rural Water Conditions: Executive Summary
570984004	National Statistical Assessment of Rural Water Conditions: Technical Summary
570983009	Nitrate Removal for Small Public Water Systems
600375030	Nitrogen in the Subsurface Environment
833B90103	NPDES Compliance Monitoring Inspector Training: Laboratory Analysis
800R94004	Office of Water Performance Evaluation Study Project: Final Report
430977005	Operations Manual: Package Treatment Plants
430977012	Operations Manual: Stabilization Ponds
600480032	Prescribed Procedures for Measurement of Radioactivity in Drinking Water
430977006	Process Control Manual: Aerobic Biological Wastewater Treatment Facilities
600387015	Processes Coefficients and Models for Simulating Toxic Organics and Heavy Metals in Surface Waters
810B92003	Quality Assurance Plan for the National Pesticide Survey of Drinking Water Wells: Analytical Method 2, Chlorinated Pesticides
810B92001	Quality Assurance Project Plan for the National Pesticide Survey of Drinking Water Wells
810B92009	Quality Assurance Project Plan for the National Pesticide Survey of Drinking Water Wells Referee Analyses for Analytical Method 2, Organochlorine Pesticides; Analytical Method 4, Carbamates; Method 5, Methylcarbamates; Method 7, Fumigants; and Method 9, Nitrate/Nitrite
810B92010	Quality Assurance Project Plan for the National Pesticide Survey of Drinking Water Wells: Analytical Method 1
810B92002	Quality Assurance Project Plan for the National Pesticide Survey of Drinking Water Wells: Analytical Method 1, Nitrogen/Phosphorous Pesticides, and Analytical Method 3, Chlorinated Acid Herbicides
810B92011	Quality Assurance Project Plan for the National Pesticide Survey of Drinking Water Wells: Analytical Method 3
810B92005	Quality Assurance Project Plan for the National Pesticide Survey of Drinking Water Wells: Analytical Method 5, Methyl Carbamates
810B92006	Quality Assurance Project Plan for the National Pesticide Survey of Drinking Water Wells: Analytical Method 6
810B92012	Quality Assurance Project Plan for the National Pesticide Survey of Drinking Water Wells: Analytical Method 6, Ethylene Thiourea
810B92007	Quality Assurance Project Plan for the National Pesticide Survey of Drinking Water Wells: Analytical Method 7, Fumigants
810B92008	Quality Assurance Project Plan for the National Pesticide Survey of Drinking Water Wells: Analytical Method 9, Nitrate and Nitrite

(*continued overleaf*)

Table 8.2. (*continued*)

EPA No.	Method
810B92015	Quality Assurance Project Plan for the National Pesticide Survey of Drinking Water Wells: Survey Statistics, Data Collection, and Processing
810B92014	Quality Assurance Project Plan for the National Survey of Pesticides in Drinking Water Wells: Well Sampling, Data Collection, and Processing
816F04026	Radionuclides and Arsenic Rules: Web Cast Training Sessions, Summer/Fall 2004
600285002	Recommended Practices for On-Line Measurement of Residual Chlorine in Wastewaters
814B96006	Reprints of EPA Methods for Chemical Analyses under the Information Collection Rule
821R01028	Results of the Interlaboratory Method Validation Study Results for Determination of *Cryptosporidium* and *Giardia* Using U.S. EPA Method 1623
430177003	Self-Monitoring Procedures: Basic Parameters for Municipal Effluents [Student Reference Manual]
822R03008	Six-Year Review: Chemical Contaminants, Health Effects Technical Support Document
816R01021	State Implementation Guidance for the Lead and Copper Rule Minor Revisions
812B94001	State Reporting Guidance for Unregulated Contaminant Monitoring
815R00002	Supplement A to the Unregulated Contaminant Monitoring Regulation Analytical Methods and Quality Control Manual
440486037	Technical Support Manual: Waterbody Surveys and Assessments for Conducting Use Attainability Analyses
440486038	Technical Support Manual: Waterbody Surveys and Assessments for Conducting Use Attainability Analyses, Volume 2: Estuarine Systems
440486039	Technical Support Manual: Waterbody Surveys and Assessments for Conducting Use Attainability Analyses, Volume 3: Lake Systems
440183079c	Test Methods: Methods for Non-conventional Pesticides Chemicals Analysis of Industrial and Municipal Wastewater
600482057	Test Method: Method for Organic Chemical Analysis of Municipal and Industrial Wastewater
815R00006	Unregulated Contaminant Monitoring Regulation Analytical Methods and Quality Control Manual
600R92070	User's Guide and Data Dictionary for Kenai Lakes Investigation Project
570990002	Your Drinking Water from Source to Tap: EPA Regulations and Guidance

gas chromatography–mass spectrometry. These methods may not be suitable for all applications, so the analyst must verify the method performance prior to use.

In all future approaches there is a need to reduce a comprehensive list of potential petroleum hydrocarbons to a manageable size. Depending on how

Table 8.3. Analytical Methods for the Analysis of Contaminants in Water

	Method	Title
100.1	Asbestos by Transmission Electron Microscopy	Analytical Method for the Determination of Asbestos Fibers in Water (EPA/600/4-83-043)
100.2	Asbestos by Transmission Electron Microscopy	Determination of Asbestos Structures Over 10 μm in Length in Drinking Water (EPA/600R-94/134)
150.1	pH, Electrometric	Methods for Chemical Analysis of Water and Wastes (EPA/600/4-79/020)
150.2	pH, Electrometric (Continuous Monitoring)	Methods for Chemical Analysis of Water and Wastes (EPA/600/4-79/020)
180.1 Rev 2.0	Turbidity, Nephelometric	Methods for the Determination of Inorganic Substances in Environmental Samples (EPA/600/R-93/100)
200.7 Rev 4.4	Metals and Trace Elements by ICP/Atomic Emission Spectrometry	Methods for the Determination of Metals in Environmental Samples, Supplement 1 (EPA/600/R-94/111)
200.8 Rev 5.4	Trace Elements by ICP/Mass Spectrometry	Methods for the Determination of Metals in Environmental Samples, Supplement 1 (EPA/600/R-94/111)
200.9 Rev 2.2	Trace Elements by Stabilized Temperature Graphite Furnace AA Spectrometry	Methods for the Determination of Metals in Environmental Samples, Supplement 1 (EPA/600/R-94/111)
245.1 Rev 3.0	Mercury by Cold Vapor AA Spectrometry [Manual]	Methods for the Determination of Metals in Environmental Samples, Supplement 1 (EPA/600/R-94/111)
245.2	Mercury by Cold Vapor AA Spectrometry [Automated]	Methods for Chemical Analysis of Water and Wastes (EPA/600/4-79/020)
300.0 Rev 2.1	Inorganic Anions by Ion Chromatography	Methods for the Determination of Inorganic Substances in Environmental Samples (EPA/600/R-93/100)
300.1 Rev 1.0	Determination of Inorganic Anions in Drinking Water by Ion Chromatography	Methods for the Determination of Organic and Inorganic Compounds in Drinking Water, Volume 1 (EPA 815-R-00-014)
335.4 Rev 1.0	Total Cyanide by Semi-automated Colorimetry	Methods for the Determination of Inorganic Substances in Environmental Samples (EPA/600/R-93/100)

(*continued overleaf*)

Table 8.3. (*continued*)

Method		Title
353.2 Rev 2.0	Nitrate–Nitrite by Automated Colorimetry	Methods for the Determination of Inorganic Substances in Environmental Samples (EPA/600/R-93/100)
365.1 Rev 2.0	Phosphorus by Automated Colorimetry. (Method for ortho-Phosphate)	Methods for the Determination of Inorganic Substances in Environmental Samples (EPA/600/R-93/100)
375.2 Rev 2.0	Sulfate by Automated Colorimetry	Methods for the Determination of Inorganic Substances in Environmental Samples (EPA/600/R-93/100)
502.2 Rev 2.1	VOCs by Purge and Trap Capillary GC with Photoionization and Electrolytic Conductivity Detectors in Series	Methods for the Determination of Organic Compounds in Drinking Water, Supplement III (EPA/600/R-95-131)
504.1 Rev 1.1	EDB, DBCP, and 1,2,3-Trichloropropane by microextraction and GC	Methods for the Determination of Organic Compounds in Drinking Water, Supplement III (EPA/600/R-95-131)
505 Rev 2.1	Organohalide Pesticides and PCBs by Microextraction and GC	Methods for the Determination of Organic Compounds in Drinking Water, Supplement III (EPA/600/R-95-131)
506 Rev 1.1	Phthalate and Adipate Esters by Liquid–Liquid or Liquid–Solid Extraction by GC with a Photoionization Detector	Methods for the Determination of Organic Compounds in Drinking Water, Supplement III (EPA/600/R-95-131)
507 Rev 2.1	Nitrogen- and Phosphorus-Containing Pesticides by GC with a Nitrogen–Phosphorus Detector	Methods for the Determination of Organic Compounds in Drinking Water, Supplement III (EPA/600/R-95-131)
508 Rev 3.1	Chlorinated Pesticides by GC with an Electron Capture Detector	Methods for the Determination of Organic Compounds in Drinking Water, Supplement III (EPA/600/R-95-131)
508A Rev 1.0	Screening for PCBs by Perchlorination and GC	Methods for the Determination of Organic Compounds in Drinking Water (EPA/600/4-88-039), December 1988, Revised July 1991

Table 8.3. (*continued*)

Method		Title
508.1 Rev 2.0	Chlorinated Pesticides, Herbicides, and Organohalides by Liquid–Solid Extraction and GC with an Electron Capture Detector	Methods for the Determination of Organic Compounds in Drinking Water, Supplement III (EPA/600/R-95-131)
515.1 Rev 4.0	Chlorinated Acids by GC with an Electron Capture Detector	Methods for the Determination of Organic Compounds in Drinking Water (EPA/600/4-88-039), December 1988, Revised July 1991
515.2 Rev 1.1	Chlorinated Acids Using Liquid–Solid Extraction and GC with an Electron Capture Detector	Methods for the Determination of Organic Compounds in Drinking Water, Supplement III (EPA/600/R-95-131)
515.3 Rev 1.0	Chlorinated Acids Using Liquid–Liquid Extraction, Derivatization, and GC with Electron Capture Detection	Methods for the Determination of Organic and Inorganic Compounds in Drinking Water, Volume 1 (EPA 815-R-00-014)
515.4 Rev 1.0	Chlorinated Acids Using Liquid–Liquid Microextraction, Derivatization, and Fast Gas Chromatography with Electron Capture Detection	Determination of Chlorinated Acids in Drinking Water by Liquid–Liquid Microextraction, Derivatization, and Fast Gas Chromatography with Electron Capture Detection
524.2 Rev 4.1	Purgeable Organic Compounds by Capillary Column GC/Mass Spectrometry	Methods for the Determination of Organic Compounds in Drinking Water, Supplement III (EPA/600/R-95-131)
525.2 Rev 2.0	Organic Compounds by Liquid–Solid Extraction and Capillary Column GC/Mass Spectrometry	Methods for the Determination of Organic Compounds in Drinking Water, Supplement III (EPA/600/R-95-131)
531.1 Rev 3.1	n-Methylcarbamoyloximes and n-Methylcarbamates by HPLC with Post Column Derivatization	Methods for the Determination of Organic Compounds in Drinking Water, Supplement III (EPA/600/R-95-131)
531.2 Rev 1.0	n-Methylcarbamoyloximes and n-Methylcarbamates by Direct Aqueous Injection HPLC with Post Column Derivatization	Measurement of n-Methyl-carbamoyloximes and n-Methyl-carbamates in Water by Direct Aqueous Injection HPLC with Post Column Derivatization
547	Glyphosphate by HPLC, Post Column Derivatization, and Fluorescence Detector	Methods for the Determination of Organic Compounds in Drinking Water, Supplement I (EPA/600/4-90/020)

(*continued overleaf*)

Table 8.3. (*continued*)

Method		Title
548.1 Rev 1.0	Endothall by Ion Exchange Extraction, Acidic Methanol Methylation, and GC/Mass Spectrometry	Methods for the Determination of Organic Compounds in Drinking Water, Supplement II (EPA/600/R-92/129)
549.2 Rev 1.0	Diquat and Paraquat by Liquid–Solid Extraction and HPLC with a Photodiode Array UV Detector	Methods for the Determination of Organic and Inorganic Compounds in Drinking Water, Volume 1 (EPA 815-R-00-014)
550	Polycyclic Aromatic Hydrocarbons (PAHs) by Liquid–Liquid Extraction and HPLC with Coupled UV and Fluorescence Detection	Methods for the Determination of Organic Compounds in Drinking Water, Supplement I (EPA/600/4-90/020)
550.1	Polycyclic Aromatic Hydrocarbons (PAHs) by Liquid–Solid Extraction and HPLC with Coupled UV and Fluorescence Detection	Methods for the Determination of Organic Compounds in Drinking Water, Supplement I (EPA/600/4-90/020)
551.1 Rev 1.0	Chlorinated Disinfection By-Products and Chlorinated Solvents by Liquid–Liquid Extraction and GC with an Electron Capture Detector	Methods for the Determination of Organic Compounds in Drinking Water, Supplement III (EPA/600/R-95-131)
552.1 Rev 1.0	Haloacetic Acids and Dalapon by Ion Exchange Liquid–Solid Extraction and GC with Electron Capture Detector	Methods for the Determination of Organic Compounds in Drinking Water, Supplement II (EPA/600/R-92/129)
552.2 Rev 1.0	Haloacetic Acids and Dalapon by Liquid–Liquid Extraction, Derivatization, and GC with Electron Capture Detector	Methods for the Determination of Organic Compounds in Drinking Water, Supplement III (EPA/600/R-95-131)
555 Rev 1.0	Chlorinated Acids by HPLC with a Photodiode Array Ultraviolet Detector	Methods for the Determination of Organic Compounds in Drinking Water, Supplement II (EPA/600/R-92/129)

Table 8.4. Test Methods for Secondary Drinking Water Contaminants

Contaminant	EPA	ASTM[a]	AWWA[b]
Aluminum	200.7, 200.8, 200.9		3120B, 3113B, 3111D
Chloride	300.0	D4327, D512	4110B, 4500-Cl-D, 4500-Cl-B
Color			2120B
Copper	200.7, 200.8, 200.9	D1688, D1688	3120B, 3111B, 3113B
Fluoride	300.0	D4327, D1179	4110B, 4500-F-B,D, 4500-F-C, 4500-F-E
Foaming agents			5540C
Iron	200.7, 200.9		3120B, 3111B, 3113B
Manganese	200.7, 200.8, 200.9		3120B, 3111B, 3113B
Odor			2150B
pH	150.1, 150.2	D1293	4500-H+B
Silver	200.7, 200.8, 200.9		3120B, 3111B, 3113B
Sulfate	300.0, 375.2	D4327, D516	4110B, 4500-SO4F, 4500-SO4C,D, 4500-SO4E
Total dissolved solids			2540C
Zinc	200.7, 200.8		3120B, 3111B

[a] American Society for Testing and Materials, *Annual Book of ASTM Standards, 2004*, West Conshohocken, PA.
[b] AWWA, *Standard Methods for the Examination of Water and Wastewater*, American Water Works Association, Washington, DC.

conservative the approach is, methods that have been used select (1) the most toxic among the total petroleum hydrocarbons (indicator approach), (2) one or more representative compounds (surrogate approach, but independent of relative mix of compounds), or (3) representative compounds for fractions of similar petroleum hydrocarbons. The *fraction* approach is the most demanding in information gathering and because of that would appear to be the most rigorous approach to date.

8.2.3. Semivolatile and Nonvolatile Hydrocarbons

As mentioned above, the most usual analytical finish for hydrocarbon determination is gas chromatography. Depending on the degree of resolution and level of information required, a number of instrument configurations may be employed. The most common requirement is determination of total petroleum hydrocarbons, and this will often consist largely of diesel range organic compounds. For this purpose, the most normal procedure is gas chromatography–flame ionization (EPA 8015B).

Because of the nature of the analytes (boiling point 170 to 430°C), higher oven temperatures are required for chromatography of this fraction compared to gasoline range organic compounds. Commonly, fused silica capillary columns are used and the sample is generally introduced by direct injection. Temperatures of the injector and detector are maintained at 200°C (390°F) and 340°C (645°F), respectively, throughout the run, and the column temperature is ramped from 45°C to 275°C. GC/FID may be used simply to fingerprint the components of a hydrocarbon pollution episode (Bruce and Schmidt, 1994), the strategy being most successful if the pollutant has only recently entered the soil environment.

Most frequently, however, some attempt is made to quantify the hydrocarbon fractions represented (Whittaker et al., 1995). It is possible to employ both external and internal standards in these determinations. When internal standards are used, they are generally compounds, such as hexafluoro-2-propanol, hexafluoro-2-methyl-2-propanol, or 2-chloroacrylonitrile. As regards the determination of diesel range organic compounds, regulatory authorities vary in terms of the prescribed range. Typically, the DRO range is considered to begin at C_{10} to C_{12} and end at C_{24} to C_{28}. Whatever the range, total petroleum hydrocarbons is taken as the sum of the area within that region of the chromatogram.

More sophisticated detection methods for gas chromatography are also employed in the analysis of hydrocarbons: gas chromatography–mass spectrometry (EPA 8270C) and gas chromatography–Fourier transform infrared spectroscopy (EPA 8410). These procedures have a significant advantage in providing better characterization of the contaminants and thus are of particular use where some environmental modification of the hydrocarbons has taken place subsequent to soil deposition.

A superior approach to determination of total petroleum hydrocarbons in soil is the summation of areas for specific ranges of hydrocarbons. This allows a better profiling of the contaminants and also confers the ability to trace the source of the pollutant. Typical ranges for the hydrocarbon profiles are n-C_{10} to n-C_{14}, n-C_{15} to n-C_{20}, n-C_{21} to n-C_{26}, and n-C_{27} to n-C_{36}. However, one must be cautious in the application of statistical methods to the determination, insofar as such methods are only as good as the information and assumptions used. Recall: Garbage in, garbage out!

One of the major problems associated with profiling of hydrocarbons at contaminated sites is the phenomenon of *weathering*, which relates to a change in composition of hydrocarbons with time, through the action of volatilization, leaching, chemical reaction (usually, oxidation but can be a reaction with soil constituents), and biotransformation.

For volatile organic compounds, the most significant process is through volatilization, resulting in a decrease of the overall concentration with time. On the other hand, the higher-molecular-weight hydrocarbons are more prone to (chemical) modification through other processes, and it becomes necessary to identify the products of the various transformations. In addition, it is useful to obtain some index of overall weathering.

Such information cannot readily be obtained from simple gas chromatography–flame ionization profiles and gas chromatography–mass spectrometry has been used for such analyses. Electron impact ionization (EI) and chemical ionization (CI) procedures are available. The former procedure produces predominantly fragment ions, whereas the latter produces predominantly parent ions. With complex high-molecular-weight samples, chemical ionization can or will produce ambiguous results, since many of the analytes have identical parent ion peaks. Thus, gas chromatography–electron impact mass spectrometry has been the method of choice for analysis of most hydrocarbon studies (Altgelt and Boduszynski, 1994). The availability of this *piggyback* method (i.e., GC/MS/MS) has further enhanced the ability to examine environmental hydrocarbon samples for particular components.

Of particular significance in the study of petroleum weathering are the *biomarker* molecules (e.g., pristane, phytane, the hopanes and steranes). Historically, the biomarkers have been employed as crude oil signatures in prospecting and characterization. More recently, such molecules have also been employed in the environmental field, both for the determination of pollutant source and estimation of the degree of weathering.

The biomarker molecules are particularly resistant to microbial attack, and thus the ratio of other hydrocarbon components to the biomarker will decrease as the crude oil is biodegraded (Wang et al., 1994). In the case of an ongoing oil discharge into the soil, this ratio will be highest nearest the source and will decrease with increasing distance from the source. Thus, the ratio may be used to locate the source of the contaminant (Whittaker et al., 1995).

In a similar manner, expression of biodegradable hydrocarbons as a ratio to high-molecular-weight polynuclear aromatic hydrocarbons should have potential for fingerprinting purposes. The failure of some attempts to use PAHs for this purpose arises from the poor choice of molecules for comparison. Low-molecular-weight PAHs such as naphthalene or phenanthrene are often selected because of their abundance and relative ease of measurement, but these molecules are also the most prone to biodegradation as well as other forms of attenuation (Sadler and Connell, 2002).

There are indications that approved methods used for oil spill assessments, including the method for total petroleum hydrocarbons (EPA 418.1), the methods for semivolatile priority pollutant organics (EPA 625, 8270), and the methods for volatile organic priority pollutant methods (EPA 602, 1624, 8240) are all inadequate for generating scientifically defensible information for natural resource damage assessment. These general organic chemical methods are deficient in chemical selectivity (types of constituents analyzed) and sensitivity (detection limits); deficiencies in these two areas lead to an inability to interpret the environmental significance of the data in a scientifically defensible manner.

8.3. ASSESSMENT OF THE METHODS

Generally, measurement of the total petroleum hydrocarbon in an ecosystem is performed by the standard method (EPA 418.1) or some modification thereof. However, many other methods exist is which the data are also claimed to be representative of the total petroleum hydrocarbons in the ecosystem. In fact, many methods for determining total petroleum hydrocarbons are prone to the following:

1. Producing false negatives (reporting *nondetected* when considerable petroleum hydrocarbons were really present).
2. Underestimating the extent of petroleum hydrocarbons present (true of virtually every total petroleum hydrocarbons methodology).
3. Underestimating the overall risk from petroleum hydrocarbons due to missing significant amounts of some of the compounds of most concern (e.g., polynuclear aromatic hydrocarbons).
4. Producing misleading data related to soil hot spots versus areas of less concern due to differing moisture concentrations of otherwise similar samples.
5. Producing misleading results because an inappropriate (not close enough to the unknown being sampled) standard (oil) was used in calibration.
6. Producing soil or sediment data which cannot be compared directly with other total petroleum hydrocarbons data or guidelines because one is expressed in dry weight and the other in wet weight.
7. Producing relatively accurate dry-weight values for heavy petroleum hydrocarbons but questionable dry weight values for lighter, more volatile compounds. (*Note:* Different labs dry the samples in different ways and a sample with lots of lighter-fraction hydrocarbons is more prone to hydrocarbon loss; the variable loss of volatile hydrocarbons in a drying step is therefore an additional area of lab and data variability).
8. Producing data that cannot be compared directly with other total petroleum hydrocarbons data or guidelines because one data set is the result of a Soxhlet extraction method and the other reflects a sonication or other alternative extraction method.
9. Producing misleading data related to heavy fraction hydrocarbons (again, such as the heavier PAHs) due to loss of the heavier compounds on filter paper.

10. Producing data prone to faulty interpretation of the environmental significance of the results (100 ppm of total petroleum hydrocarbons from one type of oil may be practically nontoxic, whereas 100 ppm of total petroleum hydrocarbons from a different type of oil may be very toxic).

Another complication with total petroleum hydrocarbons values is that petroleum-derived inputs vary considerably in composition; it is essential to bear this in mind when quantifying them in general terms such as an *oil* or the *total petroleum hydrocarbons* measurement. Petroleum is complex, containing many thousands of compounds, ranging from gases to residues boiling at about 400°C.

Furthermore, since different combinations of petroleum hydrocarbons typically contribute to "total petroleum hydrocarbons" at different sites, the fate characteristics are also typically different at different sites, even if the total petroleum hydrocarbons concentration is the same. Different methods used to generate total petroleum hydrocarbon concentrations, or similar simple screening measures of petroleum contamination, all produce very different result.

It is not surprising that the data produced as *total petroleum hydrocarbons* (EPA 418.1) suffer from several shortcomings as an index of potential groundwater contamination or health risk. In fact, it does not actually measure the total petroleum hydrocarbons in the sample but rather, measures a specific range of hydrocarbon compounds. This is caused by limitations of the extraction process (solvents used and the concentration steps) and the reference standards used for instrumental analysis. The method specifically states that it does not accurately measure the lighter fractions of gasoline [benzene–toluene-ethylbenzene–xylenes fraction (BTEX)], which should include the benzene–toluene–ethylbenzene–xylenes fraction. Further, the method was originally a method for water samples that has been modified for solids, and it is subject to bias.

The total petroleum hydrocarbons represents a summation of all the hydrocarbon compounds that may be present (and detected) in a soil sample. Because of differences in product composition between, for example, gasoline and diesel, or fresh versus weathered fuels, the types of compounds present at one site may be completely different from those present at another.

Accordingly, the total petroleum hydrocarbons at a gasoline spill site will be comprised of mostly C_6 to C_{12} compounds, while total petroleum hydrocarbons at an older site where the fuel has weathered will likely measure mostly C_8 to C_{12} compounds. Because of this inherent variability in the method and the analyte, it is currently not possible to directly relate potential environmental or health risks with concentrations of total petroleum hydrocarbons. The relative mobility or toxicity of contaminants represented by total petroleum hydrocarbons analyses at one site may be completely different from that of another site (e.g., C_6 to C_{12} compared to C_{10} to C_{25}). There is no easy way to determine if total petroleum hydrocarbons from the former site will represent the same level of risk as an equal measure of the total petroleum hydrocarbons from the latter. For these reasons it is clear that the total petroleum hydrocarbons value offers limited benefits as an indicator measure for cleanup criteria. Its current widespread use as a soil cleanup criterion is a function of a lack of understanding of its proper application and limitations, and its historical use as a simple and inexpensive indicator of general levels of contamination.

When sampling in the environment, it is often impossible to determine which chemical mixtures are causing a total petroleum hydrocarbons reading, which is one of the major weaknesses of the method. At minimum, before using contaminants data from diverse sources, efforts should be made to determine that field collection methods, detection limits, and quality control techniques were acceptable and comparable. This will help the analysts compare the analysis in the concentration range with the benchmark or regulatory criteria concentrations should be very precise and accurate.

Indeed, it must be remembered that quality control field and lab blanks and duplicates will not help in the data quality assurance goal as well as intended if one is using a method prone to false negatives. Methods may be prone to false negatives due to the use of detection limits that are too high, the loss of contaminants through inappropriate handling, or the use of inappropriate methods. The use of inappropriate methods prone to false negatives (or false positives) is particularly common related to total petroleum hydrocarbons and other general scans related to oil products. This is one reason that more rigorous analyses are often recommended as alternatives to total petroleum hydrocarbons analyses.

In interpreting the data for the total petroleum hydrocarbons in a sample, the amount of moisture cannot be ignored because moisture blocks the extraction of petroleum hydrocarbons by another hydrocarbon (Freon). Potentially, sulfur or phthalate compounds also interfere with total petroleum hydrocarbons analyses. This is similar to the problem of strong interferences from phthalate esters or chlorinated solvents when one is using electron capture methods to look for chlorinated compounds such as polycholorbiphenyls or pesticides.

Too much reliance on the determination of benzene–toluene–xylenes (BTX) or benzene–toluene–ethylbenzene–xylenes (BTEX) to measure gasoline or diesel contamination may be unaware that more modern gasoline and diesel are better refined and contain fewer of such compounds. It must be remembered that the use of benzene–toluene–xylene data started as a measure of the more hazardous compounds in gasoline. Modern gasoline and diesel has a higher percentage of straight-chain alkanes, nonvolatiles, not as many aromatics, lots of long-chain aliphatic compounds, and fewer benzene–toluene–xylene compounds. In addition, determination of the benzene–toluene–xylene concentration is not appropriate for aged gasoline characterized by loss of benzene–toluene–xylene compounds over time. Thus, the problem with many analyses for benzene–toluene–xylene as related to petroleum hydrocarbons is the danger of producing false negatives. For example, the test for benzene–toluene–xylene may indicate no contamination when significant contamination is present.

Like *total petroleum hydrocarbons, total recoverable petroleum hydrocarbons* (TRPH) is defined methodologically and concentrations given as total petroleum hydrocarbons, or TRPH alone, does not produce much valuable information. To be able to understand the significance of the concentration, the method employed for the determination must be clearly identified (e.g., EPA 8015 for gasoline, EPA 8016 for diesel, EPA 418.1 for total recoverable petroleum hydrocarbons). The data must not be used or interpreted as though various total petroleum hydrocarbons methods were the same as various total recoverable petroleum hydrocarbon methods. When comparing data with soil guideline levels, it is necessary to ascertain which laboratory analysis was done to measure compliance with the current specific guideline.

Additional problems with total petroleum hydrocarbons methods (including method 418.1) include the following:

1. Most methods used to determine the total petroleum hydrocarbons in a sample are inadequate for unknowns because the methods are only as good as the calibration standards. With unknown chemicals present, the precise standards cannot be selected and employing an incorrect calibration standard can lead to erroneous data.
2. Some of the methods that have been used for determination of the total petroleum hydrocarbons also extract vegetable and animal oils that are also present in the sample.
3. The methodology related to volatility can be extremely variable: For example, low-boiling oils are more susceptible to ambient (and extraction) conditions. The time for evaporation of the oils is a variable, and the temperature and heating period used to calculate dry weight is also a variable. It is preferable to calculate wet weight total petroleum hydrocarbons values first and then to measure percentage moisture very carefully in a manner that minimizes losses.

The ASTM method for total petroleum hydrocarbons (ASTM book) is similar to the standard EPA method (EPA 418.1) and calls for extraction with Freon. The estimated variability of the test method is questionable and may leave room for serious errors in the calculation of total petroleum hydrocarbons.

Since the determination of the total petroleum hydrocarbons in a sample is subject to many questions, the bias must be defined, and alternative reliable and meaningful methods must be sought. For example, *negative bias* may result when samples are analyzed because of (1) poor extraction efficiency of the solvent (Freon, EPA 481, or n-hexane, EPA 1664) for high-molecular-weight hydrocarbons, (2) loss of volatile hydrocarbons during extract concentration (Speight, 1999, 2004), (3) differences in molar absorbtivity between the calibration standard and product type because of the presence of unknown compound types, (4) fractionation of soluble low-infrared-active aromatic hydrocarbons in groundwater during water washout, (5) removal of five- to six-ring alkylated aromatics during the silica cleanup procedure (the efficiency of silica gel fractionation varies depending on the nature of the solute) (Speight, 1999), and (6) preferential biodegradation of n-alkanes.

In addition, *positive bias* is often introduced as a result of (1) product differences in molar absorbtivity, (2) partitioning of soluble aromatics from the bulk product because of oil washout, (3) measurement of naturally occurring saturated hydrocarbons that exhibit a high molar absorbtivity (e.g., plant waxes, n-C_{25}, n-C_{27}, n-C_{29}, and n-C_{31} alkanes), and (4) infrared dispersion of clay particles.

Thus, and to reaffirm earlier statements, there is no single analytical method that is perfect or even adequate for all cases to determine the amount of total petroleum hydrocarbons in a sample. Different analytical methods have different capabilities, and (this is where the environmental analysts plays an important role) it is up to the analysts to demonstrate that the method applied at specific sites was appropriate.

REFERENCES

Altgelt, K. H., and Boduszynski, M. M. 1994. In *Comparison and Analysis of Heavy Petroleum Fractions*. Marcel Dekker, New York, p. 257.

Apitz, S. E., Borbridge, L. M., Theriault, G. A., and Lieberman, S. H. 1992. *Analysis*, 20: 461.

Askari M. D., Masakarinec M. P., Smith S., Beam P. M., and Travis C. C. 1996. *Anal. Chem.*, 68: 3431.

ASTM. 1995. *Risk-Based Corrective Action Guidance*. American Society for Testing and Materials, West Conshohocken, PA.

ASTM. 2004. *Annual Book of Standards*. American Society for Testing and Materials. West Conshohocken, PA.

Bruce L. G., and Schmidt G. W. 1994. *Am. Assoc. Pet. Geol. Bull.*, 78: 1692.

CFR. 2004. *Code of Federal Regulations*, U.S. Government, Washington, DC. The *Code of Federal Regulations* (CFR) is the codification of the general and permanent rules published in the *Federal Register* by the executive departments and agencies of the federal government. It is divided into 50 titles that represent broad areas subject to federal regulation. Each volume of the CFR is updated once each calendar year and is issued on a quarterly basis.

Dean, J. R. 2003. *Methods for Environmental Trace Analysis*. Wiley, Hoboken, NJ.

Driscoll, J. N., Hanby, J., and Panaro, J. 1992. In *Hydrocarbon Contaminated Soils*, Vol. II, P. T. Kostecki, E. J. Calabrese, and M. Bonazountas (Eds.). Lewis Publishers, Boca Raton, FL, p. 153.

Einhorn, I. N., Sears, S. F., Hickey, J. C., Vielenave, J. H., and Moore, G. S. 1992. *Hydrocarbon Contam. Soils*, 2: 89.

EPA. 2004. Environmental Protection Agency, Washington, DC. Web site: http://www.epa.gov.

Irwin, R. J., VanMouwerik, M., Stevens, Seese, M. D., and W. Basham, W. 1997. *Environmental Contaminants Encyclopedia*. National Park Service, Water Resources Division, Fort Collins, CO.

Kahrs, L. E., Horzempa, L. M., and Peterson, D. M. 1999. *Contam. Soils*, 4: 221–232.

Kasper, K. D., Twoney, D. M., and Dinsmore, D. 1991. *Ground Water Manage.* 8: 673.

Löhmannsröben, H. -G., Roch, T., and Schulyzr, R. H. 1999. *Polycycl. Aromat. Compounds*, 13: 165.

Miller, M. (Ed.). 2000. *Encyclopedia of Analytical Chemistry*. Wiley, New York.

Sadler, R., and Connell, D. 2002. In *Environmental Protection and Risk Assessment of Organic Contaminants*, R. Kookana, R. Sadler, N. Sethunathan, and R. Naidu (Eds.). Science Publisher, Enfield, NH, p. 27.

Sadler, R., and Connell, D. 2003. In *Proceedings, 5th National Workshop on the Assessment of Site Contamination*. NEPC Service Corporation, Adelaide, Australia, p. 133.

Siegrist, R. L., and Jenssen, P. D. 1990. *Environ. Sci. Technol.*, 24: 1387.

Speight, J. G. 1999. *The Chemistry and Technology of Petroleum*, 3rd ed. Marcel Dekker, New York.

Speight, J. G. 2001. *Handbook of Petroleum Analysis*. Wiley, New York.

Speight, J. G. (Ed.). 2004. *Lange's Handbook of Chemistry*, 16th ed. McGraw-Hill, New York.

Wang, Z., Fingas, M., and Sergy, Z. 1994. *Environ. Sci. Technol.*, 28: 1733.

Weisman, W. 1998. *Analysis of Petroleum Hydrocarbons in Environmental Media*. Total Petroleum Hydrocarbons Criteria Working Group Series, Vol. 1. Amherst Scientific Publishers, Amherst, MA.(See also Vol. 2, *Composition of Petroleum Mixtures*, 1998; Vol. 3, *Selection of Representation Total Petroleum Hydrocarbons Fractions Based on Fate and Transport Considerations*, 1997; Vol. 4, *Development of Fraction-Specific Reference Doses and Reference Concentrations for Total Petroleum Hydrocarbons*, 1997; and Vol. 5, *Human Health Risk-Based Evaluation of Petroleum Contaminated Sites, Implementation of the Working Group Approach*, 1999.)

West, O. R., Siegrist, R. L., Mitchell, T. J., and Jenkins, R. A. 1995. *Environ. Sci. Technol.*, 29: 647.

Whittaker, M., Pollard, S. J. T., and Fallick, T. E. 1995. *Environ. Technol.*, 16: 1009.

Xie, G., Barcelona, M., and Fang, J. 1999. *Anal. Chem.*, 71: 1899.

CHAPTER 9

ANALYSIS OF GASEOUS EFFLUENTS

In terms of waste definition, there are three basic approaches (as it pertains to petroleum, petroleum products, and nonpetroleum chemicals) to defining petroleum or a petroleum product as hazardous: (1) a qualitative description of the waste by origin, type, and constituents; (2) classification by characteristics based on testing procedures; and (3) classification as a result of the concentration of specific chemical substances (Chapter 1). In addition, there are recommended protocols that must occur as a prelude to cleanup of emissions and the mitigation of future releases (Table 9.1).

Petroleum is capable of producing gaseous pollutant chemicals (Guthrie, 1967; Rawlinson and Ward, 1973; Francis and Peters, 1980; Hoffman, 1983; Moustafa, 1996; Speight, 1993, 1999). Gaseous emissions are often characterized by chemical species identification: for example, *inorganic gases* such as sulfur dioxide (SO_2), nitrogen oxides (NO_x), and carbon monoxide (CO), or *organic gases* such as chloroform ($CHCl_3$) and formaldehyde (HCHO). The rate of release or concentrating in the exhaust airstream (in parts per million or comparable units) along with the type of gaseous emission greatly predetermines the applicable control technology.

The three principal greenhouse gases that are products of refining are carbon dioxide, nitrous oxide, and methane (Fogg and Sangster, 2003). Carbon dioxide is the main contributor to climate change. Methane is generally not as abundant as carbon dioxide but is produced during refining and if emitted into the atmosphere is a powerful greenhouse gas and more effective at trapping heat. However, gaseous emissions associated with petroleum refining are more extensive than

Table 9.1. Necessary Actions for Cleanup of Emissions

1. Identification of the emissions
2. Identification of the emission sources
3. Estimation of emission rates
4. Atmospheric dispersion, transformation, and depletion mechanisms
5. Emission control methods
6. Air-quality evaluation methods
7. Effects on stratospheric ozone
8. Regulations

Environmental Analysis and Technology for the Refining Industry, by James G. Speight
Copyright © 2005 John Wiley & Sons, Inc.

carbon dioxide and methane and typically include process gases, petrochemical gases, volatile organic compounds (VOCs), carbon monoxide (CO), sulfur oxides (SO_x), nitrogen oxides (NO_x), particulates, ammonia (NH_3), and hydrogen sulfide (H_2S). These effluents may be discharged as air emissions and must be treated. However, gaseous emissions are more difficult to capture than wastewater or solid waste and thus are the largest source of untreated wastes released to the environment.

In the refining industry, as in other industries, air emissions include point and nonpoint sources. Point sources are emissions that exit stacks and flares and thus can be monitored and treated. Nonpoint sources are *fugitive emissions* that are difficult to locate and capture. Fugitive emissions occur throughout refineries and arise from, for example, thousands of valves, pumps, tanks, pressure relief valves, and flanges. Although individual leaks are typically small, the sum of all fugitive leaks at a refinery can be one of its largest emission sources.

The numerous process heaters used in refineries to heat process streams or to generate steam (boilers) for heating or steam stripping can be potential sources of sulfur oxides (SO_2, and SO_3), nitrogen oxides (NO and NO_2), carbon monoxide (CO), particulates, and hydrocarbons emissions. When operating properly and when burning cleaner fuels such as refinery fuel gas, fuel oil, or natural gas, these emissions are relatively low. If, however, combustion is not complete, or heaters are fired with refinery fuel pitch or residuals, emissions can be significant.

In addition to the corrosion of equipment of acid gases, the escape into the atmosphere of sulfur-containing gases can eventually lead to the formation of the constituents of acid rain [i.e., the oxides of sulfur (SO_2 and SO_3)]. Similarly, the nitrogen-containing gases can also lead to nitrous and nitric acids (through formation of the oxides NO_x, where $x = 1$ or 2), which are the other major contributors to acid rain. The release of carbon dioxide and hydrocarbons as constituents of refinery effluents can also influence the behavior and integrity of the ozone layer.

The processes that have been developed to accomplish gas purification vary from a simple once-through wash operation to complex multistep recycling systems. In many cases, the process complexities arise because of the need for recovery of the materials used to remove the contaminants or even recovery of the contaminants in the original, or altered, form (Kohl and Riesenfeld, 1979; Speight, 1993, and references cited therein).

The majority of gas streams exiting each refinery process are collected and sent to the gas treatment and sulfur recovery units to recover the refinery fuel gas and sulfur. Emissions from a sulfur recovery unit typically contain some hydrogen sulfide (H_2S), sulfur oxides, and nitrogen oxides. Other emission sources from refinery processes arise from periodic regeneration of catalysts. These processes generate streams that may contain relatively high levels of carbon monoxide, particulates, and volatile organic compounds (VOCs). Before being discharged to the atmosphere, such off-gas streams may be treated first through a carbon monoxide boiler to burn carbon monoxide and any volatile organic compounds, and then through an electrostatic precipitator or cyclone separator to remove particulates.

Analysts need consistent, reliable, and credible methodologies to produce analytical data about gaseous emissions (Patnaik, 2004). To fulfill this need in this book, this chapter is devoted to descriptions of the various analytical methods that can be applied to identify gaseous emissions from a refinery (ASTM, 2004; IP, 2001). Each gas is, in turn, referenced by its name rather than the generic term *petroleum gas* (ASTM D4150). However, the composition of each gas varies, and recognition of this is essential before testing protocols are applied.

9.1. GASEOUS PRODUCTS

9.1.1. Liquefied Petroleum Gas

Liquefied petroleum gas (LPG) is a mixture of the gaseous hydrocarbons propane [$CH_3CH_2CH_3$, boiling point: $-42°C$ ($-44°F$)] and butane [$CH_3CH_2CH_2CH_3$, boiling point: $0°C$ ($32°F$)] that are produced during natural gas refining, petroleum stabilization, and petroleum refining (Austin, 1984; Speight, 1999; Ritter, 2000). Propane and butane can be derived from natural gas or from refinery operations, but in the latter case, substantial proportions of the corresponding olefins will be present and need to be separated. The hydrocarbons are normally liquefied under pressure for transportation and storage.

The presence of propylene and butylenes in liquefied petroleum gas used as fuel gas is not critical. The vapor pressures of these olefins are slightly higher than those of propane and butane, and the flame speed is substantially higher, but this may be an advantage since the flame speeds of propane and butane are slow. However, one issue that often limits the amount of olefins in liquefied petroleum gas is the propensity of the olefins to form soot as well as the presence of mechanically entrained water (which may be limited further by specifications) (ASTM D1835). The presence of water in liquefied petroleum gas (or in natural gas) is undesirable since it can produce hydrates that will cause, for example, line blockage due to the formation of hydrates under conditions where the water *dew point* is attained (ASTM D1142). If the amount of water is above acceptable levels, the addition of a small amount of methanol will counteract any such effect. Another component of liquefied petroleum gas is propylene ($CH_3CH=CH_2$), which has a significantly lower octane number (ASTM D2623) than propane, so there is a limit to the amount of this component that can be tolerated in the mixture. Analysis by gas chromatography is possible (ASTM D5504, D6228; IP 405).

Liquefied petroleum gas and liquefied natural gas can share the facility of being stored and transported as a liquid and then vaporized and used as a gas. To achieve this, liquefied petroleum gas must be maintained at moderate pressure but at ambient temperature. The liquefied natural gas can be at ambient pressure but must be maintained at a temperature of roughly -1 to $60°C$ (30 to $140°F$). In fact, in some applications it is actually economical and convenient to use liquefied petroleum gas in the liquid phase. In such cases, certain aspects of gas composition (or quality, such as the ratio of propane to butane and the presence of traces of heavier hydrocarbons, water, and other extraneous materials) may be of less importance than use of the gas in the vapor phase.

For normal (gaseous) use, the contaminants of liquefied petroleum gas are controlled at a level at which they do not corrode fittings and appliances or impede the flow of the gas. For example, hydrogen sulfide (H_2S) and carbonyl sulfide (COS) should be absent, although to the level required for adequate odorization, organic sulfur compounds are permissible (ASTM D5305). In fact, *stenching* is a normal requirement in liquefied petroleum gas; dimethyl sulfide (CH_3SCH_3) and ethyl mercaptan (C_2H_5SH) are commonly used at a concentration of up to 50 ppm. Natural gas is treated similarly, possibly with a wider range of volatile sulfur compounds.

9.1.2. Natural Gas

Natural gas is found in petroleum reservoirs as free gas (*associated gas*) or in solution with petroleum in the reservoir (*dissolved gas*) or in reservoirs that contain only gaseous constituents and no (or little) petroleum (*unassociated gas*) (Austin, 1984; Speight, 1999; Cranmore and Stanton, 2000). The hydrocarbon content varies from mixtures of methane and ethane with very few other constituents (*dry* gas) to mixtures containing all of the hydrocarbons from methane to pentane and even hexane (C_6H_{14}) and heptane (C_7H_{16}) (*wet* gas) (Speight, 1999). In both cases, some carbon dioxide (CO_2) and inert gases, including helium (He), are present together with hydrogen sulfide (H_2S) and a small quantity of organic sulfur.

Although the major constituent of natural gas is methane, there are components such as carbon dioxide (CO), hydrogen sulfide (H_2S), and mercaptans (thiols; R–SH), as well as trace amounts of sundry other emissions. The fact that methane has a foreseen and valuable end use makes it a desirable product, but in several other situations it is considered a pollutant, having been identified as one of several greenhouse gases.

Carbon dioxide (ASTM D1137, D1945, D4984) in excess of 3% is normally removed for reasons of corrosion prevention (ASTM D1838). Hydrogen sulfide (ASTM D2420, D2385, D2725, D4084, D4810; IP 103, 272) is also removed and the odor of the gas must not be objectionable (ASTM D6273), so mercaptan content (ASTM D1988, D2385; IP 272) is important. A simple lead acetate test (ASTM D2420, D4084) is available for detecting the presence of hydrogen sulfide and is an additional safeguard that hydrogen sulfide not be present (ASTM D1835). The odor of the gases must not be objectionable. Methyl mercaptan, if present, produces a transitory yellow stain on the lead acetate paper that fades completely in less than 5 minutes. Other sulfur compounds (ASTM D5504, D6228) present in liquefied petroleum gas do not interfere.

In the lead acetate test (ASTM D2420), the vaporized gas is passed over moist lead acetate paper under controlled conditions. Hydrogen sulfide reacts with lead acetate to form lead sulfide, resulting in a stain on the paper varying in color from yellow to black, depending on the amount of hydrogen sulfide present. Other pollutants can be determined by gas chromatography (ASTM D5504, D6228; IP 318).

The total sulfur content (ASTM D1072, D2784, D3031) is normally acceptably low, frequently so low that it needs augmentation by means of alkyl sulfides, mercaptans, or thiophenes in order to maintain an acceptable safe level of odor.

The hydrocarbon dew point is reduced to such a level that retrograde condensation (i.e., condensation resulting from pressure drop) cannot occur under the worst conditions likely to be experienced in the gas transmission system. Similarly, the water dew point is reduced to a level sufficient to preclude formation of C_1 to C_4 hydrates in the system.

The natural gas after appropriate treatment for acid gas reduction, odorization, and hydrocarbon and moisture dew point adjustment (ASTM D1142) would then be sold within prescribed limits of pressure, calorific value, and possibly, the Wobbe Index [(calorific value)/(specific gravity)].

9.1.3. Refinery Gas

Refinery gas (*process gas*) is the noncondensable gas that is obtained during distillation of crude oil or treatment (cracking, thermal decomposition) of petroleum (Gary and Handwerk; 1975; Austin, 1984; Speight, 1999; Robinson and Faulkner, 2000; Speight and Ozum, 2002). There are also components of the gaseous products that must be removed prior to release of the gases to the atmosphere or prior to use of the gas in another part of the refinery (i.e., as a fuel gas or as a process feedstock).

Refinery gas consists mainly of hydrogen (H_2), methane (CH_4), ethane (C_2H_6), propane (C_3H_8), butane (C_4H_{10}), and olefins (RCH=CHR', where R and R' can be hydrogen or a methyl group) and may also include off-gases from petrochemical processes. Olefins such as ethylene [ethene, CH_2=CH_2, boiling point: $-104°C$ ($-155°F$)], propene [propylene, CH_3CH=CH_2, boiling point: $-47°C$ ($-53°F$)], butene [butene-1, CH_3CH_2CH=CH_2, boiling point: $-5°C$ ($23°F$)], isobutylene [$(CH_3)_2C$=CH_2, $-6°C$ ($21°F$)], *cis*- and *trans*-butene-2 [CH_3CH=$CHCH_3$, boiling point: ca. $1°C$ ($30°F$)] and butadiene [CH_2=$CHCH$=CH_2, boiling point: $-4°C$ ($24°F$)] as well as higher-boiling olefins are produced by various refining processes.

Refinery gas varies in composition and volume, depending on crude origin and on any additions to the crude made at the loading point. It is not uncommon to reinject light hydrocarbons such as propane and butane into the crude before dispatch by tanker or pipeline. This results in a higher vapor pressure of the crude, but it allows one to increase the quantity of light products obtained at the refinery. Since light ends in most petroleum markets command a premium, while in the oil field itself propane and butane may have to be reinjected or flared, the practice of *spiking* crude oil with liquefied petroleum gas is becoming fairly common.

In addition to the gases obtained by distillation of crude petroleum, more highly volatile products result from subsequent processing of naphtha and middle distillate to produce gasoline, from desulfurization processes involving hydrogen treatment of naphtha, distillate, and residual fuel; and from the coking or similar thermal treatments of vacuum gas oils and residual fuels. The most common processing step in the production of gasoline is the catalytic reforming of hydrocarbon fractions in the heptane (C_7) to decane (C_{10}) range.

In a series of processes commercialized as Platforming, Powerforming, Catforming, and Ultraforming, paraffinic and naphthenic (cyclic nonaromatic) hydrocarbons are converted into aromatics in the presence of hydrogen and a catalyst, or are isomerized to more highly branched hydrocarbons. Catalytic reforming processes thus not only result in the formation of a liquid product of higher octane number, but also produce substantial quantities of gases. The latter are rich in hydrogen but also contain hydrocarbons from methane to butanes, with a preponderance of propane ($CH_3 \cdot CH_2 \cdot CH_3$), n-butane ($CH_3 \cdot CH_2 \cdot CH_2 \cdot CH_3$), and isobutane [$(CH_3)_3CH$]. Their composition will vary in accordance with reforming severity and reformer feedstock. Since all catalytic reforming processes require substantial recycling of a hydrogen stream, it is normal to separate reformer gas into a propane ($CH_3 \cdot CH_2 \cdot CH_3$) and/or a butane [$CH_3 \cdot CH_2 \cdot CH_2 \cdot CH_3/(CH_3)_3CH$] stream, which becomes part of the refinery liquefied petroleum gas production, and a lighter gas fraction, part of which is recycled. In view of the excess of hydrogen in the gas, all products of catalytic reforming are saturated, and there are usually no olefinic gases present in either gas stream.

A second group of refining operations that contributes to gas production is that of the catalytic cracking processes. These consists of fluid-bed catalytic cracking, Thermofor catalytic cracking, and other variants in which heavy gas oils are converted into cracked gas, liquefied petroleum gas, catalytic naphtha, fuel oil, and coke by contacting the heavy hydrocarbon with the hot catalyst. Both catalytic and thermal cracking processes, the latter now being used largely for the production of chemical raw materials, result in the formation of unsaturated hydrocarbons, particularly ethylene ($CH_2=CH_2$) but also propylene (propene, $CH_3 \cdot CH=CH_2$), isobutylene [isobutene, $(CH_3)_2C=CH_2$] and the n-butenes ($CH_3 \cdot CH_2 \cdot CH=CH_2$ and $CH_3 \cdot CH=CH \cdot CH_3$) in addition to hydrogen (H_2), methane (CH_4) and smaller quantities of ethane ($CH_3 \cdot CH_3$), propane ($CH_3 \cdot CH_2 \cdot CH_3$), and butanes [$CH_3 \cdot CH_2 \cdot CH_2 \cdot CH_3$, $(CH_3)_3CH$]. Diolefins such as butadiene ($CH_2=CH \cdot CH=CH_2$) and are also present.

Additional gases are produced in refineries with coking or visbreaking facilities for the processing of their heaviest crude fractions. In the visbreaking process, fuel oil is passed through externally fired tubes and undergoes liquid-phase cracking reactions, which results in the formation of lighter fuel oil components. Oil viscosity is thereby reduced, and some gases, mainly hydrogen, methane, and ethane, are formed. Substantial quantities of both gas and carbon are also formed in coking (both fluid coking and delayed coking) in addition to the middle distillate and naphtha. When coking a residual fuel oil or heavy gas oil, the feedstock is preheated and contacted with hot carbon (coke), which causes extensive cracking of the feedstock constituents of higher molecular weight to produce lower-molecular-weight products, ranging from methane, via liquefied petroleum gas(es) and naphtha, to gas oil and heating oil. Products from coking processes tend to be unsaturated, and olefinic components predominate in the tail gases from coking processes.

A further source of refinery gas is hydrocracking, a catalytic high-pressure pyrolysis process in the presence of fresh and recycled hydrogen. The feedstock is again heavy gas oil or residual fuel oil, and the process is directed primarily at the production of additional middle distillates and gasoline. Since hydrogen is to be recycled, the gases produced in this process again have to be separated into lighter and heavier streams; any surplus recycle gas and the liquefied petroleum gas from the hydrocracking process are both saturated.

Both hydrocracker and catalytic reformer tail gases are commonly used in catalytic desulfurization processes. In the latter, feedstocks ranging from light to vacuum gas oils are passed at pressures of 500 to 1000 psi (3.5 to 7.0×10^3 kPa) with hydrogen over a hydrofining catalyst. This results mainly in the conversion of organic sulfur compounds to hydrogen sulfide (Bland and Davidson, 1967; Speight, 1999):

$$[S]_{feedstock} + H_2 = H_2S + \text{hydrocarbons}$$

but also produces some light hydrocarbons by hydrocracking.

Petroleum refining also produces substantial amounts of carbon dioxide, which with hydrogen sulfide, corrode refining equipment, harm catalysts, pollute the atmosphere, and prevent the use of hydrocarbon components in petrochemical manufacture. When the amount of hydrogen sulfide is high, it may be removed from a gas stream and converted to sulfur or sulfuric acid. Some natural gases contain sufficient carbon dioxide to warrant recovery as dry ice.

Thus, refinery streams, although ostensibly hydrocarbon in nature, may contain large amounts of acid gases, such as hydrogen sulfide and carbon dioxide. Most commercial plants employ hydrogenation to convert organic sulfur compounds into hydrogen sulfide. Hydrogenation is effected by means of recycled hydrogen-containing gases or external hydrogen over a nickel molybdate or cobalt molybdate catalyst.

In summary, refinery process gas, in addition to hydrocarbons, may contain other contaminants, such as carbon oxides (CO_x, where $x = 1$ and/or 2) and sulfur oxides (SO_x, where $x = 2$ and/or 3) as well as ammonia (NH_3), mercaptans (R–SH), and carbonyl sulfide (COS).

Residual refinery gases, usually in more than one stream, which allows a degree of quality control, are treated for hydrogen sulfide removal, and gas sales are usually on a thermal content (calorific value, heating value) basis, with some adjustment for variation in the calorific value and hydrocarbon type (Rawlinson and Ward, 1973; McKetta, 1993; Speight, 1993; Johansen, 1998; Cranmore and Stanton, 2000). For fuel uses, gas as specified above presents little difficulty used as supplied. Alternatively, a gas of constant Wobbe Index, say for gas turbine use, could readily be produced by the user. Part of the combustion air would be diverted into the gas stream by a Wobbe Index controller. This would be set to supply gas at the lowest Wobbe Index of the undiluted gas.

9.1.4. Sulfur Oxides, Nitrogen Oxides, Hydrogen Sulfide, and Carbon Dioxide

Sulfur oxides, nitrogen oxides, hydrogen sulfide, and carbon dioxide are commonly produced during refining operations or during use of the refined products. For example, the most common toxic gases present in diesel exhaust include carbon monoxide, sulfur dioxide, nitric oxide, and nitrogen dioxide.

These gases are also classed as primary pollutants because they are emitted directly from the source and then react to produce secondary pollutant, such as acid rain (Speight, 1993). The emissions may include a number of biologically active substances that can pose a major health concern. These gases are classed as pollutants because (1) they may not be indigenous to the location or (2) they are ejected into the atmosphere in a greater-than-natural concentration and are, in the current context, the product of human activity. Thus, they can have a detrimental effect on the environment in part or *in toto*.

For these pollutants, the atmosphere has the ability to cleanse itself within hours especially when the effects of the pollutant are minimized by the natural constituents of the atmosphere. For example, the atmosphere might be considered to be cleaning as a result of rain. However, removal of some pollutants from the atmosphere (e.g., sulfates and nitrates) by rainfall results in acid rain that can cause serious environmental damage to ecosystems within the water and land systems.

Several methods have been developed to estimate the exposure to such emissions. Most methods are based on either ambient air quality surveys or emission modeling. Exposure to other components of diesel emissions, such as PAHs, is also higher in occupational settings than it is in ambient environments. The principles of the techniques most often used in exhaust gas analysis include infrared (NDIR and FTIR), chemiluminescence, flame ionization detector (FID and fast FID), and paramagnetic methods.

9.1.5. Particulate Matter

Particulate matter is a complex emission that is classified as either *suspended particulate matter, total suspended particulate matter*, or simply, *particulate matter*. For human health purposes, the fraction of particulate matter that has been shown to contribute to respiratory diseases is termed PM_{10} (i.e., particulate matter with sizes less than 10 μm). From a control standpoint, particulate matter can be characterized as follows: (1) particle size distribution and (2) particulate matter concentration in the emission (mg/m^3). On occasion, physical property descriptions may also be employed when there are specific control applications.

Traditionally, regulatory and compliance testing requires gravimetric determination of, for example, fuel mass emissions. Instruments utilizing collecting or in situ measurement techniques are used for the analysis of various particle parameters for nonregulatory purposes.

In collecting methods, particulate matter emissions are determined through gravimetric analysis of the particulates collected on a sampling filter. Alternatively, the sample can be analyzed using thermal mass analysis (e.g., coulometric analysis). A number of other properties (e.g., surface area or biological activity) can also be analyzed. Collecting methods, especially gravimetric analysis, are well established as the most common method of particulate matter emission determination.

The health effects of particulate matter (a complex mixture of solids and liquids) emissions are not yet well understood but are recognized as major contributors to health problems. Biological activity of particulate matter may be related to particle sizes and/or particle composition. Furthermore, it has generally been concluded that exposure to particulate matter may cause increased morbidity and mortality, such as from cardiovascular disease. Long-term exposure to particulate emissions is also associated with a small increase in the relative risk of lung cancer.

9.2. ENVIRONMENTAL EFFECTS

Air pollutants are responsible for a number of adverse environmental effects, such as photochemical smog, acid rain, death of forests, or reduced atmospheric visibility. Emissions of greenhouse gases are associated with the global warming. Certain air pollutants, including black carbon, not only contribute to global warming, but are also suspected of having an immediate effect on regional climates.

Sulfur is removed from a number of refinery process off-gas streams (sour gas) in order to meet the sulfur oxide emissions limits of the Clean Air Act and to recover salable elemental sulfur. Process off-gas streams, or sour gas, from the coker, catalytic cracking unit, hydrotreating units, and hydroprocessing units can contain high concentrations of hydrogen sulfide mixed with light refinery fuel gases. Before elemental sulfur can be recovered, the fuel gases (primarily methane and ethane) need to be separated from the hydrogen sulfide. This is typically accomplished by dissolving the hydrogen sulfide in a chemical solvent. Solvents most commonly used are amines, such as diethanolamine (DEA). Dry adsorbents such as molecular sieves, activated carbon, iron sponge, and zinc oxide are also used. In the amine solvent processes, DEA solution or another amine solvent is pumped to an absorption tower, where the gases are contacted and hydrogen sulfide is dissolved in the solution. The fuel gases are removed for use as fuel in process furnaces in other refinery operations. The amine–hydrogen sulfide solution is then heated and steam stripped to remove the hydrogen sulfide gas.

Since the Claus process by itself removes only about 90% of the hydrogen sulfide in the gas stream, the Beavon process (Speight, 1993, p. 268), SCOT (Shell Claus Off-Gas Treating) process (Speight, 1993, p. 316; Hydrocarbon Processing,

2002), or the Wellman–Lord process (Speight, 1993, p. 327) is often used to recover additional sulfur. The Claus process consists of partial combustion of the hydrogen sulfide-rich gas stream (with one-third the stoichiometric quantity of air), followed by reacting the resulting sulfur dioxide and unburned hydrogen sulfide in the presence of a bauxite catalyst to produce elemental sulfur.

In the *Beavon process*, the hydrogen sulfide in the relatively low concentration gas stream from the Claus process can be removed almost completely by absorption in a quinone solution. The dissolved hydrogen sulfide is oxidized to form a mixture of elemental sulfur and hydroquinone. The solution is injected with air or oxygen to oxidize the hydroquinone back to quinone. The solution is then filtered or centrifuged to remove the sulfur, and the quinone is then reused. The Beaven process is also effective in removing small amounts of sulfur dioxide, carbonyl sulfide, and carbon disulfide that are not affected by the Claus process. These compounds are first converted to hydrogen sulfide at elevated temperatures in a cobalt molybdate catalyst prior to being fed to the Beaven unit. Air emissions from sulfur recovery units will consist of hydrogen sulfide, sulfur oxides, and nitrogen oxides in the process tail gas as well as fugitive emissions and releases from vents.

In the *SCOT process*, the sulfur compounds in the Claus tail gas are converted to hydrogen sulfide by heating and passing it through a cobalt–molybdenum catalyst with the addition of a reducing gas. The gas is then cooled and contacted with a solution of diisopropanolamine (DIPA) that removes all but trace amounts of hydrogen sulfide. The sulfide-rich diisopropanolamine is sent to a stripper, where hydrogen sulfide gas is removed and sent to the Claus plant. The diisopropanolamine is returned to the absorption column.

In the *Wellman–Lord process*, sodium sulfite is used to capture the sulfur dioxide. The sodium bisulfite thus formed is later heated to evolve sulfur dioxide and regenerate the sulfite scrubbing material. The sulfur dioxide-rich product stream can be compressed or liquefied and oxidized to sulfuric acid, or reduced to sulfur.

Most refinery process units and equipment are manifolded into a collection unit, called the *blowdown system*. Blowdown systems provide for the safe handling and disposal of liquid and gases that are either vented automatically from the process units through pressure relief valves or that are manually drawn from units. Recirculated process streams and cooling-water streams are often purged manually to prevent the continued buildup of contaminants in the stream. Part or all of the contents of equipment can also be purged to the blowdown system prior to shutdown before normal or emergency shutdowns. Blowdown systems utilize a series of flash drums and condensers to separate the blowdown into its vapor and liquid components. The liquid is typically composed of mixtures of water and hydrocarbons containing sulfides, ammonia, and other contaminants, which are sent to the wastewater treatment plant. The gaseous component typically contains hydrocarbons, hydrogen sulfide, ammonia, mercaptans, solvents, and other constituents, and is either discharged directly to the atmosphere or is combusted in a flare. The major air emissions from blowdown systems are hydrocarbons in the case of direct discharge to the atmosphere and sulfur oxides when flared.

9.3. SAMPLING

One of the more critical aspects for the analysis of gaseous (or low-boiling) hydrocarbons is the question of volumetric measurement (ASTM D1071) and sampling (ASTM D1145, D1247, D1265). However, sampling liquefied petroleum gas from a liquid storage system is complicated by existence of two phases (gas and liquid) and the composition of the supernatant vapor phase will probably differ from the composition of the liquid phase. Furthermore, the compositions of both phases will vary as a sample (or sample) is removed from one or both phases. An accurate check of composition can only be made if samples are taken during filling of the tank or from a fully charged tank.

In general, the sampling of gaseous constituents and liquefied gases is the subject of a variety of sampling methods (ASTM D5503), such as the manual method (ASTM D1265, D4057), the floating piston cylinder method (ASTM D3700), and the automatic sampling method (ASTM D4177, D5287). Methods for the preparation of gaseous and liquid blends are also available (ASTM D4051, D4307), including the sampling and handling of fuels for volatility measurements (ASTM D5842).

Sampling of methane (CH_4) and ethane (C_2H_6) hydrocarbons is usually achieved using stainless steel cylinders, either lined or unlined. However, other containers may also be employed, depending on particular situations. For example, glass cylinder containers or polyvinyl fluoride (PVF) sampling bags may also be used but obviously cannot be subjected to pressures that are far in excess of ambient pressure. The preferred method for sampling propane (C_3H_8) and butane (C_4H_{10}) hydrocarbons is by the use of piston cylinders (ASTM D3700), although sampling these materials as gases is also acceptable in many cases. The sampling of propane and higher-boiling hydrocarbons depends on the vapor pressure of the sample (IP 410). Piston cylinders or pressurized steel cylinders are recommended for high-vapor pressure sampling, where significant amounts of low-boiling gases are present, while atmospheric sampling may be used for samples having a low vapor pressure.

9.4. ANALYSIS

To monitor a process, measurement of gaseous emissions is typically performed over the time of a process cycle or over the time of use of a particular product. Emission test cycles are repeatable sequences of operating conditions. Such timely analyses allows process monitoring as well as identification of any changes that can lead to potential leakage of the gas.

Hydrocarbon gases are amenable to analytical techniques, and there has been a stronger tendency, which remains, for the determination of both major constituents and trace constituents than is the case with the heavier hydrocarbons. The complexity of the mixtures that are evident as the boiling point of petroleum fractions and petroleum products increases make identification of many of the individual constituents difficult, if not impossible. In addition, methods have been developed for the determination of physical characteristics such as calorific value, specific gravity, and enthalpy from the analyses of mixed hydrocarbon gases, but the accuracy does suffer when compared to the data produced by methods for the direct determination of these properties.

Bulk physical property tests, such as density and heating value, as well as some compositional tests, such as the Orsat analysis and the mercuric nitrate method for the determination of unsaturation, are still used. However, the choice of a particular test is dictated by (1) the requirements of the legislation, (2) the properties of the gas under study, and (3) the selection by the analyst of a suitable suite of tests to meet the various requirements. For example, judgment by the analyst is necessary whether or not a test that is applied to liquefied petroleum gas is suitable for process gas or natural gas insofar as inference from the nonhydrocarbon constituents will be minimal.

The first and most important aspect of gaseous testing is measurement of the volume of gas (ASTM D1071). In this test method, several techniques are described and may be employed for any purpose where it is necessary to know the quantity of gaseous fuel. In addition, the thermophysical properties of methane (ASTM D3956), ethane (ASTM D3984), propane (ASTM D4362), n-butane (ASTM D4650), and isobutane (ASTM D4651) should be available for use and consultation (see also Stephenson and Malanowski, 1987).

9.4.1. Calorific Value (Heat of Combustion)

Satisfactory combustion of hydrocarbon gases depends on the matching of burner and appliance design with certain gas characteristics. Various types of test methods are available for the direct determination of calorific value (ASTM D900, D1826, D3588, D4981).

The most important of these are the *Wobbe index* [or *Wobbe number* = calorific value/(specific gravity)] and the *flame speed*, usually expressed as a factor or an arbitrary scale on which that of hydrogen is 100. This factor can be calculated from the gas analysis. In fact, calorific value and specific gravity can be calculated from compositional analysis (ASTM D3588).

The Wobbe number gives a measure of the heat input to an appliance through a given aperture at a given gas pressure. Using this as a vertical coordinate and the flame speed factor as the horizontal coordinate, a combustion diagram can be constructed for an appliance, or an entire range of appliances, with the aid of appropriate test gases. This diagram would show the area within which variations in the Wobbe index of gases may occur for a given range of appliances without resulting in either incomplete combustion, flame lift, or the lighting back of preaerated flames. This method of prediction of combustion characteristics is not sufficiently accurate to eliminate entirely the need for the practical testing of new gases.

Another important combustion criterion is the gas modulus, $M = P/W$, where P is the gas pressure and W the Wobbe number of the gas. This must remain constant if a given degree of aeration is to be maintained in a preaerated burner using air at atmospheric pressure.

9.4.2. Composition

Liquefied petroleum gas, natural gas, and refinery gas are mixtures of products or naturally occurring materials and, fortunately, are relatively simple mixtures and do not suffer the complexities of the isomeric variations of the higher-molecular-weight hydrocarbons (Drews, 1998; Speight, 1999).

Thus, because of the lower-molecular-weight constituents of these gases and their volatility, gas chromatography has been the technique of choice for fixed gas and hydrocarbon speciation, and mass spectrometry is also a method of choice for compositional analysis of low-molecular-weight hydrocarbons (ASTM D2421, D2650). More recently, piggyback methods (such as gas chromatography–mass spectrometry and other double technique methods) have been developed for the identification of gaseous and low-boiling liquid constituents of mixtures. The hydrocarbon composition is limited to the total amount of ethane, butane, or pentane as well as ethylene and total dienes.

By limiting the amount of hydrocarbons that are lower boiling than the main component, the vapor pressure control is reinforced. Tests are available for vapor pressure 100°F (38°C) (ASTM D1267) and at 113°F (45°C) (IP 161). The limitation on the amount of higher-boiling hydrocarbons supports the volatility clause. The vapor pressure and volatility specifications will often be met automatically if the hydrocarbon composition is correct.

The amount of ethylene is limited because it is necessary to restrict the amount of unsaturated components so as to avoid the formation of deposits caused by the polymerization of the olefin(s). In addition, ethylene [boiling point: −104°C (−155°F)] is more volatile than ethane [boiling point: −88°C (−127°F)], and therefore a product with a substantial proportion of ethylene will have a higher vapor pressure and volatility than one that is predominantly ethane. Butadiene is also undesirable because it may also produce polymeric products that form deposits and cause blockage of lines.

Currently, the preferred method for the analysis of liquefied petroleum gas, and indeed for most petroleum-related gases, is gas chromatography (ASTM D2163; IP 264). This technique can be used for the identification and measurement of both primary and trace constituents. However, there may be some accuracy issues that arise in the measurement of the higher-boiling constituents due to relative volatility under the conditions in which the sample is held.

Capillary column gas chromatography is an even quicker and equally accurate alternative. Mass spectrometry (ASTM D1137) is also suitable for the analysis of petroleum gases. Of the other spectroscopic techniques, infrared and ultraviolet absorption may be applied to petroleum gas analysis for some specialized applications. Gas chromatography has also largely supplanted chemical absorption methods of analysis, but again, these may have some limited specialized application. Once the composition of a mixture has been determined, it is possible to calculate various properties, such as specific gravity, vapor pressure, calorific value, and dew point.

Simple evaporation tests in conjunction with vapor pressure measurement give a further guide to composition. In these tests a liquefied petroleum gas sample is allowed to evaporate naturally from an open graduated vessel. Results are recorded on the basis of volume and temperature changes, such as the temperature recorded when 95% has evaporated or the volume left at a particular temperature (ASTM D1837).

Since dew point can be calculated from composition, direct determination of dew point for a particular liquefied petroleum gas sample is a measure of composition. It is, of course, of more direct practical value, and if small quantities of higher-molecular-weight material are present, it is preferable to use a direct measurement.

Specific gravity again can be calculated, but if it is necessary to measure it, several pieces of apparatus are available. For determining the density or specific gravity of liquefied petroleum gas in its liquid state, there are two methods, using a metal pressure pycnometer. A pressure hydrometer may be used (ASTM D1267) for the relative density, which may also be calculated from compositional analysis (ASTM D2598). Various procedures, manual and recording, for specific gravity or density in the gaseous state are given in two methods (ASTM D1070; IP 59). Calculation of the density is also possible using any one of four models, depending on the composition of the gas (ASTM D4784).

Gases such as ethane, destined for use as petrochemical feedstocks, must adhere to stringent composition controls that are dependent on the process. For example, moisture content (ASTM D1142), oxygen content (ASTM D1945), carbon dioxide content (ASTM D1945), and sulfur content (ASTM D1072) must be monitored, as they all interfere with catalyst performance in petrochemical processes. The hydrocarbon composition of natural gasoline (although not specifically a gas) for petrochemical use must undergo a compositional analysis (ASTM D2427) and a test for total sulfur (ASTM D1266; IP 107, 191).

The presence of any component substantially less volatile than the primary constituents of the liquefied petroleum gas will give rise to unsatisfactory performance. It is difficult to set limits on the amount and nature of the "residue" that will make a product unsatisfactory. Obviously, small amounts of oily material can block regulators and valves. In liquid vaporizer feed systems, even gasoline-type material could cause difficulty. The residue (ASTM D2158) is a measure of the concentration of contaminants boiling above 37.8°C (100°F) that may be present in the gas. The residue as determined by the endpoint index (EPI) endeavors to give a measure of the heavier hydrocarbons, but the relationship among EPI, hydrocarbon range, and performance is not established.

Analytical methods are available in standard form for determining volatile sulfur content and certain specific corrosive sulfur compounds that are likely to be present. Volatile sulfur determination is made by a combustion procedure

(ASTM D126; IP 107) that uses a modification of the standard wick-fed lamp. Many laboratories use rapid combustion techniques with an oxyhydrogen flame in a Wickbold or Martin–Floret burner (ASTM D2784; IP 243).

This test method (ASTM D2784; IP 243) is valid for sulfur levels of >1 μg per gram of sulfur in liquefied petroleum gas, but the samples should not contain more than 100 μg per gram of chlorine. In the test, the sample is bummed in an oxyhydrogen bummer or in a lamp in a closed system in a carbon dioxide–oxygen atmosphere. The latter is not recommended for trace quantities of sulfur, due to the inordinately long combustion times needed. The sulfur oxides produced are absorbed and oxidized to sulfuric acid in a hydrogen peroxide solution. The sulfate ions are then determined by either titrating with barium perchlorate solution using a thorinmethylene blue mixed indicator, or by precipitating as barium sulfate and measuring the turbidity of the precipitate with a photometer.

Trace hydrocarbons that may be regarded as contaminants may be determined by the gas chromatographic methods already discussed. Heavier hydrocarbons in small amounts may not be removed completely from the column. If accurate information is required about the nature and amount of heavy ends, temperature programming or a concentration procedure may be used.

Analytical methods for determining traces of various other impurities, such as chlorides (ASTM D2384), are known to be in use. The presence of acetylenes in refinery gases, although unlikely, must still be considered. Acetylenes can be determined using a chemical test method, while carbonyls are determined by the classical hydroxylamine hydrochloride reaction (ASTM D1089).

The determination of traces of higher-boiling hydrocarbons and oily matter involves use of a method for residue that involves a preliminary weathering. The residue after weathering is dissolved in a solvent and the solution is applied to a filter paper. The presence of residue is indicated by the formation of an oil stain. The procedure is taken further by combining the oil stain observation with other observed values to calculate an endpoint index (ASTM D2158). The method is not very precise, and work is proceeding in several laboratories to develop a better method for the determination of residue in the form of oily matter.

In liquefied petroleum gas where the composition is such that the hydrocarbon dew point is known to be low, a dew point method will detect the presence of traces of water (ASTM D1142). Propane, isobutane [boiling point: $-12°C$ ($11°F$)], and butane generally constitute this sample type and are used for heating, motor fuels, and as chemical feedstocks (ASTM D2597, D2504, D2505). Important constituents of gas not accounted for in these analyses are moisture (water) and hydrogen sulfide, as well as other sulfur compounds (ASTM D1142, D1988, D5454, D4888, D5504, D6228).

Procedures for the determination of hydrogen, helium, oxygen, nitrogen, carbon monoxide, carbon dioxide, methane, ethene, ethane, propane, butanes, pentanes, and hexanes-plus in natural and reformed gases by packed column gas chromatography are available (ASTM D1945, D1946). These compositional analyses are used to calculate many other properties of gases, such as density, heating value, and compressibility. The first five components listed are determined using a molecular sieve column (argon carrier gas), and the remaining components are determined using polydimethylsiloxane partition or porous polymer columns. The hexanes-plus analysis is accomplished by backflushing the column after the elution of pentane or by the use of a backflushed pre-column.

Olefins (ethylene, propylene, butylenes, and pentylenes) that occur in refinery (process) gas have specific characteristics and require specific testing protocols (ASTM D5234, D5273, D5274). Thus, hydrocarbon analysis of ethylene is accomplished by two methods (ASTM D2505, D6159) one of which (ASTM D6159) uses wide-bore (0.57-mm) capillary columns, including a Al_2O_3/KCl PLOT column. Another method (ASTM D2504) is recommended for determination of noncondensable gases, and yet another (ASTM D2505) is used for the determination of carbon dioxide.

Hydrocarbon impurities in propylene can be determined by gas chromatographic methods (ASTM D2712, D2163), and another test is available for determination of traces of methanol in propylene (ASTM Test Method D4864). A gas chromatographic method (ASTM D5303) is available for the determination of trace amounts of carbonyl sulfide in propylene using a flame photometric detector. Also, sulfur in petroleum gas can be determined by oxidative microcoulometry (ASTM D3246).

Commercial butylenes, high-purity butylenes, and butane–butylene mixtures are analyzed for hydrocarbon constituents (ASTM D4424), and hydrocarbon impurities in 1,7-butadiene can also be determined by gas chromatography (ASTM D2593). The presence of butadiene dimer and styrene is determined in butadiene by gas chromatography (ASTM D2426). Carbonyls in C_4 hydrocarbons are determined by a titrimetric technique (ASTM D4423) and by use of a peroxide method (ASTM D5799).

In general, gas chromatography will undoubtedly continue to be the method of choice for characterization of light hydrocarbon materials. New and improved detection devices and techniques, such as chemiluminescence, atomic emission, and mass spectroscopy, will enhance selectivity, detection limits, and analytical productivity. Laboratory automation through autosampling, computer control, and data handling will provide improved precision and productivity, as well as simplified method operation.

Compositional analysis can be used to calculate calorific value, specific gravity, and compressibility factor (ASTM D3588).

Mercury in natural gas is also measured by atomic fluorescence spectroscopy (ASTM D6350) and by atomic absorption spectroscopy (ASTM D5954).

9.4.3. Density

The density of light hydrocarbons can be determined by several methods (ASTM D1070), including a hydrometer method (ASTM D1298) and a pressure hydrometer method (ASTM D1657; IP 235). The specific gravity (relative density) (ASTM D1070, D1657) by itself has little significance, compared to its use for higher-molecular-weight liquid petroleum products, and can give an indication of quality characteristics only when combined with values for volatility and vapor pressure. It is important for stock quantity calculations and is used in connection with transport and storage.

9.4.4. Sulfur

The manufacturing processes for liquefied petroleum gas are designed so that the majority, if not all, of the sulfur compounds are removed. The total sulfur level is therefore considerably lower than for other petroleum fuels, and a maximum limit for sulfur content helps to define the product more completely. The sulfur compounds that are primarily responsible for corrosion are hydrogen sulfide, carbonyl sulfide, and sometimes, elemental sulfur. Hydrogen sulfide and mercaptans have distinctive unpleasant odors.

A control of the total sulfur content, hydrogen sulfide, and mercaptans ensures that the product is not corrosive or nauseating. Stipulating a satisfactory copper strip test further ensures the control of the corrosion. Total sulfur in gas can be determined by combustion (ASTM D1072), by the lamp method (ASTM D1266), or by hydrogenation (ASTM D3031, D4468). Trace total organic and bound nitrogen is determined (ASTM D4629). The current test method for heavy residues in liquefied petroleum gas (ASTM D2158,) involves evaporation of a liquefied petroleum gas sample, measuring the volume of residue and observing the residue for oil stain on a piece of filter paper.

Corrosive sulfur compounds can be detected by their effect on copper and the form in which the general copper strip corrosion test (ASTM D1838) for petroleum products is applied to liquefied petroleum gas. Hydrogen sulfide can be detected by its action on moist lead acetate paper, and a procedure is also used as a measure of sulfur compounds. The method follows the principle of the standard Doctor test.

9.4.5. Volatility and Vapor Pressure

The vaporization and combustion characteristics of liquefied petroleum gas are defined for normal applications by volatility, vapor pressure, and to a lesser extent, specific gravity. Volatility is expressed in terms of the temperature at which 95% of the sample is evaporated and presents a measure of the least volatile component present (ASTM D1837). Vapor pressure (IP 410) is therefore a measure of the most extreme low-temperature conditions under which initial vaporization can take place. By setting limits to vapor pressure and volatility jointly, the specification serves to ensure essentially single-component products for the butane and propane grades (ASTM D1267, D2598; IP 410). By combining vapor pressure and volatility limits with specific gravity for propane–butane mixtures, two-component systems are essentially ensured.

The residue (ASTM D1025, D2158; IP 317) (i.e. nonvolatile matter) is a measure of the concentration of contaminants boiling above 37.8°C (100°F) that may be present in the gas.

For natural gasoline, the primary criteria are volatility (vapor pressure and knock performance). Determination of the vapor pressure (ASTM D323, D4953, D5190, D5191) and distillation profile (ASTM D216; IP 191) are essential. Knock performance is determined is determined by rating in knock test engines by both the motor method (ASTM D2700; IP 236) and the research method (ASTM D2699; IP 237). The knock characteristics of liquefied petroleum gases can also be determined (IP 238).

Other considerations for natural gasoline are copper corrosion (ASTM D130; IP 154, IP 411) and specific gravity (ASTM D1298; IP 160), the latter determination being necessary for measurement and transportation.

REFERENCES

ASTM. 2004. *Annual Book of ASTM Standards*. American Society for Testing and Materials, West Conshohocken, PA.

Austin, G. T. 1984. *Shreve's Chemical Process Industries*, 5th ed. McGraw-Hill, New York, Chap. 6.

Cranmore, R. E., and Stanton, E. 2000. In *Modern Petroleum Technology*, Vol. 1, *Upstream*, R. A. Dawe (Ed.). Wiley, New York, Chap. 9.

Drews, A. W. 1998. In *Manual on Hydrocarbon Analysis*, 6th ed, A. W. Drews (Ed.). American Society for Testing and Materials, West Conshohocken, PA, Introduction.

Fogg, P. G. T., and Sangster, J. M. (Eds.). 2003. *Chemicals in the Atmosphere: Solubility, Sources, and Reactivity*. Wiley, Hoboken, NJ.

Francis, W., and Peters, M. C. 1980. *Fuels and Fuel Technology: A Summarized Manual*. Pergamon Press, Elmsford, NY, Sec. C.

Guthrie, V. B. 1967. In *Petroleum Processing Handbook*, W. F. Bland and R. L. Davidson (Eds.). McGraw-Hill, New York, Sec. 11.

Hoffman, H. L. 1983. In *Riegel's Handbook of Industrial Chemistry*, 8th ed, J. A. Kent (Ed.). Van Nostrand Reinhold, New York, Chap. 14.

Hydrocarbon Processing. 2002. Gas Processing Handbook: A Special Report, 81(5): 118.

IP. 2001. *IP Standard Methods 2001*. Institute of Petroleum, London.

Johansen, N. G. 1998. In *Manual on Hydrocarbon Analysis*, 6th ed, A. W. Drews (Ed.). American Society for Testing and Materials, West Conshohocken, PA, Chap. 1.

McKetta, J. J. 1993. In *Chemical Processing Handbook*, J. J. McKetta (Ed.). Marcel Dekker, New York, p. 59.

Moustafa, S. M. A. 1996. In *The Engineering Handbook*, R. C. Dorf (Ed.). CRC Press, Boca Raton, FL, Chap. 64.

Patnaik, P. (Ed.). 2004. *Dean's Analytical Chemistry Handbook*, 2nd ed. McGraw-Hill, New York.

Rawlinson, D., and Ward, E. R. 1973. In *Criteria for Quality of Petroleum Products*. J. P. Allinson (Ed.). Wiley, New York, Chap. 3.

Ritter, T. J. 2000. In *Modern Petroleum Technology*, Vol. 2, *Downstream*, A. G. Lucas (Ed.). Wiley, New York, Chap. 23.

Robinson, J. D., and Faulkner, R. P. 2000. In *Modern Petroleum Technology*, Vol. 2, *Downstream*, A. G. Lucas (Ed.). Wiley, New York, Chap. 1.

Speight, J. G. 1993. *Gas Processing: Environmental Aspects and Methods.* Butterworth-Heinemann, Oxford.

Speight, J. G. 1999. *The Chemistry and Technology of Petroleum,* 3rd ed. Marcel Dekker, New York.

Speight, J. G., and Ozum, B. 2002. *Petroleum Refining Processes.* Marcel Dekker, New York.

Stephenson, R. M., and Malanowski, S. 1987. *Handbook of the Thermodynamics of Organic Compounds.* Elsevier, New York.

CHAPTER 10

ANALYSIS OF LIQUID EFFLUENTS

In terms of waste definition, there are three basic approaches (as it pertains to petroleum, petroleum products, and nonpetroleum chemicals) to defining petroleum or a petroleum product as hazardous: (1) a qualitative description of the waste by origin, type, and constituents; (2) classification by characteristics based on testing procedures; and (3) classification as a result of the concentration of specific chemical substances (Chapter 1).

A wide variety of liquid products are produced from petroleum, that varying from high-volatile naphtha to low-volatile lubricating oil (Guthrie, 1967; Speight, 1999). The liquid products are often characterized by a variety of techniques including measurement of physical properties and fractionation into group types (Chapter 7).

The purpose of this chapter is to present examples of the various liquid effluents and the methods of analysis that can be applied to the analysis of various liquid effluents from refinery processes (ASTM, 2004; IP, 2001; Mushrush and Speight, 1995). The examples chosen are those at the lower end of the petroleum product boiling range (naphtha) (Section 1.3.1) and at the higher end of the petroleum product boiling range (residual fuel oil) (Section 1.3.1). The environmental behavior of products that are intermediate in boiling range between these two can be predicted, with some degree of caution since behavior is very much dependent on composition. Nevertheless, similar tests can be applied from which reasonable deduction about behavior can be made.

The impact of the release of liquid products on the environment can, in part, be predicted from knowledge of the properties of the released liquid. Each part of an ocular liquid product from petroleum has its own set of unique analytical characteristics (Speight, 1999, 2002). Since these are well documented, there is no need for repetition here. The decision is to include the properties of the lowest-boiling liquid product (naphtha) and a high-boiling liquid product (fuel oil). For the properties of each product (as determined by analysis) a reasonable estimate can be made of other liquid products, but the relationship may not be linear and is subject to the type of crude oil and the distillation range of the product.

Nevertheless, reference is made to the various test methods dedicated to these products which can be applied to the products boiling in the intermediate range. In the light of the various tests available for composition, such tests will be deemed necessary depending on the environmental situation and the requirements of the legislation as well as at the discretion of the analyst.

Environmental Analysis and Technology for the Refining Industry, by James G. Speight
Copyright © 2005 John Wiley & Sons, Inc.

10.1. NAPHTHA

Naphtha is a liquid petroleum product that boils from about 30°C (86°F) to approximately 200°C (392°F), although there are different grades of naphtha within this extensive boiling range that have different boiling ranges (Guthrie, 1967; Goodfellow, 1973; Weissermel and Arpe, 1978; Francis and Peters, 1980; Hoffman, 1983; Austin, 1984; Chenier, 1992; Speight, 1999; Hori, 2000). The term *petroleum solvent* is often used synonymously with *naphtha*.

On a chemical basis, naphtha is difficult to define precisely because it can contain varying amounts of the constituents (paraffins, naphthenes, aromatics, and olefins) in different proportions, in addition to the potential isomers of the paraffins that exist in the naphtha boiling range (Tables 10.1 and 10.2). Naphtha is also represented as having a boiling range and carbon number similar to those of gasoline, being a precursor to gasoline.

The *petroleum ether* solvents are a specific-boiling-range naphtha, as is *ligroin*. Thus, the term *petroleum solvent* describes a special liquid hydrocarbon fraction obtained from naphtha and used in industrial processes and formulations (Weissermel and Arpe, 1978). These fractions are also referred to as *industrial naphtha*. Other solvents include *white spirit*, which is subdivided into *industrial spirit* [distilling between 30 and 200°C (86 to 392°F)] and *white spirit* [light oil with a distillation range of 135 to 200°C (275 to 392°F)]. The special value of naphtha as a solvent lies in its stability and purity.

Naphtha is produced by any one of several methods, which include (1) fractionation of straight-run, cracked, and reforming distillates, or even fractionation of crude petroleum; (2) solvent extraction; (3) hydrogenation of cracked distillates; (4) polymerization of unsaturated compounds (olefins); and

Table 10.1. General Summary of Product Types and Distillation Range

Product	Lower Carbon Limit	Upper Carbon Limit	Lower Boiling Point		Upper Boiling Point	
			°C	°F	°C	°F
Refinery gas	C_1	C_4	−161	−259	−1	31
Liquefied petroleum gas	C_3	C_4	−42	−44	−1	31
Naphtha	C_5	C_{17}	36	97	302	575
Gasoline	C_4	C_{12}	−1	31	216	421
Kerosene/diesel fuel	C_8	C_{18}	126	302	258	575
Aviation turbine fuel	C_8	C_{16}	126	302	287	548
Fuel oil	C_{12}	$>C_{20}$	216	>343	421	>649
Lubricating oil	$>C_{20}$		>343	>649		
Wax	C_{17}	$>C_{20}$	302	575	>343	>649
Asphalt	$>C_{20}$		>343	>649		
Coke	$>C_{50}$[a]		>1000[a]	>1832[a]		

[a]Carbon number and boiling point are difficult to assess; inserted for illustrative purposes only.

Table 10.2. Increase in the Number of Isomers with Carbon Number

Number of Carbon Atoms	Number of Isomers
1	1
2	1
3	1
4	2
5	3
6	5
7	9
8	18
9	35
10	75
15	4,347
20	366,319
25	36,797,588
30	4,111,846,763
40	62,491,178,805,831

(5) alkylation processes. In fact, the naphtha may be a combination of product streams from more than one of these processes.

The more common method of naphtha preparation is distillation. Depending on the design of the distillation unit, either one or two naphtha steams may be produced: (1) a single naphtha with an endpoint of about 205°C (400°F), similar to straight-run gasoline; or (2) the same fraction divided into a light naphtha and a heavy naphtha. The endpoint of the light naphtha is varied to suit subsequent subdivision of the naphtha into narrower boiling fractions and may be on the order of 120°C (250°F).

Sulfur compounds are most commonly removed or converted to a harmless form by chemical treatment with lye, Doctor solution, copper chloride, or similar treating agents (Speight, 1999). Hydrorefining processes (Speight, 1999) are also often used in place of chemical treatments. When used as a solvent, naphtha is selected for its low sulfur content; and the usual treatment processes remove only sulfur compounds. Naphtha, with its small aromatic content, has a slight odor, but the aromatics increase the solvent power of the naphtha and there is no need to remove aromatics unless odor-free naphtha is specified.

The variety of applications emphasizes the versatility of naphtha. For example, naphtha is used in paint, printing ink, and polish manufacturing and in the rubber and adhesive industries, as well as in the preparation of edible oils, perfumes, glues, and fats. Further uses are found in the dry-cleaning, leather, and fur industries and in the pesticide field. The characteristics that determine the suitability of naphtha for a particular use are volatility, solvent properties (dissolving power), purity, and odor (generally the lack thereof).

To meet the demands of a variety of uses, certain basic naphtha grades are produced, which are identified by boiling range. The complete range of naphtha solvents may be divided, for convenience, into four general categories:

1. Special-boiling-point spirits having an overall distillation range within the limits 30 to 165°C (86 to 329°F)
2. Pure aromatic compounds, such as benzene, toluene, ethylbenzene, xylenes, or mixtures (BTEX) thereof
3. White spirit, also known as mineral spirit and naphtha, usually boiling within 150 to 210°C (302 to 410°F)
4. High-boiling petroleum fractions, boiling within the limits 160 to 325°C (320 to 617°F)

Since the end use dictates the required composition of naphtha, most grades are available in both high- and low-solvency categories, and the various text methods can have a major significance in some applications and less significance in others. Hence the application and significance of tests must be considered in light of the proposed end use. Odor is particularly important, since unlike most other petroleum liquids, many of the products manufactured containing naphtha are used in confined spaces, in factory workshops, and in the home.

On the other hand, and at the other end of the spectrum of petroleum liquids, *fuel oil* is applied not only to distillate products (*distillate fuel oil*, Chapter 9) but also to residual material that is distinguished from distillate-type fuel oil by boiling range and hence is referred to as *residual fuel oil* (ASTM D396). Thus, residual fuel oil is fuel oil that is manufactured from the distillation residuum and includes all residual fuel oils, including fuel oil obtained by visbreaking as well as by blending residual products from other operations (Gruse and Stevens, 1960; Guthrie, 1967; Kite and Pegg, 1973; Weissermel and Arpe, 1978; Francis and Peters, 1980; Hoffman, 1983; Austin, 1984; Chenier, 1992; Hoffman and McKetta, 1993; Hemighaus, 1998; Warne, 1998; Speight, 1999; Charlot and Claus, 2000; Heinrich and Duée, 2000). The various grades of heavy fuel oils are produced to meet rigid specifications, to assure suitability for their intended purpose.

Detailed analysis of residual products, such as residual fuel oil, is more complex than the analysis of lower-molecular-weight liquid products. As with other products, there are a variety of physical property measurements that are required to determine that residual fuel oil meets specifications. But the range of molecular types present in petroleum products increases significantly with an increase in the molecular weight (i.e., an increase in the number of carbon atoms per molecule). Therefore, characterization measurements or studies cannot, and do not, focus on the identification of specific molecular structures. The focus tends to be on molecular classes (paraffins, naphthenes, aromatics, polycyclic compounds, and polar compounds).

Several tests that are usually applied to the lower-molecular-weight colorless (or light-colored) products are not applied to residual fuel oil. For example, test methods such as those designed for determination of the aniline point (or mixed aniline point) (ASTM D611; IP 2) and the cloud point (ASTM D2500, D5771, D5772, D5773) can suffer from visibility effects due to the color of the fuel oil.

Because of the high standards set for naphtha (McCann, 1998), it is essential to employ the correct techniques when taking samples for testing (ASTM D270, D4057; IP 51). Mishandling, or the slightest trace of contaminant, can give rise to misleading results. Special care is necessary to ensure that containers are scrupulously clean and free from odor. Samples should be taken with the minimum of disturbance so as to avoid loss of volatile components; in the case of low-boiling naphtha, it may be necessary to chill the sample. Samples awaiting examination should be kept in a cool dark place so as to ensure that they do not lose volatile constituents or discolor and develop odors due to oxidation.

The physical properties of naphtha depend on the hydrocarbon types present: In general, the aromatic hydrocarbons have the highest solvent power and the straight-chain aliphatic compounds have the lowest. Solvent properties can be assessed by estimating the amount of the various hydrocarbon types present. This method provides an indication of the solvent power of the naphtha on the basis that aromatic constituents and naphthenic constituents provide dissolving ability that paraffinic constituents do not. Another method for assessing the solvent properties of naphtha measures the performance of the fraction when used as a solvent under specified conditions, such as by the Kauri-butanol test method (ASTM D1133). Another method involves measurement of the surface tension, from which the solubility parameter is calculated, and provides an indication of dissolving power and compatibility. Such calculations have been used to determine the yield of asphaltenes from petroleum by the use of various solvents (Mitchell and Speight, 1973; Speight, 1999, 2001). A similar principle is applied to determine the amount of insoluble material in lubricating oil using n-pentane (ASTM D893, D4055).

Insoluble constituents in lubricating oil can cause wear that can lead to equipment failure. Pentane-insoluble materials can include oil-insoluble materials and some oil-insoluble resinous matter originating from oil or additive degradation, or both. Toluene-insoluble constituents arise from external contamination, fuel carbon, and highly carbonized materials from degradation of fuel, oil, and additives, or engine wear and corrosion materials. A significant change in pentane- or toluene-insoluble constituents indicates a change in oil properties that could lead to machinery failure. The insoluble constituents measured can also assist in evaluating the performance characteristics of used oil or in determining the cause of equipment failure.

Thus, one test (ASTM D893) covers the determination of pentane- and toluene-insoluble constituents in used lubricating oils using pentane dilution and centrifugation as the method of separation. The other test (ASTM D4055) uses pentane dilution followed by membrane filtration to remove insoluble constituents larger than 0.8 μm.

10.1.1. Composition

The number of potential hydrocarbon isomers in the naphtha boiling range (Tables 10.1 and 10.2) renders complete speciation of individual hydrocarbons impossible for the naphtha distillation range, and methods are used that identify the hydrocarbon types as chemical groups rather than as individual constituents.

The data from the density (specific gravity) test method (ASTM D1298; IP 160) provides a means of identification of a grade of naphtha but is not a guarantee of composition and can only be used to indicate evaluate product composition or quality when used in conjunction with the data from other test methods. Density data are used primarily to convert naphtha volume to a weight basis, a requirement in many of the industries concerned. For the necessary temperature corrections and also for volume corrections, the appropriate sections of the petroleum measurement tables (ASTM D1250; IP 200) are used.

The first level of compositional information is group-type totals, as deduced by adsorption chromatography (ASTM D1319; IP 156) to give volume percent saturates, olefins, and aromatics in materials that boil below 315°C (600°F). In this test method a small amount of sample is introduced into a glass adsorption column packed with activated silica gel, of which a small layer contains a mixture of fluorescent dyes. When the sample has been adsorbed on the gel, alcohol is added to desorb the sample down the column, and the hydrocarbon constituents are separated according to their affinities into three types (aromatics, olefins, and saturates). The fluorescent dyes also react selectively with the hydrocarbon types and make the boundary zones visible under ultraviolet light. The volume percentage of each hydrocarbon type is calculated from the length of each zone in the column.

Other test methods are available. Benzene content and other aromatics may be estimated by spectrophotometric analysis (ASTM D1017) and by gas–liquid chromatography (ASTM D2267, D2600; IP 262). However, two test methods based on the adsorption concept (ASTM D2007, D2549) are used for classifying oil samples of initial boiling point of at least 200°C (392°F) into the hydrocarbon types of polar compounds, aromatics, and saturates, and recovery of representative fractions of these types. Such methods are unsuitable for the majority of naphtha samples because of volatility constraints.

An indication of naphtha composition may also be obtained from the determination of aniline point (ASTM D1012; IP 2), freezing point (ASTM D852, D1015, D1493), cloud point (ASTM D2500), and solidification point (ASTM D1493). Although refinery treatment should ensure no alkalinity and acidity (ASTM D847, D1093, D1613, D2896; IP 1) and that no olefins are present, the relevant tests using bromine number (ASTM D875, D1159; IP 130), bromine index (ASTM D2710), and flame ionization absorption (ASTM D1319; IP 156) are necessary to ensure low levels (at the maximum) of hydrogen sulfide (ASTM D853) as well as sulfur compounds in general (ASTM D130, D849, D1266, D2324, D3120, D4045, D6212; IP 107, 154), and especially, corrosive sulfur compounds such as are determined by the doctor test method (ASTM D4952; IP 30).

Since aromatic content is a key property of low-boiling distillates such as naphtha and gasoline because the aromatic constituents influence a variety of properties, including boiling range (ASTM D86; IP 123), viscosity (ASTM D88, D445, D2161, IP 71), stability (ASTM D525; IP 40), and compatibility (ASTM

D1133) with a variety of solutes. Existing methods use physical measurements and need suitable standards. Tests such as aniline point (ASTM D611) and Kauributanol number (ASTM D1133) are of a somewhat empirical nature and can serve a useful function as control tests. Naphtha composition, however, is monitored mainly by gas chromatography, and although most of the methods may have been developed for gasoline (ASTM D2427, D6296), the applicability of the methods to naphtha is sound.

A multidimensional gas chromatographic method (ASTM D5443) provides for the determination of paraffins, naphthenes, and aromatics by carbon number in low-olefinic hydrocarbon streams having final boiling points lower than 200°C (392°F). In the method, the sample is injected into a gas chromatographic system that contains a series of columns and switching values. First, a polar column retains polar aromatic compounds, binaphthenes, and high-boiling paraffins and naphthenes. The eluent from this column goes through a platinum column that hydrogenates olefins and then to a molecular sieve column that performs carbon number separation based on the molecular structure, that is, naphthenes and paraffins. The fraction remaining on the polar column is further divided into three separate fractions, which are then separated on a nonpolar column by boiling point. A flame ionization detector detects eluting compounds.

In another method (ASTM D4420) for the determination of the amount of aromatic constituents, a two-column chromatographic system connected to a dual-filament thermal conductivity detector (or two single-filament detectors) is used. The sample is injected into the column containing a polar liquid phase. The nonaromatics are directed to the reference side of the detector and vented to the atmosphere as they elute. The column is backflushed immediately before the elution of benzene, and the aromatic portion is directed into the second column, containing a nonpolar liquid phase. The aromatic components elute in the order of their boiling points and are detected on the analytical side of the detector. Quantitation is achieved by utilizing peak factors obtained from the analysis of a sample having a known aromatic content.

Other methods for the determination of aromatics in naphtha include a method (ASTM D5580) using a flame ionization detector and methods that use combinations of gas chromatography and Fourier transform infrared spectroscopy (GC-FTIR) (ASTM D5986) and gas chromatography and mass spectrometry (GC-MS) (ASTM D5769).

Hydrocarbon composition is also determined by mass spectrometry, a technique that has seen wide use for hydrocarbon-type analysis of naphtha and gasoline (ASTM D2789) as well as for the identification of hydrocarbon constituents in higher-boiling naphtha fractions (ASTM D2425).

One method (ASTM D6379, IP 436) is used to determine the mono- and diaromatic hydrocarbon contents in distillates boiling in the range 50 to 300°C (122 to 572°F). In the method the sample is diluted with an equal volume of a hydrocarbon such as heptane, and a fixed volume of this solution is injected into a high-performance liquid chromatograph fitted with a polar column where aromatic hydrocarbons are separated from nonaromatic hydrocarbons. The separation of aromatic constituents appears as distinct bands according to ring structure, and a refractive index detector is used to identify the components as they elute from the column. The peak areas of the aromatic constituents are compared with those obtained from previously run calibration standards to calculate the % w/w monoaromatic hydrocarbon constituents and diaromatic hydrocarbon constituents in the sample. Compounds containing sulfur, nitrogen, and oxygen could possibly interfere with the performance of the test. Monoalkenes do not interfere, but if present, conjugated di- and polyalkenes may interfere with test performance.

Another method (ASTM D2425) provides more compositional detail (in terms of molecular species) than chromatographic analysis and the hydrocarbon types are classified in terms of a z-series, in which z (in the empirical formula C_nH_{2n+z}) is a measure of the hydrogen deficiency of the compound. This method requires that the sample be separated into saturated and aromatic fractions before mass spectrometric analysis (ASTM D2549), and the separation is applicable to some fractions but not to others. For example, the method is applicable to high-boiling naphtha but not to low-boiling naphtha, since it is impossible to evaporate the solvent used in the separation without also losing the lower-boiling constituents of the naphtha under investigation.

The percentage of aromatic hydrogen atoms and aromatic carbon atoms can be determined by high-resolution nuclear magnetic resonance spectroscopy (ASTM D5292) that gives the mole percent of aromatic hydrogen or carbon atoms. Proton (hydrogen) magnetic resonance spectra are obtained on sample solutions in either chloroform or carbon tetrachloride using a continuous wave or pulse Fourier transform high-resolution magnetic resonance spectrometer. Carbon magnetic resonance spectra are obtained on the sample solution in chloroform-d using a pulse Fourier transform high-resolution magnetic resonance.

The data obtained by this method (ASTM D5292) can be used to evaluate changes in aromatic contents in naphtha as well as kerosene, gas oil, mineral oil, and lubricating oil. However, results from this test are not equivalent to mass or volume percent aromatics determined by the chromatographic methods since the chromatographic methods determine the percent by weight or percent by volume of molecules that have one or more aromatic rings, and alkyl substituents on the rings will contribute to the percentage of aromatics determined by chromatographic techniques.

Low-resolution nuclear magnetic resonance spectroscopy can also be used to determine percent by weight hydrogen in jet fuel (ASTM D3701) and in light distillate, middle distillate, and gas-oil (ASTM D4808). As noted above, chromatographic methods are not applicable to naphtha, where losses can occur by evaporation.

The nature of the uses found for naphtha demands compatibility with the many other materials employed in formulation, with waxes, pigments, resins, and so on; thus, the solvent properties of a given fraction must be carefully measured and controlled. For most purposes, volatility is important, and because of the wide use of naphtha in industrial and recovery plants, information on some other fundamental characteristics is required for plant design.

Although the focus of many tests is analysis of the hydrocarbon constituents of naphtha and other petroleum fractions, heteroatoms compounds that contain sulfur and nitrogen atoms cannot be ignored, and methods for their determination are available. The combination of gas chromatography with element-selective detection gives information about the distribution of the element. In addition, many individual heteroatomic compounds can be determined.

Nitrogen compounds in middle distillates can be detected selectively by chemiluminescence. Individual nitrogen compounds can be detected down to 100 ppb nitrogen. Gas chromatography with either sulfur chemiluminescence detection or atomic emission detection has been used for sulfur-selective detection.

Estimates of the purity of these products were determined in laboratories using a variety of procedures, such as freezing point, flame ionization absorbance, ultraviolet absorbance, gas chromatography, and capillary gas chromatography (ASTM D850, D852, D853, D848, D849, D1015, D1016, D1078, D1319, D2008, D22368, D2306, D2360, D5917; IP 156).

Gas chromatography (GC) has become a primary technique for determining hydrocarbon impurities in individual aromatic hydrocarbons and the composition of mixed aromatic hydrocarbons. Although a measure of purity by gas chromatography is often sufficient, gas chromatography is not capable of measuring absolute purity; not all possible impurities will pass through a GC column, and not all those that do will be measured by the detector. Despite some shortcomings, gas chromatography is a standard, widely used technique and is the basis of many current test methods for aromatic hydrocarbons (ASTM D2306 D2360, D3054, D3750, D3797, D3798, D4492, D4534, D4735, D5060, D5135, D5713, D5917, D6144). When classes of hydrocarbons such as olefins need to be measured, techniques such as bromine index are used (ASTM D1492, D5776).

Impurities other than hydrocarbons are of concern in the petroleum industry. For example, many catalytic processes are sensitive to sulfur contaminants. Consequently, there is also a series of methods to determine trace concentrations of sulfur-containing compounds (ASTM D1685, D3961, D4045, D4735).

Chloride-containing impurities are determined by various test methods (ASTM D5194, D5808, D6069) that have sensitivity to 1 mg/kg, reflecting the needs of industry to determine very low levels of these contaminants.

Water is a contaminant in naphtha and should be measured using the Karl Fischer method (ASTM E-203, D1364, D1744, D4377, D4928, D6304), by distillation (ASTM D4006), or by centrifuging (ASTM D96) and excluded by relevant drying methods.

Tests should also be carried out for sediment if the naphtha has been subjected to events (such as oxidation) that could lead to sediment formation and instability of the naphtha and resulting products. Test methods are available for the determination of sediment by extraction (ASTM D473, IP 285) or by membrane filtration (ASTM D4807; IP 286) and the determination of simultaneously sediment with water by centrifugation (ASTM D96, D1796, D2709, D4007; IP 373, 374).

10.1.2. Density (Specific Gravity)

Density (the mass of liquid per unit volume at 15°C) and the related terms *specific gravity* (the ratio of the mass of a given volume of liquid at 15°C to the mass of an equal volume of pure water at the same temperature) and *relative density* (same as *specific gravity*) are important properties of petroleum products as they are a part of product sales specifications, although playing only a minor role in studies of product composition. Usually, a hydrometer, pycnometer, or digital density meter is used for determination in all these standards.

The determination of density (specific gravity) (ASTM D287, D891, D941, D1217, D1298, D1555, D1657, D2935, D4052, D5002; IP 160, 235, 365) provides a check on the uniformity of the naphtha and permits calculation of the weight per gallon. The temperature at which the determination is carried out and for which the calculations are to be made should also be known (ASTM D1086). However, the methods are subject to vapor pressure constraints and are used with appropriate precautions to prevent vapor loss during sample handling and density measurement. In addition, some test methods should not be applied if the samples are so dark in color that the absence of air bubbles in the sample cell cannot be established with certainty. The presence of such bubbles can have serious consequences for the reliability of the test data.

10.1.3. Evaporation Rate

Environmentally, the evaporation rate is an important property of naphtha, and although there is a significant relation between distillation range and evaporation rate, the relationship is not straightforward. A simple procedure for determining the evaporation rate involves use of at least a pair of weighed shallow containers, each containing a weighed amount of naphtha. The cover-free containers are placed in a temperature- and humidity-controlled draft-free area. The containers are reweighed at intervals until the samples have evaporated completely or left a residue that does not evaporate further (ASTM D381, D1353; IP 131).

The evaporation rate can be derived either (1) by a plot of time versus weight using a solvent having a known evaporation rate for comparison or (2) from the distillation profile (ASTM D86; IP 123). Although the results obtained on the naphtha provide a useful guide, it is, wherever possible, better to carry out a performance test on the final product when assessing environmental effects.

10.1.4. Flash Point

The flash point is the lowest temperature at atmospheric pressure (760 mmHg, 101.3 kPa) at which application of a test flame will cause the vapor of a sample to ignite under specified test conditions. The sample is deemed to have reached the flash point when a large flame appears and propagates itself instantaneously over the surface of the sample. The flash point data is used in shipping and safety regulations to define *flammable* and *combustible* materials. Flash point data can also indicate the possible presence of highly volatile and flammable constituents in a relatively nonvolatile or nonflammable material.

Of the available test methods, the most common method of determining the flash point confines the vapor (in a closed cup) until the instant the flame is applied (ASTM D56, D93, D3828, D6450; IP 34, IP 94, IP 303). An alternative method that does not confine the vapor (open cup method; ASTM D92, D1310; IP 36) gives slightly higher values of the flash point.

Another test (ASTM E-659) is available that can be used as a complement to the flash point test and involves determination of the autoignition temperature. However, the flash point should not be confused with the autoignition temperature, which measures spontaneous combustion with no external source of ignition.

10.1.5. Odor and Color

In general, the paraffinic hydrocarbons possess the mildest odor and the aromatics the strongest, the odor level (ASTM D268, D1296; IP 89) being related to chemical character and volatility of the constituents. Odors due to the presence of sulfur compounds or unsaturated constituents are excluded by specification; and apart from certain high-boiling aromatic fractions, that are usually excluded by volatility from the majority of the naphtha fractions, which may be pale yellow in color, although naphtha is usually colorless (water white). Measurement of color (ASTM D156, D848, D1209, D1555, D5386; IP 17) provides a rapid method of checking the degree of freedom from contamination. Observation of the test for residue on evaporation (ASTM D381, D1353; IP 131) provides a further guard against adventitious contamination.

10.1.6. Volatility

One of the most important physical parameters that also relates to the evaporation rate is the boiling-range distribution (ASTM D86, D1078, D2887, D2892; IP 123). The significance of the distillation test is the indication of volatility, which dictates the evaporation rate, which is an important property for naphtha used in coatings and similar applications, where the premise is that the naphtha evaporates over time, leaving the coating applied to the surface. In the basic test method (ASTM D86; IP 123) a 100-mL sample is distilled (manually or automatically) under prescribed conditions. The boiling profile is derived from temperatures and volumes of condensate recorded at regular intervals.

The determination of the boiling-range distribution of distillates such as naphtha and gasoline by gas chromatography (ASTM D3710) not only helps identify the constituents but also facilitates on-line controls at the refinery. This test method is designed to measure the entire boiling range of naphtha with either high or low Reid vapor pressure (ASTM D323; IP 69). In the method, the sample is injected into a gas chromatographic column that separates hydrocarbons in boiling-point order. The column temperature is raised at a reproducible rate, and the area under the chromatogram is recorded throughout the run. Calibration is performed using a known mixture of hydrocarbons covering the expected boiling of the sample.

Another method is described as a method for determining the carbon number distribution (ASTM D2887; IP 321), and the data derived by this test method are essentially equivalent to those obtained by the test method for true boiling-point distillation (ASTM D2892). The sample is introduced into a gas chromatographic column that separates hydrocarbons in boiling-point order. The column temperature is raised at a reproducible rate and the area under the chromatogram is recorded throughout the run. Boiling temperatures are assigned to the time axis from a calibration curve, obtained under the same conditions by running a known mixture of hydrocarbons covering the boiling range expected in the sample. From these data, the boiling-range distribution may be obtained. However, this test method is limited to samples having a boiling range greater than 55°C (130°F) and having a vapor pressure (ASTM D323, D4953, D5190, D5191, D5482, D6377, D6378; IP 69, 394) sufficiently low to permit sampling at ambient temperature.

In studying the environmental effects of naphtha, it is necessary to relate volatility to the fire hazard associated with its use, storage, and transport, and also with the handling of the products arising from the process. This is normally based on the characterization of the solvent by flash point limits (ASTM D56, D93; IP 34, 170)

10.2. FUEL OIL

The significance of the measured properties of residual fuel oil is dependent to a large extent on the ultimate uses of the fuel oil. Such uses include steam generation for various processes, as well as electrical power generation and propulsion. Corrosion, ash deposition, atmospheric pollution, and product contamination are side effects of the use of residual fuel oil, and in particular cases, properties such as vanadium, sodium, and sulfur contents may be significant.

The character of fuel oil generally renders the usual test methods for *total petroleum hydrocarbons* (Chapters 7 and 8) ineffective since high proportions of the fuel oil (specifically, residual fuel oil) are insoluble in the usual solvents employed for the test. In particular, the asphaltene constituents are insoluble in hydrocarbon solvents and are only soluble in aromatic solvents and chlorinated hydrocarbons (chloroform, methylene dichloride, and the like). Residua and asphalt (Chapter 10) have high proportions of asphaltene constituents, which render any test for *total petroleum hydrocarbons* meaningless unless a suitable solvent is employed in the test method.

Testing residual fuel oil does not suffer from the issues that are associated with sample volatility but the test methods are often sensitive to the presence of gas bubbles in the fuel oil. An air release test is available for application to lubricating oil (ASTM D3427; IP 313) and may be applied, with modification, to residual fuel oil. However, with dark-colored samples, it may be difficult to determine whether all air bubbles have been eliminated. And, as with the analysis and testing of other petroleum products, the importance of correct sampling of fuel oil cannot be over emphasized, because no proper assessment of quality may be made unless the data are obtained on truly representative samples (ASTM D270; IP 51).

10.2.1. Asphaltene Content

The asphaltene fraction (ASTM D893, D2006, D2007, D3279, D4124, D6560; IP 143) is the highest-molecular-weight and most complex fraction in petroleum (Speight, 1994). The asphaltene content gives an indication of the amount of coke that can be expected during exposure to thermal conditions (Speight, 1999, 2001; Speight and Ozum, 2002).

In any of the methods for determination of the asphaltene content (Speight et al., 1984), the residual fuel oil is mixed with a large excess (usually >30 volumes hydrocarbon per volume of sample) of low-boiling hydrocarbon such as *n*-pentane or *n*-heptane. For an extremely viscous sample, a solvent such as toluene may be used prior to addition of the low-boiling hydrocarbon, but an additional amount of the hydrocarbon (usually >30 volumes hydrocarbon per volume of solvent) must be added to compensate for the presence of the solvent. After a specified time, the insoluble material (the asphaltene fraction) is separated (by filtration) and dried. The yield is reported as a percentage (% w/w) of the original sample. Furthermore, different hydrocarbons (such as *n*-pentane or *n*-heptane) give different yields of the asphaltene fraction, and if the presence of the solvent is not compensated by use of additional hydrocarbon, the yield will be erroneous. In addition, if the hydrocarbon is not present in a large excess, the yields of the asphaltene fraction will vary and will be erroneous (Speight et al., 1984; Speight, 1999).

Another method, not specifically described as an asphaltene separation method, is designed to remove pentane-insoluble constituents by membrane filtration (ASTM D4055). In the method, a sample of oil is mixed with pentane in a volumetric flask, and the oil solution is filtered through a 0.8-μm membrane filter. The flask, funnel, and filter are washed with pentane to transfer any particulates completely onto the filter, after which the filter (with particulates) is dried and weighed to give the pentane insoluble constituents as a percent by weight of the sample. Particulates can also be determined by membrane filtration (ASTM D2276, D5452, D6217; IP 415).

The *precipitation number* is often equated with the asphaltene content, but several issues remain obvious in its rejection for this purpose. For example, the method to determine the precipitation number (ASTM D91) advocates the use of naphtha for use with black oil or lubricating oil and the amount of insoluble material (as a % v/v of the sample) is the precipitating number. In the test, 10 mL of sample is mixed with 90 mL of ASTM precipitation naphtha (which may or may nor have a constant chemical composition) in a graduated centrifuge cone and centrifuged for 10 minutes at 600 to 700 rpm. The volume of material on the bottom of the centrifuge cone is noted until repeat centrifugation gives a value within 0.1 mL (the precipitation number). Obviously, this can be substantially different as to asphaltene content.

If the residual fuel oil is produced by a thermal process such as visbreaking, it may also be necessary to determine if toluene-insoluble material is present by the methods, or modifications thereof, used to determine the toluene insoluble of tar and pitch (ASTM D4072, D4312). In the methods, a sample is digested at 95°C (203°F) for 25 minutes and then extracted with hot toluene in an alundum thimble. The extraction time is 18 hours (ASTM D4072) or 3 hours (ASTM D4312). The insoluble matter is dried and weighed.

10.2.2. Composition

The composition of residual fuel oils is often reported in the form of four or five major fractions as deduced by adsorption chromatography (Figure 10.1). In the case of cracked feedstocks, thermal decomposition products (carbenes and carboids) may also be present. *Column chromatography* is used for several types of hydrocarbon analyses that involve fractionation of viscous oils (ASTM D2007, D2549), including residual fuel oil. The former method (ASTM D2007) advocates the use of adsorption on clay and clay–silica gel followed by elution of the clay with pentane to separate saturates, elution of clay with acetone–toluene to separate polar compounds; and elution of the silica gel fraction with toluene to separate aromatic compounds. The latter method (ASTM D2549) uses adsorption on a bauxite–silica gel column. Saturates are eluted with pentane; aromatics are eluted with ether, chloroform, and ethanol.

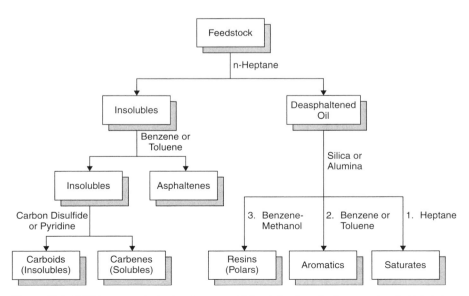

Figure 10.1. SARA-type analysis showing two additional fractions (carbenes and carboids) that are produced by thermal treatment of petroleum and petroleum products.

Correlative methods are derived relationships between fundamental chemical properties of a substance and measured physical or chemical properties. They provide information about oil from readily measured properties (ASTM D2140, D2501, D2502, D3238). One method (ASTM D2501) describes the calculation of the viscosity–gravity coefficient (VGC), a parameter derived from kinematic viscosity and density that has been found to relate to the saturate/aromatic composition. Correlations between the viscosity-gravity coefficient (or molecular weight and density) and refractive index to calculate carbon type composition in percent of aromatic, naphthenic, and paraffinic carbon atoms are employed to estimate of the number of aromatic and naphthenic rings present (ASTM D2140, D3238). Another method (ASTM D2502) permits estimation of molecular weight from kinematic viscosity measurements at 38 and 99°C (100 and 210°F) (ASTM D445). It is applicable to samples with molecular weights in the range 250 to 700 but should not be applied indiscriminately for oils that represent extremes of composition for which different constants are derived (Moschopedis et al., 1976).

A major use for *gas chromatography* for hydrocarbon analysis has been simulated distillation, as discussed previously. Other gas chromatographic methods have been developed for contaminant analysis (ASTM D3524, D4291). The aromatic content of fuel oil is a key property that can affect a variety of other properties, including viscosity, stability, and compatibility with other fuel oil or blending stock. Existing methods for this work use physical measurements and need suitable standards. Thus, methods have been standardized using nuclear magnetic resonance (NMR) for hydrocarbon characterization (ASTM D4808, D5291, D5292). The nuclear magnetic resonance method is simpler and more precise. Procedures are described that cover light distillates with a 15 to 260°C boiling range of middle distillates and gas oils with boiling ranges of 200 to 370°C (390 to 700°F) and 370 to 510°C (700 to 950°F), and residuum boiling above 510°C (950°F). One of the methods (ASTM D5292) is applicable to a wide range of hydrocarbon oils that are completely soluble in chloroform and carbon tetrachloride at ambient temperature. The data obtained by this method can be used to evaluate changes in aromatic contents of hydrocarbon oils due to process changes.

High-ionizing voltage mass spectrometry (ASTM D2786, D3239) is also employed for compositional analysis of residual fuel oil. These methods require preliminary separation using elution chromatography (ASTM D2549). A third method (ASTM D2425) may be applicable to some residual fuel oil samples in the lower-molecular-weight range.

10.2.3. Density (Specific Gravity)

Density or specific gravity (relative density) is used whenever conversions must be made between mass (weight) and volume measurements. This property is often used in combination with other test results to predict oil quality, and several methods are available for measurement of density (or specific gravity). However, the density (specific gravity) (ASTM D1298; IP 160) is probably of least importance in determining fuel oil performance, but it is used in product control, in weight–volume relationships and in the calculation of calorific value (heating value).

Two of the methods (ASTM D287, D1298) use an immersed hydrometer for measurement of density. The former method (ASTM D287) provides the results as API gravity. Two other methods (ASTM D1480, D1481) use a pycnometer to measure density or specific gravity and have the advantage of requiring a smaller sample size and can be used at higher temperatures than is normal provided that the vapor pressure of the liquid does not exceed specific limits at the temperature of the test. Two other test methods (ASTM D4052, D5002) measure density with a digital density analyzer. This device determines density by analysis of the change in oscillating frequency of a sample tube when filled with the test sample.

Another test method (ASTM D4052) covers the determination of the density or specific gravity of viscous oil, such as residual fuel oil, that are liquids at test temperatures between 15 and 35°C (59 and 95°F). However, application of the method is restricted to liquids with vapor pressures below 600 mmHg and viscosity below 15,000 cSt at the temperature of test. In addition (and this is crucial for residual fuel oil), this test method should not be applied to samples so dark in color that the absence of air bubbles in the sample cell cannot be established with certainty.

10.2.4. Elemental Analysis

Elemental analysis of fuel oil often plays a more major role that it may appear to do in lower-boiling products. Aromaticity (through the atomic hydrogen/carbon ratio), sulfur content, nitrogen content, oxygen content, and metals content are all important features that can influence the use of residual fuel oil.

Carbon content and *hydrogen content* can be determined simultaneously by the method designated for coal and coke (ASTM D3178) or by the method designated for municipal solid waste (ASTM E777). However, as with any analytical method, the method chosen for the analysis may be subject to the peculiarities or character of the feedstock under investigation and should be assessed in terms of accuracy and reproducibility. There methods that are designated for elemental analysis are:

1. *Carbon* and *hydrogen content* (ASTM D1018, D3178, D3343, D3701, D5291, E777; IP 338)
2. *Nitrogen content* (ASTM D3179, D3228, D3431, E148, E258, D5291, E778)
3. *Oxygen content* (ASTM E385)
4. *Sulfur content* (ASTM D124, D129, D139, D1266, D1552, D1757, D2622, D2785, D3120, D3177, D4045 D4294, E443; IP 30, 61, 103, 104, 107, 154, 155, 243)

The hydrogen content of fuel oil can also be measured by low-resolution magnetic resonance spectroscopy (ASTM D3701, D4808). The method is claimed to provide a simple and more precise alternative to existing test methods, specifically combustion techniques (ASTM D5291), for determining the hydrogen content of a variety of petroleum-related materials.

Nitrogen occurs in residua, and therefore in residual fuel oil, and causes serious environmental problems as a result, especially when the levels exceed 0.5% by weight, as happens often in residua. In addition to the chemical character of the nitrogen, the amount of nitrogen in a feedstock determines the severity of the process, the hydrogen requirements, and to some extent, the sediment formation and deposition.

The determination of nitrogen in petroleum products is performed regularly by the Kjeldahl method (ASTM D3228), the Dumas method, and the microcoulometric (ASTM D3431) method. The chemiluminescence method is the most recent technique applied to nitrogen analysis for petroleum and is used to determine the amount of chemically bound nitrogen in liquid samples.

In the method, the samples are introduced to the oxygen-rich atmosphere of a pyrolysis tube maintained at 975°C (1785°F). Nitrogen in the sample is converted to nitric oxide during combustion, and the combustion products are dried by passage through magnesium perchlorate [$Mg(ClO_4)_2$] before entering the reaction chamber of a chemiluminescence detector. In the detector, ozone reacts with the nitric oxide to form excited nitrogen dioxide:

$$NO + O_3 = NO_2^* + O_2$$

Photoemission occurs as the excited nitrogen dioxide reverts to the ground state:

$$NO_2^* = NO_2 + h\nu$$

The emitted light is monitored by a photomultiplier tube to yield a measure of the nitrogen content of the sample. Quantitation is based on comparison with the response for carbazole in toluene standards.

Oxygen is one of the five (C, H, N, O, and S) major elements in fuel oil but rarely exceeds 1.5% by weight unless oxidation has occurred during transportation and storage. Many petroleum products do not specify a particular oxygen content, but if the oxygen compounds are present as acidic compounds such a phenols (Ar–OH) and naphthenic acids (cycloalkyl–COOH), they are controlled in different specifications by a variety of tests. The *total acidity* (ASTM D974; IP 139, 273) is determined for many products, especially fuels and fuel oil. Oxygen-containing impurities in the form of *gum* are determined by the *existent gum* test method (ASTM D381; IP 131) and *potential gum* test method (ASTM D873; IP 138). Elemental analysis of the gum can then provide its composition with some indication of the elements (other than carbon and hydrogen) that played a predominant role in its formation.

Being the third most common element (after carbon and hydrogen) in petroleum product, *sulfur* has been analyzed extensively. Analytical methods range from elemental analyses to functional group (sulfur-type) analyses to structural characterization to molecular speciation (Speight, 2001). Of the methods specified for the determination of sulfur (Speight, 2001), the method applied to the corrosion effect of sulfur is extremely important for liquid fuels. In this method (ASTM D1266; IP 154), fuel corrosivity is assessed by the action of the fuel on a copper strip (the *copper strip test*), which helps determine any discoloration of the copper due to the presence of corrosive compounds. The copper strip is immersed in the fuel and heated at 100°C (212°F) for two hours in a bomb. A test using silver as the test metal (IP 227) has also been published. Mercaptans are usually the corrosive reference for sulfur compounds, and metal discoloration is due to formation of the metal sulfide. Thus, mercaptan sulfur is an important property of potential fuels. In addition to the copper strip test, the mercaptan sulfur (R–SH) content (ASTM D1219; IP 104) provides valuable information. As an alternative to determining the mercaptan content, a negative result in the *Doctor test* (ASTM D484; IP 30) may also be acceptable for the qualitative absence of mercaptans. The copper strip method (ASTM D130, D849, D4048; IP 154) may also be employed to determine the presence of corrosive sulfur compounds in residual fuel oil.

The Doctor test measures the amount of sulfur available to react with metallic surfaces at the temperature of the test. The rates of reaction depend on metal type, temperature, and time. In the test, a sample is treated with copper powder at 149°C or 300°F. The copper powder is filtered from the mixture. Active sulfur is calculated from the difference between the sulfur contents of the sample (ASTM D129) before and after treatment with copper.

The determination of sulfur in liquid products by x-ray fluorescence (ASTM D2622; IP 336) has become an extremely well used method over the past two decades. This method can be used to determine the amount of sulfur in homogeneous liquid petroleum hydrocarbons over the range 0.1 to 6.0% by weight. Samples with a sulfur content above this range may be determined after dilution in toluene. The method utilizes the principle that when a sample is irradiated with a ^{55}Fe source, fluorescent x-rays result. The sulfur K_α fluorescence and a background correction at adjacent wavelengths are counted. A calibration of the instrument, wherein the integration time for counting is adjusted such that the signal displayed for the background-corrected radiation equals the concentration of the calibration standard, gives a direct readout of the weight percent sulfur in the sample. Interfering elements include aluminum, silicon, phosphorus, chlorine, argon, and potassium. Generally, the amounts of these elements are insufficient to affect sulfur x-ray counts in samples covered by this method. Atmospheric argon is eliminated by a helium purge.

It is also possible to determine nitrogen and sulfur simultaneously by chemiluminescence and fluorescence. An aliquot of the sample undergoes high-temperature oxidation in a combustion tube maintained at 1050°C (1920°F). Oxidation of the sample converts the chemically bound nitrogen to nitric oxide (NO) and sulfur to sulfur dioxide (SO_2). In a nitrogen detector, ozone reacts with the nitric oxide to form excited nitrogen dioxide (NO_2). As the nitrogen dioxide reverts to its ground state, chemiluminescence occurs and this light emission is monitored by a photomultiplier tube. The light emitted is proportional to the amount of nitrogen in the sample. In the sulfur detector, the sulfur dioxide is exposed to ultraviolet radiation and produces a fluorescent emission. This light emission is proportional to the amount of sulfur and is also measured by a photomultiplier tube. Quantitation is determined by comparison to the responses given by standards containing carbazole and dimethyl sulfoxide in xylene.

Oxidative microcoulometry has become a widely accepted technique for the determination of low concentrations of sulfur in petroleum and petroleum products (ASTM D3120). The method involves combustion of the sample in an oxygen-rich atmosphere followed by microcoulometric generation of a triiodide ion to consume the resulting sulfur dioxide. It is intended to distinguish the technique from reductive microcoulometry, which converts sulfur in the sample to hydrogen sulfide that is titrated with coulometrically generated silver ion.

The bomb method for sulfur determination (ASTM D129) uses sample combustion in oxygen and conversion of the sulfur to barium sulfate, which is determined by mass. This method is suitable for samples containing 0.1 to 5.0% w/w sulfur and can be used for most low-volatility petroleum products. Elements that produce residues insoluble in hydrochloric acid interfere with this method; this includes aluminum, calcium, iron, lead, and silicon, plus minerals such as asbestos, mica, and silica, and an alternative method (ASTM D1552) is preferred. This method describes three procedures: the sample is first pyrolyzed in either an induction furnace or a resistance furnace; the sulfur is then converted to sulfur dioxide, and the sulfur dioxide is either titrated with potassium iodate–starch reagent or is analyzed by infrared spectroscopy. This method is generally suitable for samples containing from 0.06 to 8.0% w/w sulfur that distill at temperatures above 177°C (351°F).

A variety of *miscellaneous elements* can also occur in a residual fuel oil fraction. For example, *chlorine* is present as a chlorinated hydrocarbon and can be determined (ASTM D808, D1317, D6160). A rapid test method suitable for analysis of samples by nontechnical personnel is also available (ASTM D5384) and uses a commercial test kit where the oil sample is reacted with metallic sodium to convert organic halogens to halide, which is titrated with mercuric nitrate using diphenyl carbazone indicator. Iodides and bromides are reported as chloride.

Phosphorus is a common component of additives and appears most commonly as a zinc dialkyl dithiophosphate or triaryl phosphate ester, but other forms also occur. Two wet chemical methods are available, one of which (ASTM D1091) describes an oxidation procedure that converts phosphorus to aqueous orthophosphate anion. This is then determined by mass as magnesium pyrophosphate or photochemically as molybdivanadophosphoric acid. In an alternative test (ASTM D4047), samples are oxidized to phosphate with zinc oxide, dissolved in acid, precipitated as quinoline phosphomolybdate, treated with excess standard alkali, and back-titrated with standard acid. Both of these methods are used primarily for referee samples. Phosphorus is most commonly determined using x-ray fluorescence (ASTM D4927) or ICP (ASTM D4951).

10.2.5. Flash Point

As for all petroleum products, considerations of safety in storage and transportation and, more particularly, contamination by more volatile products are required. This is usually accommodated by the Pensky–Martens flash point test (ASTM D93; IP 34). For the fuel oil, a minimum flash point of 55°C (131°F) or 66°C (150°F) is included in most specifications.

10.2.6. Metals Content

The analysis for metal constituents in residual fuel oil can be accomplished by several instrumental techniques: inductively coupled argon plasma (ICAP) spectrometry, atomic absorption (AA) spectrometry, or x-ray fluorescence (XRF) spectrometry. Each technique has limitations in terms of sample preparation, sensitivity, sampling, time for analysis, and overall ease of use. Thus, a variety of tests (ASTM D482, D1026, D1262, D1318, D1368, D1548, D1549, D2547 D2599, D2788, D3340, D3341, D3605) either directly or as the constituents of combustion ash have been designated to determine metals in petroleum products based on a variety of techniques. At the time of writing, a specific test for the determination of metals in whole feeds has not been designated. However, this task can be accomplished by combustion of the sample so that only inorganic ash remains (ASTM D482). The ash can then be digested with an acid and the solution examined for metal species by atomic absorption (AA) spectroscopy (IP 288, 285) or by inductively coupled argon plasma (ICP) spectrometry (ASTM C-1109, C-1111).

Atomic absorption provides very high sensitivity but requires careful subsampling, extensive sample preparation, and detailed sample-matrix corrections. X-ray fluorescence requires little in terms of sample preparation but suffers from low sensitivity and the application of major matrix corrections. Inductively coupled argon plasma spectrometry provides high sensitivity and few matrix corrections but requires a considerable amount of sample preparation, depending on the process stream to be analyzed.

In the inductively coupled argon plasma emission spectrometer (ICAP) method, the nickel, iron, and vanadium contents of gas oil samples range from 0.1 to 100 mg/kg. Thus, a 10-g sample of gas oil is charred with sulfuric acid and subsequently combusted to leave the ash residue. The resulting sulfates are then converted to their corresponding chloride salts to ensure complete solubility. A barium internal standard is added to the sample before analysis. In addition, use of the ICAP method for the analysis of nickel, vanadium, and iron present counteracts the two basic issues arising from metals analysis. The most serious issue is the fact that these metals are partly or totally in the form of volatile chemically stable porphyrin complexes, and extreme conditions are needed to destroy the complexes without losing the metal through volatilization of the complex. The second issue is that the alternative, direct aspiration of the sample, introduces large quantities of carbon into the plasma. This carbon causes marked and somewhat variable background changes in all direct measurement techniques.

Finally, the analytical method should be selected depending on the sensitivity required, the compatibility of the sample matrix with the specific analysis technique, and the availability of facilities. Sample preparation, if it is required, can present problems. Significant losses can occur, especially in the case of organometallic complexes, and contamination of environmental sample is of serious concern. The precision of the analysis depends on the metal itself, the method used, and the standard used for calibration of the instrument.

10.2.7. Pour Point and Viscosity

The pour point (ASTM D97, D5949, D5950, D5853, D5985; IP 15) is the lowest temperature at which oil will flow under prescribed conditions. The test method for determining the solidification point (ASTM D1493) might also be applied to residual fuel oil.

The pour point test is still included in many specifications but not in some (ASTM D396; BS 2869) for residual fuel oil for assessing the mobility characteristics of residual fuel oil (ASTM D3245). Pour point procedures involving various preheat treatments prior to the pour point determination, and the use of viscosity at low temperatures have been proposed. The fluidity test (ASTM D1659) is one such procedure, as is the pumping temperature test (ASTM D3829), and another test, based on viscosity measurements (IP 230), is also available.

Viscosity is an important property of residual fuel oils, as it provides information on the ease (or otherwise) with which a fuel can be transferred from storage tank to burner system under prevailing temperature and pressure conditions. Viscosity data also indicate the degree to which a fuel oil needs to be preheated to obtain the correct atomizing temperature for efficient combustion. Most residual fuel oils function best when the burner input viscosity lies within a certain specified range.

The Saybolt Universal and Saybolt Furol viscometers are widely used in the United States and the Engler in Europe. In the United States, viscosities on the lighter fuel grades are determined using the Saybolt Universal instrument at 38°C (100°F); for the heaviest fuels the Saybolt Furol viscometer is used at 50°C (122°F). Similarly, in Europe, the Engler viscometer is used at temperatures of 20°C (68°F), 50°C (122°F), and in some instances at 100°C (212°F). Use of these empirical procedures for fuel oils is being superseded by kinematic system (ASTM D396; BS 2869) specifications for fuel oils.

The determination of residual fuel oil viscosities is complicated by the fact that fuel oils containing significant quantities of wax do not behave as simple Newtonian liquids, in which the rate of shear is directly proportional to the shearing stress applied. At temperatures in the region of 38°C (100°F) these fuels tend to deposit wax from solution, with a resulting adverse effect on the accuracy of the viscosity result unless the test temperature is raised sufficiently high for all wax to remain in solution. Although the present reference test temperature of 50°C (122°F) is adequate for use with the majority of residual fuel oils, there is a growing trend in favor of a higher temperature [82°C (180°F)], particularly in view of the availability of waxier fuel oils.

10.2.8. Stability

The problem of instability in residual fuel oil may manifest itself either as waxy sludge deposited on the soil or as fouling coastlines. Asphaltene-type deposition may, however, result from the mixing of fuels of different origin and treatment, each of which may be perfectly satisfactory when used alone. For example, straight-run fuel oils from the same crude oil are normally stable and mutually compatible, whereas fuel oils produced from thermal cracking and visbreaking operations may be stable but can be unstable or incompatible if blended with straight-run fuels, and vice versa (ASTM D1661).

Another procedure for predicting the stability of residual fuel oil involves the use of a spot test to show compatibility or cleanliness of the blended fuel oil (ASTM D2781, D4740). The former method (ASTM D2781) covers two spot test procedures for rating a residual fuel with respect to its compatibility with a specific distillate fuel. Procedure A indicates the degree of asphaltene deposition that may be expected in blending the components and is used when wax deposition is not considered a fuel application problem. Procedure B indicates the degree of wax and asphalt deposition in the mixture at room temperature. The latter method (ASTM D4740) is applicable to fuel oils with viscosities up to 50 cSt at 100°C (212°F) to identify fuels or blends that could result in excessive centrifuge loading, strainer plugging, tank sludge formation, or similar operating problems. In the method, a drop of the preheated sample is put on a test paper and placed in an oven at 100°C. After 1 hour, the test paper is removed from the oven and the resulting spot is examined for evidence of suspended solids and rated for cleanliness using the procedure described in the method. In a parallel procedure for determining *compatibility*, a blend composed of equal volumes of the fuel oil sample and the blend stock is tested and rated in the same way as just described for the *cleanliness* procedure.

For oxidative stability, an important effect after a spill, a test method (ASTM D4636) is available to determine resistance to oxidation and corrosion degradation and their tendency to corrode various metals. The test method consists of one standard and two alternative procedures. In the method, a large glass tube containing an oil sample and metal specimens is placed in a constant-temperature bath (usually from 100 to 360°C) and heated for a specific number of hours while air is passed through the oil to provide agitation and a source of oxygen. The corrosiveness of the oil is determined by the loss in metal mass and microscopic examination of the sample metal surface(s). Oil samples are withdrawn from the test oil and checked for changes in viscosity and acid number as a result of the oxidation reactions. At the end of the test, the amount of sludge present in the oil remaining in the same tube is determined by centrifugation. Also, the quantity of oil lost during the test is determined gravimetrically. Metals used in the basic and alternative tests are aluminum, bronze, cadmium, copper, magnesium, silver, steel, and titanium. Other metals may also be specified, as determined by the history and storage of the fuel oil.

10.3. WASTEWATERS

Wastewaters from petroleum refining consist of cooling water, process water, stormwater, and sanitary sewage water. A large portion of water used in petroleum refining is used for cooling. Most cooling water is recycled over and over. Cooling water typically does not come into direct contact with process oil streams and therefore contains fewer contaminants than does process wastewater. However, it may contain some oil contamination, due to leaks in the process equipment. Water used in processing operations accounts for a significant portion of the total wastewater. Process wastewater arises from desalting crude oil, steam stripping operations, pump gland cooling, product fractionator reflux drum drains, and boiler blowdown. Because process water often comes into direct contact with oil, it is usually highly contaminated. Stormwater (i.e., surface water runoff) is intermittent and will contain constituents from spills to the surface, leaks in equipment, and any materials that may have collected in drains. Runoff surface water also includes water coming from crude and product storage tank roof drains.

Wastewaters are treated in on-site wastewater treatment facilities and then discharged to publicly owned treatment works (POTWs) or discharged to surface waters under National Pollution Discharge Elimination System (NPDES) permits. Petroleum refineries typically utilize primary and secondary wastewater treatment. Primary wastewater treatment consists of the separation of oil, water, and solids in two stages. During the first stage, an API separator, corrugated plate interceptor, or other separator design is used. Wastewater moves very slowly through the separator, allowing free oil to float to the surface and be skimmed off, and solids to settle to the bottom and be scraped off to a sludge-collecting hopper. The second stage utilizes physical or chemical methods to separate emulsified oils from the wastewater. Physical methods may include the use of a series of settling ponds with a long retention time, or the use of dissolved air flotation (DAF). In DAF, air is bubbled through the wastewater, and both oil and suspended solids are skimmed off the top. Chemicals such as ferric hydroxide or aluminum hydroxide can be used to coagulate impurities into a froth or sludge that can be skimmed more easily off the top. Some wastes associated with the primary treatment of wastewater at petroleum refineries may be considered hazardous and include API separator sludge, primary treatment sludge, sludge from other gravitational separation techniques, float from DAF units, and wastes from settling ponds.

After primary treatment, the wastewater can be discharged to a publicly owned treatment works (POTW) or undergo secondary treatment before being discharged directly to surface waters under a National Pollution Discharge Elimination System (NPDES) permit. In secondary treatment, dissolved oil and other organic pollutants may be consumed by microorganisms biologically. Biological treatment may require the addition of oxygen through a number of different techniques, including activated sludge units, trickling filters, and rotating biological contactors. Secondary treatment generates biomass waste that is typically treated anaerobically and then dewatered.

Some refineries employ an additional stage of wastewater treatment called *polishing* to meet discharge limits. The polishing step can involve the use of activated carbon, anthracite coal, or sand to filter out any remaining impurities, such as biomass, silt, trace metals, and other inorganic chemicals, as well as any remaining organic chemicals.

Certain refinery wastewater streams are treated separately, prior to the wastewater treatment plant, to remove contaminants that would not easily be treated after mixing with other wastewater. One such waste stream is the sour water drained from distillation reflux drums. Sour water contains dissolved hydrogen sulfide and other organic sulfur compounds and ammonia, which are stripped in a tower with gas or steam before being discharged to the wastewater treatment plant.

Wastewater treatment plants are a significant source of refinery air emissions and solid wastes. Air releases arise from fugitive emissions from the numerous tanks, ponds, and sewer system drains. Solid wastes are generated in the form of sludge from a number of treatment units.

Many refineries unintentionally release, or have released, liquid hydrocarbons to groundwater and surface waters. At some refineries contaminated groundwater has migrated off-site and resulted in continuous "seeps" to surface waters. Although the actual volume of hydrocarbons released in such a manner is relatively small, there is the potential to contaminate large volumes of groundwater and surface water, possibly posing a substantial risk to human health and the environment.

REFERENCES

ASTM. 2004. *Annual Book of ASTM Standards*. American Society for Testing and Materials, West Conshohocken, PA.

Austin, G. T. 1984. *Shreve's Chemical Process Industries*, 5th ed. McGraw-Hill, New York, Chap. 37.

Chenier, P. J. 1992. *Survey of Industrial Chemistry*, 2nd rev. ed. VCH Publishers, New York, Chaps. 7 and 8.

Charlot, J.-C., and Claus, G. 2000. In *Modern Petroleum Technology*, Vol. 2, *Downstream*, A. G. Lucas (Ed.). Wiley, New York, Chap. 21.

Francis, W., and Peters, M. C. 1980. *Fuels and Fuel Technology: A Summarized Manual*. Pergamon Press, Elmsford, NY, Sec. B.

Goodfellow, A. J. 1973. In *Criteria for Quality of Petroleum Products*. J. P. Allinson (Ed.). Wiley, New York, Chap. 4.

Gruse, W. A., and Stevens, D. R. 1960. *Chemical Technology of Petroleum*. McGraw-Hill, New York, Chap. 11.

Guthrie, V. B. 1967. In *Petroleum Processing Handbook*, W. F. Bland and R. L. Davidson (Eds.). McGraw-Hill, New York, Sec. 11.

Heinrich, H., and Duée, D. 2000. In *Modern Petroleum Technology*, Vol. 2, *Downstream*, A. G. Lucas (Ed.). Wiley, New York, Chap. 10.

Hemighaus, G. 1998. In *Manual on Hydrocarbon Analysis*, 6th ed., A. W. Drews (Ed.). American Society for Testing and Materials, West Conshohocken, PA, Chap. 3.

Hoffman, H. L. 1983. In *Riegel's Handbook of Industrial Chemistry*, 8th ed., J. A. Kent (Ed.). Van Nostrand Reinhold, New York, Chap. 14.

Hoffman, H. L., and McKetta, J. J. 1993. In *Chemical Processing Handbook*, J. J. McKetta (Ed.). Marcel Dekker, New York, p. 851.

Hori, Y. 2000. In *Modern Petroleum Technology*, Vol. 2, *Downstream*, A. G. Lucas (Ed.). Wiley, New York, Chap. 2.

IP. 2001. *IP Standard Methods 2001*. Institute of Petroleum, London.

Kite, W. H., Jr., and Pegg, R. E. 1973. In *Criteria for Quality of Petroleum Products*, J. P. Allinson (Ed.). Wiley, New York, Chap. 7.

McCann, J. M. 1998. In *Manual on Hydrocarbon Analysis*, 6th ed., A. W. Drews (Ed.). American Society for Testing and Materials, West Conshohocken, PA, Chap. 2.

Mitchell, D. L., and Speight, J. G. 1973. *Fuel*, 52: 149.

Moschopedis, S. E., Fryer, J. F., and Speight, J. G. 1976. *Fuel*, 55: 227.

Mushrush, G. W., and Speight, J. G. 1995. *Petroleum Products: Instability and Incompatibility*. Taylor & Francis, New York.

Speight, J. G. 1994. In *Asphaltenes and Asphalts*, I. *Developments in Petroleum Science*, 40, T. F. Yen and G. V. Chilingarian (Eds.). Elsevier, Amsterdam, Chap. 2.

Speight, J. G. 1999. *The Chemistry and Technology of Petroleum*, 3rd ed. Marcel Dekker, New York.

Speight, J. G. 2001. *Handbook of Petroleum Analysis*. Wiley, New York.

Speight, J. G. 2002. *Handbook of Petroleum Product Analysis*. Wiley, Hoboken, NJ.

Speight, J. G., and Ozum, B. 2002. *Petroleum Refining Processes*. Marcel Dekker, New York.

Speight, J. G., Long, R. B., and Trowbridge, T. D. 1984. *Fuel*, 63: 616.

Warne, T. M. 1998. In *Manual on Hydrocarbon Analysis*, 6th ed., A. W. Drews (Ed.). American Society for Testing and Materials, West Conshohocken, PA, Chap. 3.

Weissermel, K., and Arpe, H.-J. 1978. *Industrial Organic Chemistry*. Verlag Chemie, New York, Chap. 13.

CHAPTER
11
ANALYSIS OF SOLID EFFLUENTS

In terms of waste definition, there are three basic approaches (as it pertains to petroleum, petroleum products, and nonpetroleum chemicals) to defining petroleum or a petroleum product as hazardous: (1) a qualitative description of the waste by origin, type, and constituents; (2) classification by characteristics based on testing procedures; and (3) classification as a result of the concentration of specific chemical substances (Chapter 1; EPA, 2004).

Solid effluents are generated from many of the refining processes, petroleum handling operations, and wastewater treatment. Both hazardous and nonhazardous wastes are generated, treated, and disposed. Refinery wastes are typically in the form of sludge (including sludge from wastewater treatment), spent process catalysts, filter clay, and incinerator ash. Treatment of these wastes includes incineration, land treating off-site, land filling on-site, land filling off-site, chemical fixation, neutralization, and other treatment methods.

A significant portion of the nonpetroleum product outputs of refineries is transported off-site and sold as by-products. These outputs include sulfur, acetic acid, phosphoric acid, and recovered metals. Metals from catalysts and from the crude oil that have deposited on the catalyst during the production often are recovered by third-party recovery facilities.

Storage tanks are used throughout the refining process to store crude oil and intermediate process feeds for cooling and further processing. Finished petroleum products are also kept in storage tanks before transport off-site. Storage tank bottoms are mixtures of iron rust from corrosion, sand, water, and emulsified oil and wax, which accumulate at the bottom of tanks. Liquid tank bottoms (primarily water and oil emulsions) are periodically drawn off to prevent their continued buildup. Tank bottom liquids and sludge are also removed during periodic cleaning of tanks for inspection. Tank bottoms may contain amounts of tetraethyl- or tetramethyllead (although this is increasingly rare, due to the phase-out of leaded products), other metals, and phenols. Solids generated from leaded gasoline storage tank bottoms are listed as a RCRA hazardous waste.

11.1. RESIDUA AND ASPHALT

The importance of residua and asphalt to the environmental analyst arises from spillage or leakage in the refinery or on the road. In either case, the properties of

---

Environmental Analysis and Technology for the Refining Industry, by James G. Speight
Copyright © 2005 John Wiley & Sons, Inc.

these materials are detrimental to the ecosystem in which the release occurred. As with other petroleum products, knowledge of the properties of residua and asphalt can help determine the potential cleanup methods and may even allow regulators to trace the product to the refinery where it was produced. In addition, the character of residua and asphalt render the usual test methods for *total petroleum hydrocarbons* (Chapters 7 and 8) ineffective, since high proportions of asphalt and residua are insoluble in the usual solvents employed for the test. Application of the test methods for *total petroleum hydrocarbons* to fuel oil (Chapter 10) is subject to similar limitations.

Residua are the dark-colored nearly solid or solid products of petroleum refining that are produced by atmospheric and vacuum distillation (Figure 11.1; Chapter 3). Asphalt is usually produced from a residuum and is a dark brown to black cementitious material obtained from petroleum processing that contains very high-molecular-weight molecular polar species called *asphaltenes* that are soluble in carbon disulfide, pyridine, aromatic hydrocarbons, and chlorinated hydrocarbons (Chapter 3) (Gruse and Stevens, 1960; Guthrie, 1967; Broome and Wadelin, 1973; Weissermel and Arpe, 1978; Hoffman, 1983; Austin, 1984; Chenier, 1992; Hoffman and McKetta, 1993).

Residua and asphalt derive their characteristics from the nature of their crude oil precursor, the distillation process being a concentration process in which most of the heteroatoms and polynuclear aromatic constituents of the feedstock are concentrated in the residuum (Speight, 2001). Asphalt may be similar to its parent residuum but with some variation possible by choice of manufacturing process. In general terms, residua and asphalt are a hydrocarbonaceous material that consist constituents (containing carbon, hydrogen, nitrogen, oxygen, and sulfur) that are completely soluble in carbon disulfide (ASTM D4). Trichloroethylene or

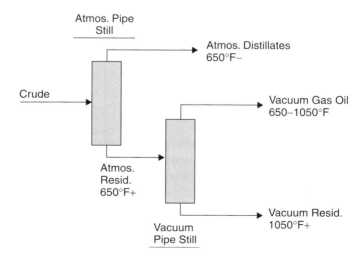

Figure 11.1. Distillation section of a refinery.

1,1,1-trichloroethane has been used in recent years as solvents for the determination of asphalt (and residua) solubility (ASTM D2042).

The residua from which asphalt are produced, once considered the garbage of a refinery, have little value and little use other than as a road oil. In fact, the development of delayed coking (once the so-called "refinery garbage can") was with the purpose of converting residua to liquids (valuable products) and coke (fuel).

Asphalt manufacture involves distilling everything possible from crude petroleum until a residuum with the desired properties is obtained. This is usually done by stages in which distillation at atmospheric pressure removes the lower-boiling fractions and yields an atmospheric residuum (*reduced crude*) that may contain higher-boiling (lubricating) oils, wax, and asphalt. Distillation of the reduced crude under vacuum removes the oils (and wax) as overhead products, and the residuum remains as a bottom (or residual) product. The majority of the polar functionalities and high-molecular-weight species in the original crude oil, which tend to be nonvolatile, concentrate in the vacuum residuum (Speight, 2000), thereby conferring desirable or undesirable properties on the residuum.

At this stage the residuum is frequently, but incorrectly, referred to as *pitch* and has a softening point (ASTM D36, D61, D2319, D3104, D3461) related to the amount of oil removed, which increases with increasing overhead removal. In character with the elevation of the softening point, the pour point is also elevated; the more oil that is distilled from the residue, the higher the softening point.

Propane deasphalting of a residuum also produces asphalt and there are differences in the properties of asphalts prepared by propane deasphalting and those prepared by vacuum distillation from the same feedstock. Propane deasphalting also has the ability to reduce a residuum even further and to produce an asphalt product having a lower-viscosity, higher-ductility, and higher-temperature susceptibility than that of other asphalts, although such properties might be anticipated to be very crude oil dependent. Propane deasphalting is conventionally applied to low-asphalt-content crude oils, which are generally different in type and source from those processed by distillation of higher-yield crude oils. In addition, the properties of asphalt can be modified by air blowing in batch and continuous processes (Figure 11.2). On the other hand, the preparation of asphalts in liquid form by blending (cutting back) asphalt with a petroleum distillate fraction is customary and is generally accomplished in tanks equipped with coils for air agitation or with a mechanical stirrer or vortex mixer.

An *asphalt emulsion* is a mixture of asphalt and an anionic agent such as the sodium or potassium salt of a fatty acid. The fatty acid is usually a mixture and may contain palmitic, stearic, linoleic, and abietic acids and/or high-molecular-weight phenols. Sodium lignate is often added to alkaline emulsions to effect better emulsion stability. Nonionic cellulose derivatives are also used to increase the viscosity of the emulsion if needed. The acid number is an indicator of its asphalt emulsification properties and reflects the presence of high-molecular-weight asphaltic or naphthenic acids. Diamines, frequently used as cationic agents, are made from the reaction of tallow acid amines with acrylonitrile, followed by hydrogenation.

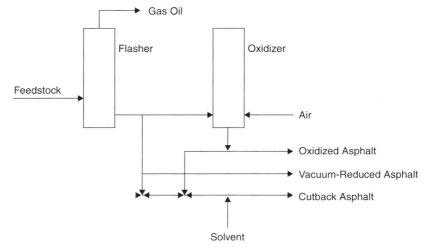

Figure 11.2. Asphalt manufacture including air blowing (Speight, 1992, 1999, and references cited therein).

The properties of asphalt emulsions (ASTM D977, D2397) allow a variety of uses. As with other petroleum products, sampling is an important precursor to asphalt analysis, and a standard method (ASTM D140) is available that provides guidance for the sampling of asphalts, liquid and semisolid, at point of manufacture, storage, or delivery.

The properties of residua and asphalt are defined by a variety of standard tests that can be used to define quality and remembering that the properties of residua vary with cut point (Speight, 2001); that is, the volume percent of the crude oil helps the refiner produce asphalt of a specific type or property (ASTM D496). Roofing and industrial asphalts are also generally specified in various grades of hardness, usually with a combination of softening point (ASTM D61, D2319, D3104, D3461) and penetration to distinguish grades (ASTM D312, D449).

The significance of a particular test is not always apparent by reading the procedure, and sometimes can only be gained through working familiarity with the test. The following tests are commonly used to characterize asphalts, but these are not the only tests used for determining the property and behavior of an asphaltic binder. As in the petroleum industry, a variety of tests are employed having evolved through local, or company, use.

11.1.1. Acid Number

The acid number is a measure of the acidity of a product and is used as a guide in the quality control of resid or asphalt properties. Since a variety of oxidation products contribute to the acid number, and the organic acids vary widely in service properties, so the test is not sufficiently accurate to predict the precise behavior of asphalt in service.

Resids and asphalt contain a small amount of organic acids and saponifiable material that is determined largely by the percentage of naphthenic (cycloparaffinic) acids of higher molecular weight that are originally present in the crude oil. With increased hardness, asphalt from a particular crude oil normally decreases in acid number as more of the naphthenic acids are removed during the distillation process. Acidic constituents may also be present as additives or as degradation products formed during service, such as oxidation products (ASTM D5770). The relative amount of these materials can be determined by titrating with bases. The acid number is used as a guide in the quality control of lubricating oil formulations. It is also sometimes used as a measure of lubricant degradation in service. Any condemning limits must be established empirically.

The *saponification number* expresses the amount of base that will react with 1 g of a sample when heated in a specific manner. Since certain elements are sometimes added to asphalt and also consume alkali and acids, the results obtained indicate the effect of these extraneous materials in addition to the saponifiable material present. In the test method (ASTM D94; IP 136), a known weight of the sample is dissolved in methyl ethyl ketone or a mixture of suitable solvents, and the mixture is heated with a known amount of standard alcoholic potassium hydroxide for between 30 and 90 minutes at 80°C (176°F). The excess alkali is titrated with standard hydrochloric acid and the saponification number is calculated.

11.1.2. Asphaltene Content

The asphaltene fraction (ASTM D2006, D2007, D3279, D4124, D6560; IP 143) is the highest molecular weight and most complex fraction in petroleum. The asphaltene content gives an indication of the amount of coke that can be expected during processing (Speight, 2001; Speight and Ozum, 2002).

In any of the methods for the determination of the asphaltene content, the crude oil or product (such as a residuum or asphalt) is mixed with a large excess (usually >30 volumes hydrocarbon per volume of sample) of low-boiling hydrocarbon such as n-pentane or n-heptane. For an extremely viscous sample, a solvent such as toluene may be used prior to the addition of the low-boiling hydrocarbon, but an additional amount of the hydrocarbon (usually >30 volumes hydrocarbon per volume of solvent) must be added to compensate for the presence of the solvent. After a specified time, the insoluble material (the asphaltene fraction) is separated (by filtration) and dried. The yield is reported as a percentage (% w/w) of the original sample. In any of these tests, different hydrocarbons (such as n-pentane or n-heptane) will give different yields of the asphaltene fraction, and if the presence of the solvent is not compensated by use of additional hydrocarbon, the yield will be erroneous. In addition, if the hydrocarbon is not present in a large excess, the yields of the asphaltene fraction will vary and will be erroneous (Speight, 2001).

The *precipitation number* is often equated to the asphaltene content, but there are several issues that remain obvious in its rejection for this purpose. For example, the method to determine the precipitation number (ASTM D91) advocates the use of naphtha for use with black oil or lubricating oil, and the amount of insoluble material (as a % v/v of the sample) is the precipitating number. In the test, 10 mL of sample is mixed with 90 mL of ASTM precipitation naphtha (which may or may nor have a constant chemical composition) in a graduated centrifuge cone and is centrifuged for 10 minutes at 600 to 700 rpm. The volume of material on the bottom of the centrifuge cone is noted until repeat centrifugation gives a value within 0.1 mL (the precipitation number). Obviously, this can be substantially different from the asphaltene content.

In another test method (ASTM D4055), pentane-insoluble materials above 0.8 μm in size can be determined. In the test method, a sample of oil is mixed with pentane in a volumetric flask, and the oil solution is filtered through a 0.8-μm membrane filter. The flask, funnel, and filter are washed with pentane to transfer the particulates completely onto the filter, which is then dried and weighed to give the yield of pentane-insoluble materials.

Another test method (ASTM D893) that was originally designed for the determination of pentane- and toluene-insoluble materials in used lubricating oils can also be applied to residua and asphalt. However, the method may need modification by first adding a solvent (such as toluene) to the residuum or asphalt before adding pentane. The pentane-insoluble constituents can include toluene-insoluble materials. A significant change in the pentane- or toluene-insoluble constituents indicates a change in properties of the resid or asphalt due to thermal or oxidative effects.

11.1.3. Carbon Disulfide Insoluble Constituents

The component of highest carbon content, the fraction termed *carboids*, consists of species that are insoluble in carbon disulfide or in pyridine. The fraction that has been called *carbenes* contains molecular species that are soluble in carbon disulfide and soluble in pyridine but which are insoluble in toluene (Figure 11.3).

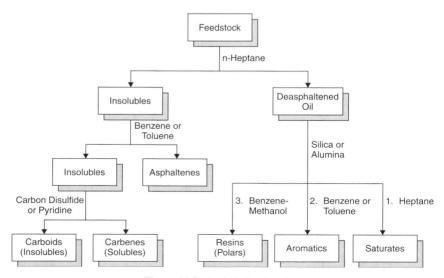

Figure 11.3. Asphalt fractionation.

The carbene and carboids fractions are generated by thermal degradation or by oxidative degradation and are not considered to be naturally occurring constituents of resids or asphalt. The test method for determining the toluene-insoluble constituents of tar and pitch (ASTM D4072, D4312) can be used to determine the amount of carbenes and carboids in resids and asphalt.

11.1.4. Composition

Determination of the composition of resids and asphalt has always presented a challenge because of the complexity and high molecular weights of the molecular constituents. The principle behind composition studies is to evaluate resids and asphalt in terms of composition and performance.

The methods employed can be arranged conveniently into a number of categories: (1) fractionation by precipitation; (2) fractionation by distillation; (3) separation by chromatographic techniques; (4) chemical analysis using spectrophotometric techniques (infrared, ultraviolet, nuclear magnetic resource, x-ray fluorescence, emission, neutron activation), titrimetric and gravimetric techniques, and elemental analysis; and (5) molecular weight analysis by mass spectrometry, vapor pressure osmometry, and size exclusion chromatography.

However, fractional separation has been the basis for most composition analysis, and the separation methods are used to produce operationally defined fractions (Figure 11.3). Three types of separation procedures are now in use: (1) chemical precipitation in which n-pentane separation of an asphaltene fraction is followed by chemical precipitation of other fractions with sulfuric acid of increasing concentration (ASTM D2006); (2) adsorption chromatography using a clay-gel procedure where after removal of the asphaltene fraction, the remaining constituents are separated by selective adsorption/desorption on an adsorbent (ASTM D2007, D4124); (3) size exclusion chromatography, in which gel permeation chromatographic (GPC) separation of constituents occurs based on their associated sizes in dilute solutions (ASTM D3593).

The fractions obtained in these schemes are defined operationally or procedurally. The amount and type of the asphaltene constituents are, for instance, defined by the solvent used for precipitating them. Fractional separation of does not provide well-defined chemical components, and the separated fractions should only be defined in terms of the particular test procedure (Speight, 1999, 2001). This is analogous to the definition of total petroleum hydrocarbons, in which the composition is defined by the method of extraction (Chapter 7). However, these fractions are generated by thermal degradation or by oxidative degradation and are not considered to be naturally occurring constituents. The test method for determining the toluene-insoluble constituents of tar and pitch (ASTM D4072, D4312) can be sued to determine the amount of carbenes and carboids in resids and asphalt.

In the methods, a sample is digested at 95°C (203°F) for 25 minutes and then extracted with hot toluene in an alundum thimble. The extraction time is 18 hours (ASTM D4072) or 3 hours (ASTM D4312). The insoluble matter is dried and weighed. Combustion will then show if the material is truly carbonaceous or if it is inorganic ash from the metallic constituents (ASTM D482, D2415, D4628, D4927, D5185, D6443; IP 4).

Another method (ASTM D893) covers the determination of pentane- and toluene-insoluble constituents in used lubricating oils and can also be applied. Pentane-insoluble constituents include oil-insoluble materials, and toluene-insoluble constituents can come from external contamination and highly carbonized materials from degradation. A significant change in pentane- or toluene-insoluble constituents indicates a change in properties that could lead to problems in further processing (for resids) or service (for asphalt).

Two test methods are used. Procedure A covers the determination of insoluble constituents without the use of coagulant in the pentane and provides an indication of materials that can readily be separated from the diluted material by centrifugation. Procedure B covers the determination of insoluble constituents that contain additives and employs a coagulant. In addition to the materials separated by using procedure A, this coagulation procedure separates some finely divided materials that may be suspended in the resid or asphalt. The results obtained by procedures A and B should not be compared since they usually give different values. The same procedure should be applied when comparing results obtained periodically when comparing results determined in different laboratories.

In procedure A, a sample is mixed with pentane and centrifuged, after which the resid or asphalt solution is decanted and the precipitate is washed twice with pentane, dried, and weighed. For toluene-insoluble constituents, a separate sample of the resid or asphalt is mixed with pentane and centrifuged. The precipitate is washed twice with pentane, once with toluene–alcohol solution, and once with toluene. The insoluble material is then dried and weighed. In procedure B, procedure A is followed except that instead of pentane, a pentane-coagulant solution is used.

Many investigations of relationships between composition and properties take into account only the concentration of the asphaltene constituents, independent of quality criterion. However, a distinction should be made between the asphaltene constituents that occur in straight-run resids and those that occur in blown asphalts. Remembering that asphaltene constituents are a solubility class rather than a distinct chemical class means that vast differences occur in the makeup of this fraction when it is produced by different processes.

11.1.5. Density (Specific Gravity)

For clarification, it is necessary to understand the basic definitions that are used: (1) *density* is the mass of liquid per unit volume at 15.6°C (60°F), (2) *relative density* is the ratio of the mass of a given volume of liquid at 15.6°C (60°F) to the mass of an equal volume of pure water at the same temperature, and (3) *specific gravity* is the same as the relative density and the terms are used interchangeably.

Density (ASTM D1298; IP 160) is an important property of petroleum products since petroleum, especially petroleum products, are usually bought and sold on that basis, or if on a volume basis, then converted to a mass basis via density measurements. This property is almost synonymously termed *density, relative density, gravity*, and *specific gravity*, all terms related to each other. Usually a hydrometer, pycnometer, or more modern digital density meter is used for the determination of density or specific gravity.

In the most commonly used method (ASTM D1298; IP 160), the sample is brought to the prescribed temperature and transferred to a cylinder at approximately the same temperature. The appropriate hydrometer is lowered into the sample and allowed to settle, and after temperature equilibrium has been reached, the hydrometer scale is read and the temperature of the sample is noted.

Although there are many methods for the determination of density due to the different nature of petroleum itself and the different products, one test method (ASTM D5002) is used for the determination of the density or relative density of petroleum that can be handled in a normal fashion as liquids at test temperatures between 15 and 35°C (59 and 95°F). This test method applies to petroleum oils with high vapor pressures provided that appropriate precautions are taken to prevent vapor loss during transfer of the sample to the density analyzer. In the method, approximately 0.7 mL of crude oil sample is introduced into an oscillating sample tube, and the change in oscillating frequency caused by the change in mass of the tube is used in conjunction with calibration data to determine the density of the sample.

Another test determines density and specific gravity by means of a digital densimeter (ASTM D4052; IP 365). In the test, a small volume (approximately 0.7 mL) of liquid sample is introduced into an oscillating sample tube, and the change in oscillating frequency caused by the change in the mass of the tube is used in conjunction with calibration data to determine the density of the sample. The test is usually applied to petroleum, petroleum distillates, and petroleum products that are liquids at temperatures between 15 and 35°C (59 and 95°F) and that have vapor pressures below 600 mmHg and viscosities below about 15,000 cSt at the temperature of test. However, the method should not be applied to samples so dark in color that the absence of air bubbles in the sample cell cannot be established with certainty.

Accurate determination of the density or specific gravity of crude oil is necessary for the conversion of measured volumes to volumes at the standard temperature of 15.56°C (60°F) (ASTM D1250; IP 200; Petroleum Measurement Tables). The specific gravity is also a factor reflecting the quality of crude oils.

The accurate determination of the API gravity of petroleum and its products (ASTM D287) is necessary for the conversion of measured volumes to volumes at the standard temperature of 60°F (15.56°C). Gravity is a factor governing the quality of crude oils. However, the gravity of a petroleum product is an uncertain indication of its quality. Correlated with other properties, gravity can be used to give approximate hydrocarbon composition and heat of combustion. This is usually accomplished though use of the API gravity, which is derived from the specific gravity:

$$\text{API gravity (deg)} = (141.5/\text{sp gr } 60°/60°\text{F}) - 131.5$$

and is also a critical measure for reflecting the quality of petroleum.

API gravity or density or relative density can be determined using one of two hydrometer methods (ASTM D287, D1298). The use of a digital analyzer (ASTM D5002) is finding increasing popularity for the measurement of density and specific gravity.

In the method (ASTM D287), the API gravity is determined using a glass hydrometer for petroleum and petroleum products that are normally handled as liquids and that have a Reid vapor pressure of 26 psi (180 kPa) or less. The API gravity is determined at 15.6°C (60°F), or converted to values at 60°F, by means of standard tables. These tables are not applicable to nonhydrocarbons or essentially pure hydrocarbons such as the aromatics.

For solid and semisolid materials, a pycnometer is generally used (ASTM D70), and a hydrometer is applicable to liquid materials (ASTM D3142). It is worthy of note at this point that the density (hence, the API gravity) of residua show pronounced changes due to the effects of temperature and pressure (Table 11.1). Therefore, isolation of the sample after leakage or spillage must also allow for equilibration to ambient conditions before measurements are made.

11.1.6. Elemental Analysis

Residua and asphalt are not composed of a single chemical species, but rather, are a complex mixture of organic molecules that vary widely in composition and are composed of carbon, hydrogen, nitrogen, oxygen, and sulfur as well as trace amounts of metals, principally vanadium and nickel. The heteroatoms, although a minor component compared to the hydrocarbon moiety, can vary in concentration over a wide range, depending on the source.

Generally, most resids and asphalt have 79 to 88% w/w carbon, 7 to 13% w/w hydrogen, trace to 8% w/w sulfur, 2 to 8% w/w oxygen, and trace to 3% w/w nitrogen. Trace metals such as iron, nickel, vanadium, calcium, titanium, magnesium, sodium, cobalt, copper, tin, and zinc occur in crude oils. Vanadium

Table 11.1. Effect of Temperature and Pressure on the Density and API Gravity of Various Residua

Source of Residuum	Property	Temperature °C	Temperature °F	Pressure: psi / atm / MPa	14.21 / 0.97 / 0.098	2843 / 193 / 19.6	5685 / 387 / 39.2	8528 / 580 / 58.8	11,371 / 774 / 78.4	14,214 / 967 / 98.0
California	Density (g/cm^3)	25	77		1.014	1.023	1.031	1.038	1.045	1.051
	API gravity				8.0	6.8	5.7	4.8	3.9	3.3
	Density (g/cm^3)	45	113		1.002	1.011	1.020	1.028	1.035	1.041
	API gravity				9.7	8.5	7.2	6.1	5.2	4.4
	Density (g/cm^3)	65	149		0.990	1.000	1.009	1.017	1.025	1.032
	API gravity				11.4	10.0	8.7	7.6	6.6	5.6
Venezuela	Density (g/cm^3)	25	77		1.024	1.032	1.040	1.048	1.054	1.061
	API gravity				6.7	5.6	4.6	3.5	2.7	1.9
	Density (g/cm^3)	45	113		1.012	1.020	1.029	1.037	1.044	1.051
	API gravity				8.3	7.2	6.0	5.0	4.0	3.1
	Density (g/cm^3)	65	149		1.000	1.009	1.018	1.027	1.034	1.041
	API gravity				10.0	8.7	7.5	6.3	5.3	4.4

and nickel are bound in organic complexes, and by virtue of the concentration (distillation) process by which asphalt is manufactured, are also found in asphalt. The catalytic behavior of vanadium has prompted studies of the relation between vanadium content and sensitivity to oxidation (viscosity ratio).

Thus, elemental analysis is still of considerable value to determine the amounts of elements, and the method chosen for the analysis may be subject to the peculiarities or character of the material under investigation and should be assessed in terms of accuracy and reproducibility. The methods that are designated for elemental analysis are:

1. *Carbon* and *hydrogen content* (ASTM D1018, D3178, D3343, D3701, D5291, E777; IP 338)
2. *Nitrogen content* (ASTM D3179, D3228, D3431, E148, ASTM E258, D5291, E778)
3. *Oxygen content* (ASTM E385)
4. *Sulfur content* (ASTM D124, D129, D139, D1266, D1552, D1757, D2622, D2785, D3120, D3177, D4045, D4294, ASTM E443; IP 30, 61, 103, 104, 107, 154, 243).

The determination of *nitrogen* has been performed regularly by the Kjeldahl method (ASTM D3228), the Dumas method, and the microcoulometric (ASTM D3431) method. The chemiluminescence method, the most recent technique applied to nitrogen analysis for petroleum, determines the amount of chemically bound nitrogen in liquid hydrocarbon samples. In the method, the samples are introduced to the oxygen-rich atmosphere of a pyrolysis tube maintained at 975°C (1785°F). Nitrogen in the sample is converted to nitric oxide during combustion, and the combustion products are dried by passage through magnesium perchlorate [$Mg(ClO_4)_2$] before entering the reaction chamber of a chemiluminescence detector.

Oxygen is one of the five (C, H, N, O, and S) major elements in resids and asphalt, although the level rarely exceeds 1.5% by weight. Many petroleum products do not specify a particular oxygen content, but if the oxygen compounds are present as acidic compounds such a phenols (Ar–OH) and naphthenic acids (cycloalkyl–COOH), they are controlled in different specifications by a variety of tests.

11.1.7. Float Test

The float test is used to determine the consistency of asphalt at a specified temperature. One test method (ASTM D139) is normally used for asphalt that is too soft for the penetration test (ASTM D5, D217, D937, D1403; IP 50, 179, 310).

11.1.8. Softening Point

The softening point of residua and asphalt is the temperature at which asphalt attains a particular degree of softness under specified conditions of test.

Resids and asphalt do not go through a solid–liquid phase change when heated and therefore do not have a true melting point. As the temperature is raised, the material gradually softens or becomes less viscous. For this reason, determination of the softening point must be made by an arbitrary but closely defined method if the test values are to be reproducible. Softening-point determination is useful in determining the consistency as one element in establishing the uniformity of shipments or sources of supply.

Several tests are available to determine the softening point (ASTM D36, D61, D2319, D3104, D3461; IP 58). In the test method (ASTM D36; IP 58), a steel ball of specified weight is laid on a layer of sample contained in a ring of specified dimensions. The softening point is the temperature, during heating under specified conditions, at which the material surrounding the ball deforms and contacts a base plate.

11.1.9. Viscosity

Viscosity is a measure of flow characteristics and is generally the most important controlling property for movement of resids from one unit to another and for selection of asphalt to meet a particular application. A number of instruments are in common use for this purpose. The vacuum capillary (ASTM D2171) is commonly used to classify paving asphalt at 60°C (140°F). Kinematic capillary instruments (ASTM D2170, D4402) are commonly used in the temperature range 60 to 135°C (140 to 275°F) for both liquid and semisolid materials in the range 30 to 100,000 cSt. Saybolt tests (ASTM D88) are also used in this temperature range and at higher temperature (ASTM E102). At lower temperatures the cone and plate instrument (ASTM D3205) has been used extensively in the viscosity range 1000 to 1,000,000 poise. Other techniques include use of the sliding plate microviscometer and the rheogoniometer.

11.1.10. Weathering

This test (ASTM D529) evaluates the relative weather resistance of asphalt used for protective-coating applications, especially for roofing. No direct measure of outdoor life or service can be obtained from this test. Methods for preparing test panels (ASTM D1669) and failure-endpoint testing (ASTM D1670) are available.

11.2. COKE

Coke does not offer the same potential environmental issues as other petroleum products (Chapter 10 and above). It is used predominantly as a refinery fuel unless other uses for the production of a high-grade coke or carbon are desired. In the former case, the constituents of the coke that will release environmentally harmful gases such as nitrogen oxides, sulfur oxides, and particulate matter should be known. In addition, stockpiling coke on a site where it awaits use or transportation can lead to leachates as a result of rainfall (or acid rainfall) which are highly detrimental. In such a case, application of the toxicity characteristic leaching procedure to coke (TCLP, EPA SW-846 Method 1311), which is designed to determine the mobility of both organic and inorganic contaminants present in materials such as coke, is warranted before stockpiling the coke in the open is warranted.

Petroleum coke is the residue left by the destructive distillation of petroleum residua in processes such as the delayed coking process (Figure 11.4). That formed in catalytic cracking operations is usually nonrecoverable, as it is often employed as fuel for the process.

Coke is a gray-to-black solid carbonaceous residue that is produced from petroleum during thermal processing; it is characterized by having a high carbon content (95%+ by weight) and a honeycomb type of appearance and is insoluble in organic solvents (ASTM D121) (Chapter 2) (Gruse and Stevens, 1960; Guthrie, 1967; Weissermel and Arpe, 1978; Hoffman, 1983; Austin, 1984; Chenier, 1992; Hoffman and McKetta, 1993; Speight, 1992, 1999; Speight and Ozum, 2002).

Coke occurs in various forms and the terminology reflects the type of coke (Tables 11.2 and 11.3) that can influence behavior in the environment. But no matter what the form, coke usually consists mainly of carbon (greater than 90% but usually greater than 95%) and has a low mineral matter content (determined as ash residue). Coke is used as a feedstock in coke ovens for the steel industry, for heating purposes, for electrode manufacture, and for production of chemicals. The two most important qualities are *green coke* and *calcined coke*. The latter category includes *catalyst coke* deposited on the catalyst during refining processes: this coke is not recoverable and is usually burned as refinery fuel.

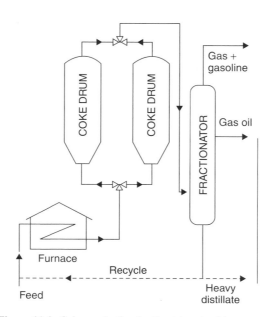

Figure 11.4. Coke production by the delayed coking process.

Table 11.2. Description of Delayed Coke Carbon Forms

Needle coke	Ribbonlike parallel-ordered anisotropic domains that can also occur as folded structures
Lenticular/granular	Lenticular anisotropic domains of various sizes that are not aligned parallel to the particle surface
Mixed layer	Ribbon and lenticular anisotropic domains of various sizes in curved and irregular layered arrangements
Sponge	Porous microstructure with walls that are generally anisotropic but with pores and walls that vary in size
Shot	Ribbon and lenticular anisotropic domains arranged in concentric patterns to form shotlike coke
Amorphous	Isotropic carbon form closely associated with parent liquor; higher in volatile matter than incipient mesophase
Incipient Mesophase	Initial stage of mesophase formation; transition stage between amorphous and mesophase
Mesophase	Nemitic liquid crystals; lower in volatile matter than incipient mesophase

Table 11.3. Description of Fluid Coke Carbon Forms

Layered	Anisotropic carbon domains aligned in concentric layers parallel to the particle surface similar to an onionlike pattern
NonLayered	Anisotropic domains are not aligned parallel to the particle surface
Aggregates	Fragments of anisotropic domains
Amorphous	Isotropic carbon form closely associated with parent liquor; higher in volatile matter than incipient mesophase
Incipient Mesophase	Initial stage of mesophase formation; transition stage between amorphous and mesophase
Mesophase	Nemitic liquid crystals; lower in volatile matter than incipient mesophase

The test methods for coke are necessary for defining the coke as a fuel (for internal use in a refinery) or for other uses, particularly those test methods where prior sale of the coke is involved. Specifications are often dictated by environmental regulations if not by the purchaser of the coke.

The test methods outlined below are the methods that are usually applied to petroleum coke but should not be thought of as the only test methods. In fact, there are many test methods for coke (ASTM, 2004, Vol. 05.06), and these should be consulted either when more detail is required or when a fuller review is required.

11.2.1. Ash

The ash content (i.e., the ash yield, which is related to the mineral matter content; see also Section 11.2.6) is one of the properties used to evaluate coke and indicates the amount of undesirable residue present. Some sample of coke may be declared to have an acceptable ash content, but this varies with the intended use of the coke.

For the test method, the preparation and sampling of the analytical sample must neither remove nor add mineral matter (ASTM D346). Improper dividing, sieving, and crushing equipment, and some muffle furnace-lining material can contaminate the coke and lead to erroneous results. In addition, a high sulfur content of the furnace gases, regardless of the source of the sulfur, can react with an alkaline ash to produce erratic results. To counteract such an effect, the furnace should be swept with air.

In the test method (ASTM D4422), a sample of petroleum coke is dried, ground, and ashed in a muffle furnace at 700 to 775°C (1292 to 1427°F). The noncarbonaceous residue is weighed and reported as the percent by weight ash. As already noted, the ash must not be understood to be the same as the mineral content of the petroleum coke.

In addition, ashing procedures can be used as a preliminary step for determination of the trace elements in coke, and by inference, in the higher-boiling fractions of the crude oil. Among the techniques used for trace-element determinations are flameless and flame atomic absorption (AA) spectrophotometry (ASTM D2788, D5863) and inductively coupled argon plasma (ICP) spectrophotometry (ASTM D5708).

Inductively coupled argon plasma emission spectrophotometry (ASTM D5708) has an advantage over atomic absorption spectrophotometry (ASTM D4628, D5863) because it can provide more complete elemental composition data than the atomic absorption method. Flame emission spectroscopy is often used successfully in conjunction with atomic absorption spectrophotometry (ASTM D3605). X-ray fluorescence spectrophotometry (ASTM D4927, D6443) is also sometimes used, but matrix effects can be a problem. The method to be used for the determination of metallic constituents is often a matter of individual preference.

11.2.2. Composition

The composition of petroleum coke varies with the source of the crude oil, but in general, large amounts of high-molecular-weight complex hydrocarbons (rich in carbon but correspondingly poor in hydrogen) make up a high proportion. The solubility of petroleum *coke* in carbon disulfide has been reported to be as high as 50 to 80%, but this is in fact a misnomer, since the coke is the insoluble, honeycomb material that is the end product of thermal processes.

Carbon and hydrogen in coke can be determined by the standard analytical procedures for coal and coke (ASTM D3178, D3179). However, in addition to carbon, hydrogen, and metallic constituents, coke contains considerable amounts of nitrogen and sulfur that must be determined prior to sale or use. These elements will appear as their oxides (NO_x, SO_x), respectively, when the coke is combusted, thereby causing serious environmental issues.

A test method (ASTM D5291) is available for simultaneous determination of carbon, hydrogen, and nitrogen in petroleum products and lubricants. There are at least three instrumental techniques available for this analysis, each based on different chemical principles. However, all involve sample combustion, components separation, and final detection.

In one of the variants of the method, a sample is combusted in an oxygen atmosphere, and the product gases are separated from each other by adsorption over chemical agents. The remaining elemental nitrogen gas is measured by a thermal conductivity cell. Carbon and hydrogen are measured separately by selective infrared cells as carbon dioxide and water. In another variant of the method, a sample is combusted in an oxygen atmosphere, and the product gases are separated from each other, and the three gases of interest are measured by gas chromatography. In the third variant of the method, a sample is combusted in an oxygen atmosphere, and the product gases are cleaned by passage over chemical agents, and the three gases of interest are separated chromatographically and measured with a thermal conductivity detector.

The nitrogen method is not applicable too samples containing <0.75% by weight nitrogen or for the analysis of volatile materials such as gasoline, gasoline oxygenate blends, or aviation turbine fuels. The details of the method should be consulted along with those given in an alternative method for the determination of carbon, hydrogen, and nitrogen in coal and coke (ASTM D3179, D5373).

A test method (ASTM D1552) is available for sulfur analysis, and the method covers three procedures applicable to samples boiling above 177°C (350°F) and containing not less than 0.06 mass % sulfur. Thus, the method is applicable to most fuel oils, lubricating oils, residua, and coke, and coke containing up to 8% by weight sulfur can be analyzed. This is particularly important for cokes that originate from heavy oil and tar sand bitumen, where the sulfur content of the coke is usually at least 5% by weight.

In the iodate detection system (ASTM D1552), the sample is burned in a stream of oxygen at a sufficiently high temperature to convert about 97% by weight of the sulfur to sulfur dioxide. The combustion products are passed into an absorber containing an acidic solution of potassium iodide and starch indicator. A faint blue color is developed in the absorber solution by the addition of standard potassium iodate solution. As combustion proceeds, bleaching the blue color, more iodate is added. The sulfur content of the sample is calculated from the amount of standard iodate consumed during the combustion.

In the infrared detection system, the sample is weighed into a special ceramic boat which is then placed into a combustion furnace at 1371°C (2500°F) in an oxygen atmosphere. Most of the sulfur present is converted to sulfur dioxide, which is then measured with an infrared detector after moisture and dust are removed by traps. The calibration factor is determined using standards approximating the material to be analyzed.

For the iodate method, chlorine in concentrations below 1 mass % does not interfere. The isoprene rubber method can tolerate somewhat higher levels. Nitrogen when present above 0.1 mass % may interfere with the iodate method; the extent depending on the types of nitrogen compounds as well as the combustion conditions. It does not interfere in the infrared method. The alkali and alkaline earth metals, zinc, potassium, and lead do not interfere with either method.

Determination of the physical composition can be achieved by any of the test methods for determining the toluene-insoluble constituents of tar and pitch (ASTM D4072, D4312). Furthermore, a variety of sample can be employed to give a gradation of soluble and insoluble fractions. The coke, of course, remains in the extraction thimble (Soxhlet apparatus) and the extracts are freed from the solvent and weight to give percent by weight yield(s).

Finally, one aspect that can pay a role in compositional studies is the sieve (screening) analysis. Like all petroleum products, sampling is, or can be, a major issue. If not performed correctly and poor sampling is the result, erroneous and very misleading data can be produced by the analytical method of choice. For this reason, reference is made to standard procedures such as the *Standard Practice for Collection and Preparation of Coke Samples for Laboratory Analysis* (ASTM D346) and the *Standards Test Method for the Sieve Analysis of Coke* (ASTM D293).

11.2.3. Density

The *density* (*specific gravity*) of coke has a strong influence on future use and can affect the characteristics of the products such as carbon and graphite. The density (specific gravity) of coke can be measured conveniently by use of a pycnometer. In the test method (ASTM D5004), the mass of the sample is determined directly and the volume is derived by determining the mass of liquid displaced when the sample is introduced into a pycnometer. Oil or other material sprayed on calcined petroleum coke to control dust will interfere. Such oil can be removed by flushing with a solvent, which must be completely removed before the density determination.

The *real density* of coke is obtained when the particle size of the specimen is smaller than 75 mm. The real density (or the particle size) exerts a direct influence on the physical and chemical properties of the carbon and graphite products that are manufactured from the coke.

In the test method (ASTM D2638), a sample is dried and ground to pass a 75-mm screen. The mass of the volume is determined directly, and the volume derived by the volume of helium displaced when the sample is introduced into a helium pycnometer. The ratio of the mass of the sample to the volume is reported as the real density.

The *vibrated bulk density* (VBD) is an indication of the porosity of calcined petroleum coke, which affects its suitability for use in pitch-bonded carbon applications. This property is strongly dependent on average particle size and range and tends to increase with decreasing coke size. In the test method (ASTM D4292), the coke is crushed, 100 g is measured after vibration, and the bulk density is calculated. The procedure is limited to particles passing through a 6.68-mm-opening sieve and retained on a 0.21-mm-opening sieve.

11.2.4. Dust Control Material

Dust control material is applied to calcined coke to help maintain a dust-free environment. It adds weight to the coke and can have a negative effect on the quality of carbon and graphite artifacts made from the treated coke. Hence, a maximum amount may be specified.

In the test method (ASTM D4930), a weighed dry representative coke sample 6.3 mm in maximum size is extracted using methylene chloride in a Soxhlet apparatus. The mass of the residue remaining after extraction and evaporation of the solvent is the mass of the dust control material. This test method is limited to those materials that are soluble in a solvent (e.g., methylene chloride) that can be used in a Soxhlet extraction type of apparatus. Toluene and methyl chloroform have also been found to give results equal to those of methylene chloride.

11.2.5. Hardness

The Hardgrove grindability index (HGI) (ASTM D5003) is used to predict the ranking in industrial-size mills used for crushing operations and is used commonly to determine the hardness of coal samples (ASTM D409) (Speight, 1994). The rankings are based on energy required and feed rate or both. With the introduction in the coal market of petroleum coke, this test method has been extended to the coke. In the current context, the Hardgrove grindability index is also used to select raw petroleum coke and coals that are compatible with each other when milled together in a blend so that segregation of the blend does not occur during particle-size reduction.

In the test method (ASTM D5003), the coke sample is crushed to produce a high yield of particles passing a No. 16 sieve and retained on a No. 30 sieve. These particles are reduced in the Hardgrove grindability machine according to the test method for coal (ASTM D409). The quantity of particles retained on a No. 200 sieve is used to calculate the Hardgrove grindability index of the sample. Both this test method and the test method for coal (ASTM D409) produce the same results on petroleum coke samples.

11.2.6. Metals

The presence and concentration of various metallic elements in petroleum coke are major factors in the suitability of the coke for various uses. In the test method (ASTM D5056), a sample of petroleum coke is ashed (thermally decomposed to leave only the ash of the inorganic constituents) at $525°C$ ($977°F$). The ash is fused with lithium tetraborate or lithium metaborate. The melt is then dissolved in dilute hydrochloric acid and the resulting solution is analyzed by atomic absorption spectroscopy to determine the metals in the sample. However, spectral interferences may occur when using wavelengths other than those recommended for analysis or when using multielement hollow cathode lamps.

This test method can be used in the commercial transfer of petroleum coke to determine whether that lot of coke meets specifications. This method can analyze raw and calcined coke for the trace elements aluminum, calcium, iron, nickel, silicon, sodium, and vanadium. The inductively coupled plasma atomic emission spectroscopy (ICPAES) method (ASTM DD 5600, D6357) is complementary to this method and can also be used for the determination of metals in petroleum coke.

In the inductively coupled plasma atomic emission spectroscopy (ICPAES) method (ASTM DD 5600), a sample of petroleum coke is ashed at 700°C (1292°F) and the ash is fused with lithium borate. The melt is dissolved in dilute hydrochloric acid, and the resulting solution is analyzed by inductively coupled plasma atomic emission spectroscopy using aqueous calibration standards. Because of the need to fuse the ash with lithium borate or other suitable salt, the fusibility of ash may need attention (ASTM D1857).

The wavelength-dispersive x-ray spectroscopy method (ASTM D6376) provides a rapid means of measuring metallic elements in coke and provides a guide for determining conformance to material specifications. A benefit of this method is that the sulfur content can also be used to evaluate potential formation of sulfur oxides, a source of atmospheric pollution. This test method specifically determines sodium, aluminum, silicon, sulfur, calcium, titanium, vanadium, manganese, iron, and nickel.

In the method, a weighed portion of a sample of coke dried at 110°C (230°F) and crushed to pass a No. 200-mesh sieve, mixed with stearic acid, and then milled and compressed into a smooth pellet. The pellet is irradiated with an x-ray beam and the characteristic x-rays of the elements analyzed are excited, separated, and detected by the spectrometer. The measured x-ray intensities are converted to elemental concentration by using a calibration equation derived from the analysis of the standard materials. The K_α spectral lines are used for all the elements determined by this test method. This test method is also applicable to the determination of additional elements provided that appropriate standards are available for use and comparison.

11.2.7. Sulfur

In addition to metallic constituents, coke contains considerable amounts of sulfur (ASTM D1552, D3177, D4239) that must be determined prior to sale or use. A test method (ASTM D1552) available for sulfur analysis covers three procedures applicable to samples boiling above 177°C (350°F) and containing not less than 0.06 mass % sulfur. Thus, the method is applicable to most fuel oils, lubricating oils, residua, and coke, and coke containing up to 8% by weight sulfur can be analyzed. This is particularly important for cokes that originate from heavy oil and tar sand bitumen, where the sulfur content of the coke is usually at least 5% by weight.

In the iodate detection system (ASTM D1552), the sample is burned in a stream of oxygen at a sufficiently high temperature to convert about 97% by weight of the sulfur to sulfur dioxide. The combustion products are passed into an absorber containing an acidic solution of potassium iodide and starch indicator. A faint blue color is developed in the absorber solution by the addition of standard potassium iodate solution. As combustion proceeds, bleaching the blue color, more iodate is added. The sulfur content of the sample is calculated from the amount of standard iodate consumed during the combustion.

In the infrared detection system, the sample is weighed into a special ceramic boat which is then placed into a combustion furnace at 1371°C (2500°F) in an oxygen atmosphere. Most of the sulfur present is converted to sulfur dioxide, which is then measured with an infrared detector after moisture and dust are removed by traps. The calibration factor is determined using standards approximating the material to be analyzed.

For the iodate method, chlorine in concentrations below 1 mass % does not interfere. The isoprene rubber method can tolerate somewhat higher levels. Nitrogen when present above 0.1 mass % may interfere with the iodate method, the extent depending on the types of nitrogen compounds as well as the combustion conditions. It does not interfere in the infrared method. The alkali and alkaline earth metals, zinc, potassium, and lead do not interfere with either method.

REFERENCES

ASTM, 2004. *Annual Book of ASTM Standards*. American Society for Testing and Materials, West Conshohocken, PA.

Austin, G. T. 1984. *Shreve's Chemical Process Industries*, 5th ed. McGraw-Hill, New York, Chap. 37.

Broome, D. C., and Wadelin, F. A. 1973. In *Criteria for Quality of Petroleum Products*, J. P. Allinson (Ed.). Wiley, New York, Chap. 10.

Chenier, P. J. 1992. *Survey of Industrial Chemistry*, 2nd rev. ed. VCH Publishers, New York, Chap. 7.

EPA. 2004. Environmental Protection Agency, Washington, DC. Web site: http://www.epa.gov.

Gruse, W. A., and Stevens, D. R. 1960. *Chemical Technology of Petroleum*. McGraw-Hill, New York, Chap. 15.

Guthrie, V. B. 1967. In *Petroleum Processing Handbook*, W. F. Bland and R. L. Davidson (Eds.). McGraw-Hill, New York, Sec. 11.

Hoffman, H. L. 1983. In *Riegel's Handbook of Industrial Chemistry*, 8th ed., J. A. Kent (Ed.). Van Nostrand Reinhold, New York, Chap. 14.

Hoffman, H. L., and McKetta, J. J. 1993. In *Chemical Processing Handbook*, J. J. McKetta (Ed.). Marcel Dekker, New York, p. 851.

IP. 2001. *IP Standard Methods 2001*. Institute of Petroleum, London.

Speight, J. G. 1992. In *Kirk–Othmer Encyclopedia of Chemical Technology*, 4th ed., Vol. 3, p. 689.

Speight, J. G. 1994. In *Asphaltenes and Asphalts*, I. *Developments in Petroleum Science*, 40, T. F. Yen and G. V. Chilingarian (Eds.), Elsevier, Amsterdam, Chap. 2.

Speight, J. G. 1999. *The Chemistry and Technology of Petroleum*, 3rd ed. Marcel Dekker, New York.

Speight, J. G. 2000. *The Desulfurization of Heavy Oils and Residua*, 2nd ed. Marcel Dekker, New York.

Speight, J. G. 2001. *Handbook of Petroleum Analysis*. Wiley, New York.

Speight, J. G., and Ozum, B. 2002. *Petroleum Refining Processes*. Marcel Dekker, New York.

Van Gooswilligen, G. 2000. In *Modern Petroleum Technology*, Vol. 2, *Downstream*, A. G. Lucas (Ed.). Wiley, New York, Chap. 32.

Weissermel, K., and Arpe, H.-J. 1978. *Industrial Organic Chemistry*. Verlag Chemie, New York, Chap. 13.

CHAPTER 12

POLLUTION PREVENTION

Having defined the process products and emission (Chapters 3 and 4), *pollution prevention* is the operational guideline for refinery operators, process engineers, process chemists, and for that matter, anyone who handles petroleum and/or petroleum products. It is in this area that environmental analysis plays a major role (EPA, 2004).

Pollution prevention is, simply, reduction or elimination of discharges or emissions to the environment. The limits of pollutants emitted to the atmosphere, the land, and water are defined by various pieces of legislation that have been put into place over the past four decades (Chapter 5) (Speight, 1996; Woodside, 1999). This includes all pollutants, such as hazardous and nonhazardous wastes and regulated and unregulated chemicals from all sources.

Pollution associated with petroleum refining typically includes volatile organic compounds (volatile organic compounds), carbon monoxide (CO), sulfur oxides (SO_x), nitrogen oxides (NO_x), particulates, ammonia (NH_3), hydrogen sulfide (H_2S), metals, spent acids, and numerous toxic organic compounds (Hydrocarbon Processing, 2003). Sulfur and metals result from the impurities in crude oil. The other wastes represent losses of feedstock and petroleum products.

These pollutants may be discharged as air emissions, wastewater, or solid waste. All of these wastes are treated. However, air emissions are more difficult to capture than wastewater or solid waste. Thus, air emissions are the largest source of untreated wastes released to the environment.

Pollution prevention can be accomplished by reducing the generation of wastes at their source (source reduction) or by using, reusing, or reclaiming wastes once they are generated (environmentally sound recycling). However, environmental analysis plays a major role in determining if emissions–effluents (air, liquid, or solid) fall within the parameters of the relevant legislation. For example, issues to be addressed are the constituents of gaseous emissions, the sulfur content of liquid fuels, and the potential for leaching contaminants (through normal rainfall or through the agency of acid rain) from solid products such as coke.

The purpose of this chapter is to present a description of the methods by which petroleum products–effluents–emissions are treated in an attempt to ensure that pollution does not occur and products–effluents–emissions fall within legislative specifications. Indeed, as already noted, environmental analysis is the major

---

Environmental Analysis and Technology for the Refining Industry, by James G. Speight
Copyright © 2005 John Wiley & Sons, Inc.

discipline by which the character of the products–effluents–emissions can be determined, and hence monitored.

12.1. REFINERY WASTES AND TREATMENT

Waste elimination is common sense and provides several obvious benefits:

1. Solves the waste disposal problems created by land bans
2. Reduces waste disposal costs
3. Reduces costs for energy, water, and raw materials
4. Reduces operating costs
5. Protects workers, the public, and the environment
6. Reduces risk of spills, accidents, and emergencies
7. Reduces vulnerability to lawsuits and improves its public image
8. Generates income from wastes that can be sold

Yet waste elimination continues to elude many companies in every sector, including the refinery section and activity from refinery waste (which is a function of their production system design). It may not matter how a refiner categorizes the waste or how the refiner chooses to pursue waste elimination. One thing remains constant: that once identified, waste can be eliminated. There are models and structures that allow a refiner to identify and eliminate waste to increase productivity, and hence cost structures, that have a direct impact on refinery operations and, more than all else, profitability. Waste elimination through identification (by judicious analysis) and treatment subscribe to the smooth operation of a refinery.

Generally, process wastes (emissions) are categorized as gaseous, liquid, and solid. This does not usually include waste from or from accidental spillage of a petroleum feedstock or product.

12.1.1. Air Emissions

Air emissions include point and nonpoint sources (Chapter 4). Point sources are emissions that exit stacks and flares and thus can be monitored and treated. Nonpoint sources are *fugitive emissions* that are difficult to locate and capture. Fugitive emissions occur throughout refineries and arise from the thousands of valves, pumps, tanks, pressure relief valves, flanges, and so on. Although individual leaks are typically small, the sum of all fugitive leaks at a refinery can be one of its largest emission sources.

The numerous process heaters used in refineries to heat process streams or to generate steam (boilers) for heating or steam stripping can be potential sources of SO_x, NO_x, CO, particulates, and hydrocarbon emissions. When operating properly and when burning cleaner fuels, such as refinery fuel gas, fuel oil, or natural gas, these emissions are relatively low. If, however, combustion is not complete, or heaters are fired with refinery fuel pitch or residuals, emissions can be significant.

The majority of gas streams exiting each refinery process contain varying amounts of refinery fuel gas, hydrogen sulfide, and ammonia. These streams are collected and sent to the gas treatment and sulfur recovery units to recover the refinery fuel gas and sulfur through a variety of add-on technologies (Speight, 1993, 1996). Emissions from the sulfur recovery unit typically contain some hydrogen sulfide, sulfur oxides, and nitrogen oxides. Other emission sources from refinery processes arise from periodic regeneration of catalysts. These processes generate streams that may contain relatively high levels of carbon monoxide, particulates, and volatile organic compounds. Before being discharged to the atmosphere, such off-gas streams may be treated first through a carbon monoxide boiler to burn carbon monoxide and any volatile organic compounds, and then through an electrostatic precipitator or cyclone separator to remove particulates.

Sulfur is removed from a number of refinery process off-gas streams (sour gas) to meet the sulfur oxide emissions limits of the Clean Air Act and to recover salable elemental sulfur. Process off-gas streams, or sour gas, from the coker, catalytic cracking unit, hydrotreating units, and hydroprocessing units can contain high concentrations of hydrogen sulfide mixed with light refinery fuel gases.

Before elemental sulfur can be recovered, the fuel gases (primarily methane and ethane) need to be separated from the hydrogen sulfide. This is typically accomplished by dissolving the hydrogen sulfide in a chemical solvent. Solvents most commonly used are amines, such as diethanolamine (DEA, $HOCH_2CH_2NHCH_2CH_2OH$). Dry adsorbents such as molecular sieves, activated carbon, iron sponge (Fe_2O_3), and zinc oxide (ZnO) are also used (Speight, 1993). In the amine solvent processes, diethanolamine solution or similar ethanolamine solution is pumped to an absorption tower, where the gases are contacted and hydrogen sulfide is dissolved in the solution. The fuel gases are removed for use as fuel in process furnaces in other refinery operations. The amine–hydrogen sulfide solution is then heated and steam stripped to remove the hydrogen sulfide gas.

Current methods for removing sulfur from the hydrogen sulfide gas streams are typically a combination of two processes, in which the primary process is the Claus process, followed by the Beaven process, the SCOT process, or the Wellman–Lord process.

In the Claus process (Figure 12.1), after separation from the gas stream using amine extraction, the hydrogen sulfide is fed to the Claus unit, where it is converted in two stages. The first stage is a thermal step, in which the hydrogen sulfide is partially oxidized with air in a reaction furnace at high temperatures [1000 to 1400°C (1830 to 2550°F)]. Sulfur is formed, but some hydrogen sulfide remains unreacted and some sulfur dioxide is produced. The second stage is a catalytic stage in which the remaining hydrogen sulfide is reacted with the sulfur dioxide at lower temperatures [200 to 350°C (390 to 660°F)] over a catalyst to produce more sulfur. The overall reaction is the conversion of hydrogen sulfide and sulfur dioxide to sulfur and water:

$$2H_2S + SO_2 \rightleftharpoons 3S + 2H_2O$$

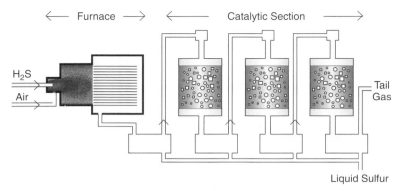

Figure 12.1. Claus process (Speight, 1993, 1999, and references cited therein).

The catalyst is necessary to ensure that the components react with reasonable speed, but unfortunately, the reaction does not always proceed to completion. For this reason, two or three stages are used, with sulfur being removed between the stages. For analysts it is valuable to know that carbon disulfide (CS_2) is a by-product from the reaction in the high-temperature furnace. The carbon disulfide can be destroyed catalytically before it enters the catalytic section proper. Generally, the Claus process may remove only about 90% of the hydrogen sulfide in the gas stream, and as already noted, other processes, such as the Beaven process, SCOT process, or Wellman–Lord process are often used to recover additional sulfur.

In the *Beaven process*, the hydrogen sulfide in the relatively low concentration gas stream from the Claus process can be almost completely removed by absorption in a quinone solution. The dissolved hydrogen sulfide is oxidized to form a mixture of elemental sulfur and hydroquinone. The solution is injected with air or oxygen to oxidize the hydroquinone back to quinone. The solution is then filtered or centrifuged to remove the sulfur, and the quinone is then reused. The Beaven process is also effective in removing small amounts of sulfur dioxide, carbonyl sulfide, and carbon disulfide that are not affected by the Claus process. These compounds are first converted to hydrogen sulfide at elevated temperatures in a cobalt molybdate catalyst prior to being fed to the Beaven unit. Air emissions from sulfur recovery units will consist of hydrogen sulfide, sulfur oxides, and nitrogen oxides in the process tail gas as well as fugitive emissions and releases from vents.

The *SCOT process* is also widely used for removing sulfur from the Claus tail gas. The sulfur compounds in the Claus tail gas are converted to hydrogen sulfide by heating and passing it through a cobalt–molybdenum catalyst with the addition of a reducing gas. The gas is then cooled and contacted with a solution of diisopropanolamine (DIPA) that removes all but trace amounts of hydrogen sulfide. The sulfide-rich diisopropanolamine is sent to a stripper, where hydrogen sulfide gas is removed and sent to the Claus plant. The diisopropanolamine is returned to the absorption column.

The *Wellman–Lord process* is divided into two main stages: (1) absorption and (2) regeneration. In the absorption section, hot flue gases are passed through a prescrubber where ash, hydrogen chloride, hydrogen fluoride, and sulfur trioxide are removed. The gases are then cooled and fed into the absorption tower. A saturated solution of sodium sulfite is then sprayed into the top of the absorber onto the flue gases; the sodium sulfite reacts with the sulfur dioxide, forming sodium bisulfite. The concentrated bisulfate solution is collected and passed to an evaporation system for regeneration. In the regeneration section, sodium bisulfite is converted, using steam, to sodium sulfite, which is recycled back to the flue gas. The remaining product, the released sulfur dioxide, is converted to elemental sulfur, sulfuric acid, or liquid sulfur dioxide.

Most refinery process units and equipment are sent into a collection unit called a *blowdown system*. Blowdown systems provide for the safe handling and disposal of liquid and gases which are either vented automatically from the process units through pressure relief valves, or are drawn from units manually. Recirculated process streams and cooling-water streams are often purged manually to prevent the continued buildup of contaminants in the stream. Part or all of the contents of equipment can also be purged to the blowdown system prior to shutdown before normal or emergency shutdowns. Blowdown systems utilize a series of flash drums and condensers to separate the blowdown into its vapor and liquid components. Typically, the liquid is composed of mixtures of water and hydrocarbons containing sulfides, ammonia, and other contaminants, which are sent to the wastewater treatment plant. The gaseous component typically contains hydrocarbons, hydrogen sulfide, ammonia, mercaptans, solvents, and other constituents, and is either discharged directly to the atmosphere or is combusted in a flare. The major air emissions from blowdown systems are hydrocarbons in the case of direct discharge to the atmosphere and sulfur oxides when flared.

12.1.2. Wastewater and Treatment

Wastewaters from petroleum refining consist of process water, cooling water, stormwater, and sanitary sewage water (Chapter 4). Water used in processing operations accounts for a significant portion of the total wastewater. Process wastewater arises from desalting crude oil, steam-stripping operations, pump gland cooling, product fractionator reflux drum drains, and boiler blowdown. Because process water often comes into direct contact with oil, it is usually highly contaminated. Most cooling water is recycled over and over. Cooling water typically does not come into direct contact with process oil streams and therefore contains less contaminants than process wastewater. However, it may contain some oil contamination, due to leaks in the process equipment. Stormwater (i.e., surface water runoff) is intermittent and will contain constituents from spills to the surface, leaks in equipment, and any materials that may have collected in drains. Runoff surface water also includes water coming from crude and product storage tank roof drains. Sewage water needs no further explanation of its origins but must be treated as opposed to discharge onto the land or into ponds.

Wastewater is treated in on-site wastewater treatment facilities and then discharged to publicly owned treatment works (POTWs) or discharged to surface waters under National Pollution Discharge Elimination System (NPDES) permits. Petroleum refineries typically utilize primary and secondary wastewater treatment.

Primary wastewater treatment consists of the separation of oil, water, and solids in two stages. During the first stage, an API separator, a corrugated plate interceptor, or other separator design is used. Wastewater moves very slowly through the separator, allowing free oil to float to the surface and be skimmed off and solids to settle to the bottom and be scraped off to a sludge collection hopper. The second stage utilizes physical or chemical methods to separate emulsified oils from the wastewater. Physical methods may include the use of a series of settling ponds with a long retention time, or the use of dissolved air flotation (DAF). In DAF, air is bubbled through the wastewater, and both oil and suspended solids are skimmed off the top. Chemicals such as ferric hydroxide or aluminum hydroxide can be used to coagulate impurities into a froth or sludge that can be more easily skimmed off the top. Some wastes associated with the primary treatment of wastewater at petroleum refineries may be considered hazardous and include API separator sludge, primary treatment sludge, sludge from other gravitational separation techniques, float from DAF units, and wastes from settling ponds.

After primary treatment, the wastewater can be discharged to a publicly owned treatment works (POTW) or undergo *secondary treatment* before being discharged directly to surface waters under a National Pollution Discharge Elimination System (NPDES) permit. In secondary treatment, microorganisms may consume dissolved oil and other organic pollutants biologically. Biological treatment may require the addition of oxygen through a number of different techniques, including activated sludge units, trickling filters, and rotating biological contactors. Secondary treatment generates biomass waste that is typically treated anaerobically and then dewatered.

Some refineries employ an additional stage of wastewater treatment called *polishing* to meet discharge limits. The polishing step can involve the use of activated carbon, anthracite coal, or sand to filter out any remaining impurities, such as biomass, silt, trace metals, and other inorganic chemicals, as well as any remaining organic chemicals.

Certain refinery wastewater streams are treated separately, prior to the wastewater treatment plant, to remove contaminants that would not easily be treated after mixing with other wastewater. One such waste stream is the sour water drained from distillation reflux drums. Sour water contains dissolved hydrogen sulfide and other organic sulfur compounds and ammonia, which are stripped in a tower with gas or steam before being discharged to the wastewater treatment plant.

Wastewater treatment plants are a significant source of refinery air emissions and solid wastes. Air releases arise from fugitive emissions from the numerous tanks, ponds, and sewer system drains. Solid wastes are generated in the form of sludge from a number of the treatment units.

Many refineries unintentionally release, or have unintentionally released in the past, liquid hydrocarbons to groundwater and surface waters. At some refineries, contaminated groundwater has migrated off-site and resulted in continuous *seeps* to surface waters. Although the actual volume of hydrocarbons released in such a manner is relatively small, there is a potential to contaminate large volumes of groundwater and surface water, possibly posing a substantial risk to human health and the environment.

12.1.3. Other Waste and Treatment

Solid wastes are generated from many of the refining processes, petroleum handling operations, and wastewater treatment (Chapter 4). Both hazardous and nonhazardous wastes are generated, treated, and disposed. Solid wastes in a refinery are typically in the form of sludge (including sludge from wastewater treatment), spent process catalysts, filter clay, and incinerator ash. Treatment of these wastes includes incineration, land treating off-site, land filling on-site, land filling off-site, chemical fixation, neutralization, and other treatment methods (Speight, 1996; Woodside, 1999).

A significant portion of the nonpetroleum product outputs of refineries is transported off-site and sold as by-products. These outputs include sulfur, acetic acid, phosphoric acid, and recovered metals. Metals from catalysts and crude oil that have deposited on the catalyst during the production often are recovered by third-party recovery facilities.

Storage tanks are used throughout the refining process to store crude oil and intermediate process feeds for cooling and further processing. Finished petroleum products are also kept in storage tanks before transport off-site. Storage tank bottoms are mixtures of iron rust from corrosion, sand, water, and emulsified oil and wax, which accumulate at the bottom of tanks. Liquid tank bottoms (primarily water and oil emulsions) are drawn off periodically to prevent their continued buildup. Tank bottom liquids and sludge are also removed during periodic cleaning of tanks for inspection. Tank bottoms may contain amounts of tetraethyl- or tetramethyllead (although this is increasingly rare, due to the phase-out of leaded products), other metals, and phenols. Solids generated from leaded gasoline storage tank bottoms are listed as a hazardous waste.

12.2. POLLUTION PREVENTION

Pollution prevention is everyone's responsibility. Preventing pollution may be a new role for production-oriented managers and workers, but their cooperation is crucial. It is the workers themselves who must make pollution prevention succeed in the workplace.

Several options have been identified that refineries can undertake to reduce pollution: pollution prevention options, recycling options, and waste treatment options. Furthermore, pollution prevention options are often presented in four categories: (1) pollution prevention options, (2) waste recycling, and (3) waste treatment. Either one or the other or any combination of the three options may be in operation in any given refinery.

12.2.1. Pollution Prevention Options

Pollution prevention options are usually subdivided into four areas: (1) good operating practices, (2) processes modification, (3) feedstock modification, and (4) product reformulation (Lo, 1991). The options described here include only the first three of these categories since product reformulation is not an option that is usually available to the environmental analyst, scientist, or engineer.

Operating Practices

Good operating practices prevent waste by better handling of feedstocks and products without making significant modifications to current production technology:

1. Specify sludge and water content for feedstock.
2. Minimize carryover to API separator.
3. Use recycled water for desalter.
4. Replace desalting with chemical treatment system.
5. Collect catalyst fines during delivery.
6. Recover coke fines.

If feedstocks are handled appropriately, they are less likely to become wastes inadvertently through spills or outdating. If products are handled appropriately, they can be managed in the most cost-effective manner.

For example, a significant portion of refinery waste arises from oily sludge found in combined process/storm sewers. Segregation of the relatively clean rainwater runoff from the process streams can reduce the quantity of oily sludge generated. Furthermore, there is a much higher potential for recovery of oil from smaller, more concentrated process streams.

Solids released to the refinery wastewater sewer system can account for a large portion of a refinery's oily sludge. Solids entering the sewer system (primarily, soil particles) become coated with oil and are deposited as oily sludge in the API oil/water separator. Because a typical sludge has a solids content of 5 to 30% by weight, preventing 1 lb of solids from entering the sewer system can eliminate 3 to 20 lb of oily sludge.

Methods used to control solids include using a street sweeper on paved areas, paving unpaved areas, planting ground cover on unpaved areas, relining sewers, cleaning solids from ditches and catch basins, and reducing heat exchanger bundle cleaning solids by using antifoulants in cooling water. Benzene and other solvents in wastewater can often be treated more easily and effectively at the point at which they are generated rather than at the wastewater treatment plant after it is mixed with other wastewater.

Process Modifications

The petroleum industry requires very large, capital-intensive process equipment. Expected lifetimes of process equipment are measured in decades. This limits economic incentives to make capital-intensive process modifications to reduce wastes generation. Reductions in waste generation can be accomplished by process modifications:

1. Add coking operations. Certain refinery hazardous wastes can then be used as coker feedstock, reducing the quantity of sludge for disposal.
2. Install secondary seals on floating roof tanks. Where appropriate, replace with fixed roofs to eliminate the collection of rainwater, contamination of crude oil or finished products, and oxidation of crude oil.
3. Where feasible:
   a. Replace clay filtration with hydrotreating.
   b. Substitute air coolers or electric heaters for water heat exchangers to reduce sludge production.
   c. Install tank agitators. This can prevent solids from settling out.
   d. Concentrate similar wastewater streams through a common dewatering system.

or by process improvements:

1. Segregate oily wastes to reduce the quantity of oily sludge generated and increase the potential for oil recovery.
2. Reuse rinse waters where possible.
3. Use optimum pressures, temperatures, and mixing ratios.
4. Sweep or vacuum streets and paved process areas to reduce solids going to sewers.
5. Use water softeners in cooling water systems to extend the useful life of the water.

The petroleum industry has made many improvements in the design and modification of processes and technologies to recover product and unconverted raw materials. In the past, they pursued this strategy to the point that the cost of further recovery could not be justified. Now the costs of end-of-pipe treatment and disposal have made source reduction a good investment. Greater reductions are possible when process engineers trained in pollution prevention plan to reduce waste at the design stage. For example, although barge loading is not a factor for all refineries, it is an important emissions source for many facilities. One of the largest sources of volatile organic carbon emissions is the fugitive emissions from loading of tanker barges. These emissions could be reduced by more than 90% by installing a vapor loss control system that consists of vapor recovery or the destruction of the volatile organic carbon emissions in a flare.

Fugitive emissions are one of the largest sources of refinery hydrocarbon emissions. A leak detection and repair (LDAR) program consists of using a portable detecting instrument to detect leaks during regularly scheduled inspections of valves, flanges, and pump seals. Older refinery boilers may also be a significant source of emissions of sulfur oxides (SO_x), nitrogen oxides (NO_x), and particulate matter. It is possible to replace a large number of old boilers with a single new cogeneration plant with emission controls.

Since storage tanks are one of the largest sources of VOC emissions, a reduction in the number of these tanks can have a significant impact. The need for certain tanks can often be eliminated through improved production planning and more continuous operations. By minimizing the number of storage tanks, tank bottom solids and decanted wastewater may also be reduced. Installing secondary seals on the tanks can significantly reduce the losses from storage tanks containing gasoline and other volatile products.

Eventually, solids entering the crude distillation unit are likely to attract more oil and produce additional emulsions and sludge. The amount of solids removed from the desalting unit should therefore be maximized. A number of techniques can be used, such as using low-shear mixing devices to mix desalter wash water and crude oil; using lower-pressure water in the desalter to avoid turbulence; and replacing the water jets used in some refineries with mud rakes, which add less turbulence when removing settled solids.

Purging or blowing down a portion of the cooling water stream to the wastewater treatment system controls the dissolved solids concentration in the recirculating cooling water. Solids in the blowdown eventually create additional sludge in the wastewater treatment plant. However, minimizing the dissolved solids content of the cooling water can lower the amount of cooling tower blowdown. A significant portion of the total dissolved solids in the cooling water can originate in the cooling-water makeup stream in the form of naturally occurring calcium carbonates. Such solids can be controlled either by selecting a source of cooling tower makeup water with smaller quantities of dissolved solids or by removing the dissolved solids from the makeup water stream. Common treatment methods include cold lime softening, reverse osmosis, or electrodialysis.

In many refineries, using high-pressure water to clean heat exchanger bundles generates and releases water and entrained solids to the refinery wastewater treatment system. Exchanger solids may then attract oil as they move through the sewer system and may also produce finer solids and stabilized emulsions that are more difficult to remove. Solids can be removed at the heat-exchanger cleaning pad by installing concrete overflow weirs around the surface drains or by covering drains with a screen. Other ways to reduce solids generation are by using antifoulants on the heat-exchanger bundles to prevent scaling and by cleaning with reusable cleaning chemicals that also allow for the easy removal of oil.

Surfactants entering the refinery wastewater streams will increase the amount of emulsions and sludge generated. Surfactants can enter the system from a number of sources, including washing unit pads with detergents; treating gasoline with an endpoint over 200°C (>392°F), thereby producing spent caustics; cleaning tank truck tank interiors; and using soaps and cleaners for miscellaneous tasks. In addition, the overuse and mixing of the organic polymers used to separate oil, water, and solids in the wastewater treatment plant can actually stabilize emulsions. The use of surfactants should be minimized by educating operators, routing surfactant sources to a point downstream of the DAF unit, and by using dry cleaning, high-pressure water, or steam to clean oil surfaces of oil and dirt.

Replacing 55-gallon drums with bulk storage facilities can minimize the chances of leaks and spills. And just as 55-gallon drums can lead to leaks, underground piping can be a source of undetected releases to the soil and groundwater. Inspecting, repairing, or replacing underground piping with surface piping can reduce or eliminate these potential sources.

Finally, open ponds used to cool, settle out solids, and store process water can be a significant source of volatile organic carbon emissions. Wastewater from coke cooling and coke volatile organic carbon removal is occasionally cooled in open ponds where volatile organic carbon easily escapes to the atmosphere. In many cases, open ponds can be replaced by closed storage tanks.

Material Substitution Options

Spent conventional degreaser solvents can be reduced or eliminated through substitution with less toxic and/or biodegradable products. In addition, chromate-containing wastes can be reduced or eliminated in cooling tower and heat exchanger sludge by replacing chromates with less toxic alternatives, such as phosphates.

Using higher-quality catalysts will lead in increased process efficiency while the required frequency of catalyst replacement can be reduced. Similarly, the replacement of ceramic catalyst support with activated alumina supports presents the opportunity for recycling the activated alumina supports with the spent alumina catalyst.

12.2.2. Recycling

Recycling is the use, reuse, or reclamation of a waste after it is generated. At present the petroleum industry is focusing on recycling and reuse as the best opportunities for pollution prevention:

1. Use phenols and caustics produced in the refining operations as chemical feeds in other applications.
2. Use oily waste sludge as feedstock in coking operations.
3. Regenerate catalysts. Extend useful life. Recover valuable metals from spent catalyst. Possibly use catalyst as a concrete admixture or as a fertilizer.

4. Maximize slop oil recovery. Agitate sludge with air and steam to recover residual oils.
5. Regenerate filtration clay. Wash clay with naphtha, dry by steam heating and feed to a burning kiln for regeneration.
6. Recover valuable product from oily sludge with solvent extraction.

Although pollution is reduced more if wastes are prevented in the first place, a next-best option for reducing pollution is to treat wastes so that they can be transformed into useful products.

Caustic substances used to absorb and remove hydrogen sulfide and phenol contaminants from intermediate and final product streams can often be recycled. Spent caustics may be salable to chemical recovery companies if concentrations of phenol or hydrogen sulfide are high enough. Process changes in the refinery may be needed to raise the concentration of phenols in the caustic to make recovery of the contaminants economical. Caustics containing phenols can also be recycled on-site by reducing the pH of the caustic until the phenols become insoluble, thereby allowing physical separation. The caustic can then be treated in the refinery wastewater system.

Oily sludge can be sent to a coking unit or the crude distillation unit, where it becomes part of the refinery products. Sludge sent to the coker can be injected into the coke drum with the quench water, injected directly into the delayed coker, or injected into the coker blowdown contactor used in separating the quenching products. Use of sludge as a feedstock has increased significantly in recent years and currently is carried out by most refineries. The quantity of sludge that can be sent to the coker is restricted by coke quality specifications that may limit the amount of sludge solids in the coke. Coking operations can be upgraded, however, to increase the amount of sludge that they can handle.

Significant quantities of catalyst fines are often present around the catalyst hoppers of fluid catalytic cracking reactors and regenerators. Coke fines are often present around the coker unit and coke storage areas. The fines can be collected and recycled before being washed to the sewers or migrating off-site via the wind. Collection techniques include dry sweeping the catalyst and coke fines and sending the solids to be recycled or disposed of as nonhazardous waste. Coke fines can also be recycled for fuel use. Another collection technique involves the use of vacuum ducts in dusty areas (and vacuum hoses for manual collection) that run to a small baghouse for collection.

An issue that always arises relates to the disposal of laboratory sample from any process control or even environmental laboratory that is associated with a refinery. Samples from such a laboratory can be recycled to the oil recovery system.

12.2.3. Treatment Options

When pollution prevention and recycling options are not economically feasibly, pollution can still be reduced by treating wastes so that they are transformed into less environmentally harmful wastes or can be disposed of in a less environmentally harmful media.

The toxicity and volume of some deoiled and dewatered sludge can be reduced further through thermal treatment. Thermal sludge treatment units use heat to vaporize the water and volatile components in the feed and leave behind a dry solid residue. The vapors are condensed for separation into hydrocarbon and water components. Noncondensable vapors are either flared or sent to the refinery amine unit for treatment and use as refinery fuel gas.

Furthermore, because oily sludge makes up a large portion of refinery solid wastes, any improvement in the recovery of oil from the sludge can significantly reduce the volume of waste. A number of technologies are currently in use to separate oil, water, and solids mechanically, including: belt filter presses, recessed chamber pressure filters, rotary vacuum filters, scroll centrifuges, disk centrifuges, shakers, thermal driers, and centrifuge–drier combinations.

Waste material such as tank bottoms from crude oil storage tanks constitute a large percentage of refinery solid waste and pose a particularly difficult disposal problem due to the presence of heavy metals. Tank bottoms are comprised of heavy hydrocarbons, solids, water, rust, and scale. Minimization of tank bottoms is carried out most cost-effectively through careful separation of the oil and water remaining in the tank bottom. Filters and centrifuges can also be used to recover the oil for recycling.

Spent clay from refinery filters often contains significant amounts of entrained hydrocarbons and must therefore be designated as hazardous waste. Backwashing spent clay with water or steam can reduce the hydrocarbon content to levels so that it can be reused or handled as a nonhazardous waste. Another method used to regenerate clay is to wash the clay with naphtha, dry it by steam heating, and then feed it to a burning kiln for regeneration. In some cases, clay filtration can be replaced entirely with hydrotreating process options.

Decant oil sludge from the fluidized-bed catalytic cracking unit can (and often does) contain significant concentrations of catalyst fines. These fines often prevent the use of decant oil as a feedstock or require treatment that generates an oily catalyst sludge. Catalyst fines in the decant oil can be minimized by using a decant oil catalyst removal system. One system incorporates high-voltage electric fields to polarize and capture catalyst particles in the oil. The amount of catalyst fines reaching the decant oil can be minimized by installing high-efficiency cyclones in the reactor to shift catalyst fines losses from the decant oil to the regenerator, where they can be collected in the electrostatic precipitator.

12.3. ADOPTION OF POLLUTION REDUCTION OPTIONS

Although numerous cases have been documented where petroleum refineries have reduced pollution and operating costs simultaneously, there are often barriers to doing so. The primary barrier to most pollution reduction projects is cost. Many pollution reduction options simply do not pay for themselves. Corporate investments typically must earn an adequate return on invested capital for the shareholders, and some pollution prevention options at some facilities may not meet the requirements set by the companies. In addition, the equipment used in the petroleum refining industry are very capital intensive and have very long lifetimes. This reduces the incentive to make process modifications to (expensive) installed equipment that is still useful. It should be noted that pollution prevention techniques are, nevertheless, often more cost-effective than pollution reduction through end-of-pipe treatment.

Of course, facility training programs that emphasize the importance of keeping solids out of sewer systems will help reduce that portion of wastewater treatment plant sludge arising from the everyday activities of refinery personnel. For example, educating personnel on how to avoid leaks and spills can reduce contaminated soil.

A systematic approach will produce better results than piecemeal efforts. An essential first step is a comprehensive waste audit:

1. List all waste generated.
2. Identify the composition of the waste and the source of each substance.
3. Identify options to reduce the generation of these substances in the production or manufacturing process.
4. Focus on wastes that are most hazardous and techniques that are implemented most easily.
5. Compare the technical and economic feasibility of the options identified.
6. Evaluate the results and schedule periodic reviews of the program so it can be adapted to reflect changes in regulations, technology, and economic feasibility.

The waste audit should systematically evaluate opportunities for improved operating procedures, process modifications, process redesign, and recycling.

Petroleum refinery wastes result from processes designed to remove naturally occurring contaminants in the crude oil, including water, sulfur, nitrogen, and heavy metals:

1. Segregate process (oily) waste streams from relatively clean rainwater runoff to reduce the quantity of oily sludge.
2. Generate an increased the potential for oil recovery. A significant portion of refinery waste comes from oily sludge found in combined process/storm sewers.
3. Conduct an inspection of petroleum refinery systems for leaks. For example, check hoses, pipes, valves, pumps, and seals. Make necessary repairs where appropriate.
4. Conserve water. Reuse rinse waters if possible. Reduce equipment-cleaning frequency where beneficial in reducing net waste generation.

5. Use correct pressures, temperatures, and mixing ratios for optimum recovery of product and reduction in waste produced.
6. Employ street sweeping or vacuuming of paved process areas to reduce solids to the sewers.
7. Pave runoff areas to reduce transfer of solids to waste systems. Use water softeners in cooling-water systems to extend the useful cycling time of the water.

Setting up a pollution prevention program does not require exotic or expensive technologies. Some of the most effective techniques are simple and inexpensive. Others require significant capital expenditures, but many provide a return on that investment.

REFERENCES

EPA. 2004. Environmental Protection Agency, Washington, DC. Web site: http://www.epa.gov.

Hydrocarbon Processing. 2003. Special Report: Plant Safety and Environment, 82(11): 37 et seq.

Lo, P. 1991. *Wastewater and Solid Waste Management*. County Sanitation District of Los Angeles County, Whittier, CA.

Speight, J. G. 1993. *Gas Processing: Environmental Aspects and Methods*, Butterworth-Heinemann, Oxford.

Speight, J. G. 1996. *Environmental Technology Handbook*. Taylor & Francis, Philadelphia.

Woodside, G. 1999. *Hazardous Materials and Hazardous Waste Management*. Wiley, New York.

GLOSSARY

This glossary provides definitions that are commonly used in reference to refining operations (processes, equipment, products) and will be of use to the reader.

ABN separation: a method of fractionation by which petroleum is separated into acidic, basic, and neutral constituents.

Accuracy: a measure of how close the test result will be to the true value of the property being measured; a relative term in the sense that systematic errors or biases can exist but be small enough to be inconsequential.

Acid catalyst: a catalyst having acidic character; alumina is an example of such a catalyst.

Acid deposition: acid rain; a form of pollution depletion in which pollutants, such as nitrogen oxides and sulfur oxides, are transferred from the atmosphere to soil or water; often referred to as *atmospheric self-cleaning*. The pollutants usually arise from the use of fossil fuels.

Acidity: the capacity of an acid to neutralize a base such as a hydroxyl ion (OH^-).

Acid number: a measure of the reactivity of petroleum with a caustic solution, given in terms of milligrams of potassium hydroxide that are neutralized by 1 g of petroleum.

Acid rain: the precipitation phenomenon that incorporates anthropogenic acids and other acidic chemicals from the atmosphere to the land and water (see also **Acid deposition**).

Acid sludge: the residue left after treating petroleum oil with sulfuric acid for the removal of impurities; a black, viscous substance containing the spent acid and impurities.

Acid treating: a process in which unfinished petroleum products, such as gasoline, kerosene, and lubricating-oil stocks, are contacted with sulfuric acid to improve their color, odor, and other properties.

Additive: a substance added to petroleum products (such as lubricating oils) to impart new or to improve existing characteristics.

Environmental Analysis and Technology for the Refining Industry, by James G. Speight
Copyright © 2005 John Wiley & Sons, Inc.

Adsorption: the transfer of a substance from a solution to the surface of a solid, resulting in a relatively high concentration of the substance at the place of contact; see also **Chromatographic adsorption**.

Air pollution: the discharge of toxic gases and particulate matter introduced into the atmosphere, principally as a result of human activity.

Air toxics: hazardous air pollutants.

Alicyclic hydrocarbon: a hydrocarbon that has a cyclic structure (e.g., cyclohexane); also collectively called *naphthenes*.

Aliphatic hydrocarbon: a hydrocarbon in which the carbon–hydrogen groupings are arranged in open chains that may be branched. The term includes *paraffins* and *olefins* and provides a distinction from *aromatics* and *naphthenes*, which have at least some of their carbon atoms arranged in closed chains or rings.

Aliquot: that quantity of material of proper size for measurement of the property of interest; test portions may be taken from the gross sample directly, but often preliminary operations such as mixing or further reduction in particle size are necessary.

Alkalinity: the capacity of a base to neutralize the hydrogen ion (H^+).

Alkanes: hydrocarbons that contain only single bonds. The chemical name indicates the number of carbon atoms and ends with the suffix *ane*.

Alkenes: hydrocarbons that contain carbon–carbon double bonds. The chemical name indicates the number of carbon atoms and ends with the suffix *ene*.

Alkylate: the product of an alkylation (*q.v.*) process.

Alkylation: in the petroleum industry, a process by which an olefin (e.g., ethylene) is combined with a branched-chain hydrocarbon (e.g., isobutane); alkylation may be accomplished as a thermal or a catalytic reaction.

Alkyl groups: a group of carbon and hydrogen atoms that branch from the main carbon chain or ring in a hydrocarbon molecule. The simplest alkyl group, a methyl group, is a carbon atom attached to three hydrogen atoms.

Alumina (Al_2O_3): used in separation methods as an adsorbent and in refining as a catalyst.

American Society for Testing and Materials (ASTM): the official organization in the United States for designing standard tests for petroleum and other industrial products.

Analyte: the chemical for which a sample is tested, or analyzed.

Analytical equivalence: the acceptability of the results obtained from various laboratories; a range of acceptable results.

Antibody: a molecule having chemically reactive sites specific for certain other molecules.

API gravity: a measure of the *lightness* or *heaviness* of petroleum that is related to density and specific gravity:

$$°API = (141.5/\text{sp gr at } 60°F) - 131.5$$

Aromatic hydrocarbon: a hydrocarbon characterized by the presence of an aromatic ring or condensed aromatic rings; benzene and substituted benzene, naphthalene and substituted naphthalene, phenanthrene and substituted phenanthrene, as well as the higher condensed ring systems; compounds that are distinct from those of aliphatic compounds (*q.v.*) or alicyclic compounds (*q.v.*).

Asphalt: the nonvolatile product obtained by distillation and further processing of an asphaltic crude oil; a manufactured product.

Asphaltene (asphaltenes): the brown to black powdery material produced by treatment of petroleum, petroleum residua, or bituminous materials with a low-boiling liquid hydrocarbon (e.g., pentane or heptane); soluble in benzene (and other aromatic solvents), carbon disulfide, and chloroform (or other chlorinated hydrocarbon solvents).

ASTM: see **American Society for Testing and Materials**.

Atmospheric equivalent boiling point (AEBP): a mathematical method of estimating the boiling point at atmospheric pressure of nonvolatile fractions of petroleum.

Atmospheric residuum: a residuum (*q.v.*), obtained by distillation of a crude oil under atmospheric pressure, which boils above 350°C (660°F).

Attainment area: a geographical area that meets NAAQS for criteria air pollutants; see also **Nonattainment area**.

Attapulgus clay: see **Fuller's earth**.

BACT: best available control technology.

Baghouse: a filter system for the removal of particulate matter from gas streams; so called because of the similarity of the filters to coal bags.

Barrel: the unit of measurement of liquids in the petroleum industry; equivalent to 42 U.S. standard gallons or 33.6 imperial gallons.

Base number: the quantity of acid, expressed in milligrams of potassium hydroxide per gram of sample, that is required to titrated a sample to a specified endpoint.

Basic nitrogen: nitrogen (in petroleum) that occurs in pyridine form.

Basic sediment and water (bs&w, bsw): the material that collects in the bottom of storage tanks, usually composed of oil, water, and foreign matter; also called **bottoms; bottom settlings**.

Baumé gravity: the specific gravity of liquids expressed as degrees on the Baumé (°B or °Bé) scale. For liquids lighter than water:

$$\text{sp gr } 60°F = \frac{140}{130 + °BJ}$$

For liquids heavier than water:

$$\text{sp gr } 60°F = \frac{145}{145 - °BJ}$$

Bbl: see **Barrel**.

Benzene: a colorless aromatic liquid hydrocarbon (C_6H_6).

Benzin: refined light naphtha used for extraction purposes.

Benzine: an obsolete term for light petroleum distillates covering the gasoline and naphtha range; see also **Ligroine (ligroin)**.

Benzol: the general term that refers to commercial or technical (not necessarily pure) benzene; also the term used for aromatic naphtha.

Biogenic: material derived from bacterial or vegetation sources.

Biological lipid: any biological fluid that is miscible with a nonpolar solvent. These materials include waxes, essential oils, and chlorophyll.

Biological oxidation: the oxidative consumption of organic matter by bacteria by which the organic matter is converted into gases.

Biomass: biological organic matter.

Bitumen: a semisolid to solid hydrocarbonaceous material found filling pores and crevices of sandstone, limestone, or argillaceous sediments.

Bituminous: containing bitumen or constituting the source of bitumen.

Bituminous rock: see **Bituminous sand**.

Bituminous sand: a formation in which the bituminous material (see **Bitumen**) is found as a filling in veins and fissures in fractured rocks or impregnating relatively shallow sand, sandstone, and limestone strata; a sandstone reservoir that is impregnated with a heavy, viscous black petroleumlike material that cannot be retrieved through a well by conventional production techniques.

Black acid(s): a mixture of the sulfonates found in acid sludge which are insoluble in naphtha, benzene, and carbon tetrachloride; very soluble in water but insoluble in 30% sulfuric acid; in the dry, oil-free state, sodium soaps are black powders.

Black oil: any of the dark-colored oils; a term now often applied to fuel oil (*q.v.*).

Boiling point: a characteristic physical property of a liquid at which the vapor pressure is equal to that of the atmosphere and the liquid is converted to a gas.

Boiling range: the range of temperature, usually determined at atmospheric pressure in standard laboratory apparatus, over which the distillation of an oil begins, proceeds, and finishes.

Bromine number: the number of grams of bromine absorbed by 100 g of oil, which indicates the percentage of double bonds in the material.

Brown acid: oil-soluble petroleum sulfonates found in acid sludge that can be recovered by extraction with naphtha solvent. Brown-acid sulfonates are somewhat similar to mahogany sulfonates but are more water soluble. In the dry, oil-free state, the sodium soaps are light-colored powders.

BS&W: see **Basic sediment and water**.

BTEX: benzene, toluene, ethylbenzene, and the xylene isomers.

Bunker fuel: heavy *residual oil*, also called bunker C, bunker C fuel oil, or bunker oil; see also **No. 6 Fuel oil**.

Burner fuel oil: any petroleum liquid suitable for combustion.

Burning oil: an illuminating oil, such as kerosene (kerosine) suitable for burning in a wick lamp.

Burning point: see **Fire point**.

C_1, C_2, C_3, C_4, C_5 fractions: a common way of representing fractions containing a preponderance of hydrocarbons having one, two, three, four, or five carbon atoms, respectively, and without reference to hydrocarbon type.

CAA: Clean Air Act; this act is the foundation of air regulations in the United States.

Carbene: the pentane- or heptane-insoluble material that is insoluble in benzene or toluene but which is soluble in carbon disulfide (or pyridine).

Carboid: the pentane- or heptane-insoluble material that is insoluble in benzene or toluene and which is also insoluble in carbon disulfide (or pyridine).

CAS: Chemical Abstract Service.

Cat cracking: see **Catalytic cracking**.

Catalyst: a chemical agent which, when added to a reaction (process), will enhance the conversion of a feedstock without being consumed in the process.

Catalyst selectivity: the relative activity of a catalyst with respect to a particular compound in a mixture, or the relative rate in competing reactions of a single reactant.

Catalyst stripping: the introduction of steam at a point where spent catalyst leaves the reactor in order to strip (i.e., remove) deposits retained on the catalyst.

Catalytic activity: the ratio of the space velocity of the catalyst under test to the space velocity required for the standard catalyst to give the same conversion as the catalyst being tested; usually multiplied by 100 before being reported.

Catalytic cracking: the conversion of high-boiling feedstocks into lower-boiling products by means of a catalyst that may be used in a fixed or fluid bed.

Catalytic reforming: rearranging hydrocarbon molecules in a gasoline-boiling-range feedstock to produce other hydrocarbons having a higher antiknock quality; isomerization of paraffins, cyclization of paraffins to naphthenes (q.v.), dehydrocyclization of paraffins to aromatics (q.v.).

Cetane index: an approximation of the cetane number (q.v.) calculated from the density (q.v.) and mid-boiling-point temperature (q.v.).

Cetane number: a number indicating the ignition quality of diesel fuel; a high cetane number represents a short ignition delay time; the ignition quality of diesel fuel can also be estimated from the following formula:

diesel index = [aniline point (°F)

$\times$ API gravity]

$\times$ 100

CFR: *Code of Federal Regulations*; Title 40 (40 CFR) contains the regulations for protection of the environment.

Characterization factor: the UOP characterization factor K, defined as the ratio of the cube root of the molal average boiling point, T_B, in degrees Rankine (°R = °F + 460), to the specific gravity at 60°/60°F:

$$K = (T_B)^{1/3}/\text{sp gr}$$

Ranges from 12.5 for paraffinic stocks to 10.0 for the highly aromatic stocks; also called the *Watson characterization factor*.

Chemical waste: any solid, liquid, or gaseous material discharged from a process that may pose substantial hazards to human health and environment.

Chromatogram: the resultant electrical output of sample components passing through a detection system following chromatographic separation. A chromatogram may also be called a *trace*.

Chromatographic adsorption: selective adsorption on materials such as activated carbon, alumina, or silica gel; liquid or gaseous mixtures of hydrocarbons are passed through the adsorbent in a stream of diluent, and certain components are adsorbed preferentially.

Chromatography: a method of separation based on selective adsorption; see also **Chromatographic adsorption**.

Clay: silicate minerals that also usually contain aluminum and have particle sizes less than 0.002 μm; used in separation methods as an adsorbent and in refining as a catalyst.

Cleanup: a preparatory step following extraction of a sample medium designed to remove components that may interfere with subsequent analytical measurements.

Cloud point: the temperature at which paraffin wax or other solid substances begin to crystallize or separate from the solution, imparting a cloudy appearance to the oil when the oil is chilled under prescribed conditions.

Coke: a gray to black solid carbonaceous material produced from petroleum during thermal processing; characterized by having a high carbon content (95%+ by weight), a honeycomb appearance, and is insoluble in organic solvents.

Coker: the processing unit in which coking takes place.

Coking: a process for the thermal conversion of petroleum in which gaseous, liquid, and solid (coke) products are formed.

Composition: the general chemical makeup of petroleum.

Confirmation column: a secondary column in chromatography that contains a stationary phase having different affinities for components in a mixture than in the primary column. Used to confirm analyses that may not be completely resolved using the primary column.

Contaminant: a substance that causes deviation from the normal composition of an environment.

Cracking: the thermal processes by which the constituents of petroleum are converted to lower-molecular-weight products; a process whereby the relative proportion of lighter or more volatile components of crude oil is increased by changing the chemical structure of the constituent hydrocarbons.

Criteria air pollutants: air pollutants or classes of pollutants regulated by the Environmental Protection Agency; the air pollutants are (including VOCs) ozone, carbon monoxide, particulate matter, nitrogen oxides, sulfur dioxide, and lead.

Crude oil: see *Petroleum*.

Cut: the *distillate* obtained between two given temperatures during a distillation process.

Cut point: the boiling-temperature division between distillation fractions of petroleum.

Cycloalkane: a class of alkanes that are in the form of a ring.

Cycloparaffin: synonymous with **Cycloalkane**.

Deasphaltened oil: the fraction of petroleum after the asphaltenes have been removed.

Deasphaltening: removal of a solid powdery asphaltene fraction from petroleum by the addition of low-boiling liquid hydrocarbons such as n-pentane or n-heptane under ambient conditions.

Deasphalting: the removal of the asphaltene fraction from petroleum by the addition of a low-boiling hydrocarbon liquid such as n-pentane or n-heptane; more correctly, the removal asphalt (tacky, semisolid) from petroleum (as occurs in a refinery asphalt plant) by the addition of liquid propane or liquid butane under pressure.

Decoking: removal of petroleum coke from equipment such as coking drums; hydraulic decoking uses high-velocity water streams.

Delayed coking: a coking process in which the thermal reaction are allowed to proceed to completion to produce gaseous, liquid, and solid (coke) products.

Density: the mass (or weight) of a unit volume of any substance at a specified temperature; see also **Specific gravity**.

Desalting: removal of mineral salts (mostly chlorides) from crude oils.

Desorption: the reverse process of adsorption whereby adsorbed matter is removed from the adsorbent; also used as the reverse of absorption.

Diesel fuel: fuel used for internal combustion in diesel engines; usually, that fraction which distills within the temperature range approximately 200 to 370°C. A general term covering oils used as fuel in diesel and other compression ignition engines.

Distillate: a product obtained by condensing the vapors evolved when a liquid is boiled and collecting the condensation in a receiver that is separate from the boiling vessel.

Distillation: a process for separating liquids with different boiling points.

Distillation curve: see **Distillation profile**.

Distillation profile: the distillation characteristics of petroleum or a petroleum product, showing the temperature and the percent distilled.

Distillation range: the difference between the temperature at the initial boiling point and at the endpoint, as obtained by the distillation test.

Domestic heating oil: see **No. 2 fuel oil**.

Effluent: any contaminating substance, usually a liquid that enters the environment via a domestic industrial, agricultural, or sewage plant outlet.

Electrical precipitation: a process using an electrical field to improve the separation of hydrocarbon reagent dispersions. May be used in chemical treating processes on a wide variety of refinery stocks.

Electric desalting: a continuous process to remove inorganic salts and other impurities from crude oil by settling out in an electrostatic field.

Electrostatic precipitators: devices used to trap fine dust particles (usually in the size range 30 to 60 μm) that operate on the principle of imparting an electric charge to particles in an incoming airstream which are then collected on an oppositely charged plate across a high-voltage field.

Eluate: the solutes, or analytes, moved through a chromatographic column; see **Elution**.

Eluent: solvent used to elute sample.

Elution: a process whereby a solute is moved through a chromatographic column by a solvent (liquid or gas), or eluent.

Emission control: the use of gas cleaning processes to reduce emissions.

Emission standard: the maximum amount of a specific pollutant permitted to be discharged from a particular source in a given environment.

EPA: Environmental Protection Agency.

Extract: the portion of a sample preferentially dissolved by the solvent and recovered by physically separating the solvent.

Fabric filters: filters made from fabric materials and used for removing particulate matter from gas streams (see **Baghouse**).

FCC: fluid catalytic cracking.

FCCU: fluid catalytic cracking unit.

Feedstock: petroleum as it is fed to the refinery; a refinery product that is used as the raw material for another process; the term is also generally applied to raw materials used in other industrial processes.

Filtration: the use of an impassable barrier to collect solids but which allows liquids to pass.

Fingerprint analysis: a direct-injection GC/FID analysis in which the detector output—the chromatogram—is compared to chromatograms of reference materials as an aid to product identification.

Fire point: the lowest temperature at which, under specified conditions in standardized apparatus, a petroleum product vaporizes sufficiently rapidly to form above its surface an air–vapor mixture that burns continuously when ignited by a small flame.

Flame ionization detector (FID): a detector for a gas chromatograph that measures anything that can burn.

Flammability range: the range of temperature over which a chemical is flammable.

Flammable: a substance that will burn readily.

Flammable liquid: a liquid having a flash point below 37.8°C (100°F).

Flammable solid: a solid that can ignite from friction or from heat remaining from its manufacture, or which may cause a serious hazard if ignited.

Flash point: the lowest temperature to which the product must be heated under specified conditions to give off sufficient vapor to form a mixture with air that can be ignited momentarily by a flame.

Flue gas: gas from the combustion of fuel, the heating value of which has been substantially spent and which is, therefore, discarded to the flue or stack.

Fluid catalytic cracking: cracking in the presence of a fluidized bed of catalyst.

Fluid coking: a continuous fluidized solids process that cracks feed thermally over heated coke particles in a reactor vessel to gas, liquid products, and coke.

Fly ash: particulate matter produced from mineral matter in coal that is converted during combustion to finely divided inorganic material which emerges from the combustor in the gases.

Fractional composition: the composition of petroleum as determined by fractionation (separation) methods.

Fractional distillation: the separation of the components of a liquid mixture by vaporizing and collecting the fractions, or cuts, which condense in different temperature ranges.

Fractionating column: a column arranged to separate various fractions of petroleum by a single distillation which may be tapped at different points along its length to separate various fractions in the order of their boiling points.

Fractionation: the separation of petroleum into the constituent fractions using solvent or adsorbent methods; chemical agents such as sulfuric acid may also be used.

Fuel oil: a general term applied to oil used for the production of power or heat. In a more restricted sense, it is applied to any petroleum product that is used as boiler fuel or in industrial furnaces. These oils are normally *residues*, but blends of distillates and *residues* are also used as fuel oil. The wider term *liquid fuel* is sometimes used, but the term *fuel oil* is preferred; also called **heating oil**; see also **No. 1 to No. 4 fuel oils**.

Fuller's earth: a clay that has high adsorptive capacity for removing color from oils; attapulgus clay is a widely used fuller's earth.

Functional group: the portion of a molecule that is characteristic of a family of compounds and determines the properties of these compounds.

Furnace oil: a distillate fuel intended primarily for use in domestic heating equipment.

Gas chromatography: an analytical technique, employing a gaseous mobile phase, which separates mixtures into their individual components.

Gas-oil: a petroleum distillate with a viscosity and *distillation range* intermediate between those of *kerosene* and *light lubricating oil.*

Gaseous pollutants: gases released into the atmosphere that act as primary or secondary pollutants.

Gasoline: fuel for the internal combustion engine that is commonly, but improperly, referred to simply as *gas*; the terms *petrol* and *benzine* are commonly used in some countries.

Gravimetric: gravimetric methods weigh a residue.

Grease: a semisolid or solid lubricant consisting of a stabilized mixture of mineral, fatty, or synthetic oil with soaps, metal salts, or other thickeners.

Greenhouse effect: warming of the Earth due to entrapment of the sun's energy by the atmosphere.

Greenhouse gases: gases that contribute to the greenhouse effect (*q.v.*).

HAP(s): hazardous air pollutant(s).

Headspace: the vapor space above a sample into which volatile molecules evaporate. Certain methods sample this vapor.

Heating oil: see **Fuel oil**.

Heavy ends: the highest-boiling portion of a petroleum fraction; see also **Light ends**.

Heavy fuel oil: fuel oil having a high density and viscosity; generally, residual fuel oil such as No. 5 and No. 6 fuel oil (*q.v.*).

Heavy oil: petroleum having an API gravity of less than 20°.

Heavy petroleum: see Fuel **oil**.

Heteroatom compounds: chemical compounds that contain nitrogen and/or oxygen and/or sulfur and/or metals bound within their molecular structure(s).

HF alkylation: an alkylation process whereby olefins (C_3, C_4, C_5) are combined with isobutane in the presence of hydrofluoric acid catalyst.

Hydraulic fluid: a fluid supplied for use in hydraulic systems. Low viscosity and low *pour point* are desirable characteristics. Hydraulic fluids may be of petroleum or nonpetroleum origin.

Hydrocarbons: molecules that consist *only* of hydrogen and carbon atoms.

Hydrocracking: a catalytic high-pressure high-temperature process for the conversion of petroleum feedstocks in the presence of fresh and recycled hydrogen; carbon–carbon bonds are cleaved in addition to the removal of heteroatomic species.

Hydrocracking catalyst: a catalyst used for hydrocracking which typically contains separate hydrogenation and cracking functions.

Hydrotreating: the removal of heteroatomic (nitrogen, oxygen, and sulfur) species by treatment of a feedstock or product at relatively low temperatures in the presence of hydrogen.

Ignitability: characteristic of liquids whose vapors are likely to ignite in the presence of an ignition source; also characteristic of nonliquids that may catch fire from friction or contact with water and that burn vigorously.

Illuminating oil: oil used for lighting purposes.

Immunoassay: portable tests that take advantage of an interaction between an antibody and a specific analyte. Immunoassay tests are semiquantitative and usually rely on color changes of varying intensities to indicate relative concentrations.

Infrared spectroscopy: an analytical technique that quantifies the vibration (stretching and bending) that occurs when a molecule absorbs (heat) energy in the infrared region of the electromagnetic spectrum.

Isomerization: the conversion of a *normal* (straight-chain) paraffin hydrocarbon into an *iso* (branched-chain) paraffin hydrocarbon having the same atomic composition.

Jet fuel: fuel meeting the required properties for use in jet engines and aircraft turbine engines.

Kauri-butanol number: A measurement of solvent strength for hydrocarbon solvents; the higher the Kauri-butanol (KB) value, the stronger the solvency; the test method (ASTM D1133) is based on the principle that Kauri resin is readily soluble in butyl alcohol but not in hydrocarbon solvents and that the resin solution will tolerate only a certain amount of dilution and is reflected as cloudiness when the resin starts to come out of solution; solvents such as toluene can be added in a greater amount (and thus have a higher KB value) than weaker solvents such as hexane.

Kerosene (kerosine): a fraction of petroleum that was initially sought as an illuminant in lamps; a precursor to diesel fuel with a *distillation* range that generally falls within the limits of 150 and 300°C; main uses are as a jet engine fuel, an illuminant, for heating purposes, and as a fuel for certain types of internal combustion engines.

K-factor: see **Characterization factor**.

LAER: lowest achievable emission rate; the required emission rate in nonattainment permits.

Light ends: the lower-boiling components of a mixture of hydrocarbons; see also **Heavy ends and Light hydrocarbons**.

Light hydrocarbons: hydrocarbons with molecular weights less than that of heptane (C_7H_{16}).

Light oil: the products distilled or processed from crude oil up to, but not including, the first lubricating-oil distillate.

Light petroleum: petroleum having an API gravity greater than 20°.

Ligroine (ligroin): a saturated petroleum naphtha boiling in the range 20 to 135°C (68 to 275°F) and suitable for general use as a solvent; also called *benzine* or *petroleum ether*.

Liquefied petroleum gas: propane, butane, or mixtures thereof, gaseous at atmospheric temperature and pressure, held in the liquid state by pressure to facilitate storage, transport, and handling.

Liquid chromatography: a chromatographic technique that employs a liquid mobile phase.

Liquid–liquid extraction: an extraction technique in which one liquid is shaken with or contacted by an extraction solvent to transfer molecules of interest into the solvent phase.

MACT: maximum achievable control technology. Applies to major sources of hazardous air pollutants.

Major source: a source that has a potential to emit for a regulated pollutant that is at or greater than an emission threshold set by regulations.

Maltenes: that fraction of petroleum that is soluble in, for example, pentane or heptane; deasphaltened oil (*q.v.*); also the term arbitrarily assigned to the pentane-soluble portion of petroleum that is relatively high boiling (>300°C, 760 mm); see also **Petrolenes**.

Mass spectrometer: an analytical technique that *fractures* organic compounds into characteristic "fragments" based on functional groups that have a specific mass-to-charge ratio.

MCL: the maximum contaminant level dictated by regulations.

MDL: see **Method detection limit**.

Method detection limit: the smallest quantity or concentration of a substance that an instrument can measure.

Microcrystalline wax: wax extracted from certain petroleum residua and having a finer and less apparent crystalline structure than paraffin wax.

Middle distillate: one of the distillates obtained between *kerosene* and *lubricating oil* fractions in the refining processes. These include *light fuel oils* and *diesel fuels*.

Mineral hydrocarbons: petroleum hydrocarbons, considered *mineral* because they come from the earth rather than from plants or animals.

Mineral oil: the older term for petroleum; the term was introduced in the nineteenth century as a means of differentiating petroleum (rock oil) from whale oil or oil from plants that, at the time, were the predominant illuminant for oil lamps.

Mobile phase: in chromatography, the phase (gaseous or liquid) responsible for moving an introduced sample through a porous medium to separate components of interest.

MSDS: material safety data sheet.

NAAQS: National Ambient Air Quality Standards; standards exist for the pollutants known as the criteria air pollutants: nitrogen oxides (NO_x), sulfur oxides (SO_x), lead, ozone, particulate matter less than 10 μm in diameter, and carbon monoxide (CO).

Naphtha: a generic term applied to refined, partly refined, or unrefined petroleum products and liquid products of natural gas, the majority of which distills below 240°C (464°F); the volatile fraction of petroleum which is used as a solvent or as a precursor to gasoline.

Naphthenes: cycloparaffins.

NESHAP: National Emissions Standards for Hazardous Air Pollutants; emission standards for specific source categories that emit or have the potential to emit one or more hazardous air pollutants; the standards are modeled on the best practices and most effective emission reduction methodologies in use at the affected facilities.

Nonattainment area: a geographical area that does not meet NAAQS for criteria air pollutants; see also **Attainment area**).

NO_x: oxides of nitrogen.

No. 1 fuel oil: very similar to kerosene (*q.v.*); used in burners where vaporization before burning is usually required and a clean flame is specified.

No. 2 fuel oil: has properties similar to those of diesel fuel and heavy jet fuel; used in burners where complete vaporization is not required before burning; see also **domestic heating oil**.

No. 4 fuel oil: a light industrial heating oil and is used where preheating is not required for handling or burning; there are two grades of No. 4 fuel oil, differing in safety (flash point) and flow (viscosity) properties.

No. 5 fuel oil: a heavy industrial fuel oil that requires preheating before burning.

No. 6 fuel oil: a heavy fuel oil and is more commonly known as *bunker C oil* when it is used to fuel ocean-going vessels; preheating is always required for burning this oil.

Olefin: synonymous with *alkene*.

Oxygenated gasoline: gasoline with added ethers or alcohols, formulated according to the Federal Clean Air Act to reduce carbon monoxide emissions during winter months.

Paraffin (alkane): one of a series of saturated aliphatic hydrocarbons, the lowest numbers of which are methane, ethane, and propane. The higher homologs are solid waxes.

Paraffin wax: the colorless, translucent, highly crystalline material obtained from the light lubricating fractions of paraffinic crude oils (wax distillates).

Particulate matter: particles in the atmosphere or on a gas stream that may be organic or inorganic and originate from a wide variety of sources and processes.

Partitioning: in chromatography, the physical act of a solute having different affinities for the stationary and mobile phases.

Partition ratios, K: the ratio of total analytical concentration of a solute in the stationary phase, CS, to its concentration in the mobile phase, CM.

Petrol: a term commonly used in some countries for **gasoline**.

Petrolatum: a semisolid product, ranging from white to yellow in color, produced during refining of residual stocks.

Petrolenes: the term applied to that part of the pentane- or heptane-soluble material that is low boiling ($<300°C$, $<570°F$, 760 mm) and can be distilled without thermal decomposition; see also **Maltenes**).

Petroleum (crude oil): naturally occurring mixture consisting essentially of many types of hydrocarbons, but also containing sulfur, nitrogen, or oxygen derivatives. Petroleum may be of paraffinic, asphaltic, or mixed base, depending on the presence of *paraffin* wax and *bitumen* in the *residue* after atmospheric distillation. Petroleum composition varies according to the geological strata of its origin.

Petroleum refinery: see **Refinery**.

Petroleum refining: a complex sequence of events that result in the production of a variety of products.

Photoionization: a gas chromatographic detection system that utilizes an ultraviolet lamp as an ionization source for analyte detection. It is usually used as a selective detector by changing the photon energy of the ionization source.

PINA analysis: a method of analysis for paraffins, isoparaffins, naphthenes, and aromatics.

PIONA analysis: a method of analysis for paraffins, isoparaffins, olefins, naphthenes, and aromatics.

Pipe still: a still in which heat is applied to the oil while being pumped through a coil or pipe arranged in a suitable firebox.

Pipestill gas: the most volatile fraction that contains most of the gases that are generally dissolved in the crude. Also known as *pipestill light ends*.

PNA: see **Polynuclear aromatic compound**.

PNA analysis: a method of analysis for paraffins, naphthenes, and aromatics.

Pollution: the introduction into the land, water, and air systems of a chemical or chemicals that are not indigenous to these systems, or introduction into the land, water, and air systems of indigenous chemicals in greater-than-natural amounts.

Polycyclic aromatic hydrocarbons (PAHs): polycyclic aromatic hydrocarbons are a suite of compounds comprised of two or more condensed aromatic rings. They are found in many petroleum mixtures, and they are introduced to the environment predominantly through natural and anthropogenic combustion processes.

Polynuclear aromatic compound: an aromatic compound having two or more fused benzene rings (e.g., naphthalene, phenanthrene).

PONA analysis: a method of analysis for paraffins (P), olefins (O), naphthenes (N), and aromatics (A).

Porphyrins: organometallic constituents of petroleum that contain vanadium or nickel; the degradation products of chlorophyll that became included in the protopetroleum.

Positive bias: a result that is incorrect and too high.

Pour point: the lowest temperature at which oil will pour or flow when it is chilled without disturbance under definite conditions.

Propane deasphalting: solvent deasphalting using propane as the solvent.

Propane dewaxing: a process for dewaxing lubricating oils in which propane serves as solvent.

PSD: prevention of significant deterioration.

PTE: potential to emit; the maximum capacity of a source to emit a pollutant, given its physical or operation design, and considering certain controls and limitations.

Purge and trap: a chromatographic sample introduction technique in volatile components that are purged from a liquid medium by bubbling gas through it. The components are then concentrated by "trapping" them on a short intermediate column, which is subsequently heated to drive the components on to the analytical column for separation.

Purge gas: typically helium or nitrogen, used to remove analytes from the sample matrix in purge-and-trap extractions.

RACT: reasonably available control technology standards; implemented in areas of nonattainment to reduce emissions of volatile organic compounds and nitrogen oxides.

Recycling: the use or reuse of chemical waste as an effective substitute for a commercial products or as an ingredient or feedstock in an industrial process.

Reduced crude: a residual product remaining after the removal, by distillation or other means, of an appreciable quantity of the more volatile components of crude oil.

Refinery: a series of integrated unit processes by which petroleum can be converted to a slate of useful (salable) products.

Refinery gas: a gas (or a gaseous mixture) produced as a result of refining operations.

Refining: the process(es) by which petroleum is distilled and/or converted by application of a physical and chemical processes to form a variety of products are generated.

Reformate: the liquid product of a reforming process.

Reforming: the conversion of hydrocarbons with low octane numbers into hydrocarbons having higher octane numbers (e.g., the conversion of an n-paraffin into an isoparaffin).

Reformulated gasoline (RFG): gasoline designed to mitigate smog production and to improve air quality by limiting the emission levels of certain chemical compounds such as benzene and other aromatic derivatives; often contains oxygenates ($q.v.$).

Residual fuel oil: obtained by blending the residual product(s) from various refining processes with suitable diluent(s) (usually, middle distillates) to obtain the required fuel oil grades.

Residual oil: see **Residuum**.

Residuum (resid; *pl.* residua): the residue obtained from petroleum after nondestructive distillation has removed all the volatile materials from crude oil [e.g., an atmospheric (345°C, 650°F+) residuum].

Resins: that portion of the maltenes (*q.v.*) that is adsorbed by a surface-active material such as clay or alumina; the fraction of deasphaltened oil that is insoluble in liquid propane but soluble in *n*-heptane.

Retention time: the time it takes for an eluate to move through a chromatographic system and reach the detector. Retention times are reproducible and can therefore be compared to a standard for analyte identification.

SARA analysis: a method of analysis for saturates, aromatics, resins, and asphaltenes.

SARA separation: see **SARA analysis**.

Saturates: paraffins and cycloparaffins (naphthenes).

Saybolt Furol viscosity: the time, in seconds (**Saybolt Furol seconds, SFS**), for 60 mL of fluid to flow through a capillary tube in a Saybolt Furol viscometer at specified temperatures between 70 and 210°F; the method is appropriate for high-viscosity oils such as transmission, gear, and heavy fuel oils.

Saybolt Universal viscosity: the time, in seconds (**Saybolt Universal seconds, SUS**), for 60 mL of fluid to flow through a capillary tube in a Saybolt Universal viscometer at a given temperature.

Separatory funnel: glassware shaped like a funnel with a stoppered rounded top and a valve at the tapered bottom, used for liquid–liquid separations.

Sludge: a semisolid to solid product that results from the storage instability and/or the thermal instability of petroleum and petroleum products.

Soap: an emulsifying agent made from sodium or potassium salts of fatty acids.

Solvent: a liquid in which certain kinds of molecules dissolve. Although they typically are liquids with low boiling points, they may include high-boiling liquids, supercritical fluids, or gases.

Solvent extraction: a process for separating liquids by mixing the stream with a solvent that is immiscible with part of the waste but that will extract certain components of the waste stream.

Sonication: a physical technique employing ultrasound to intensely vibrate a sample media in extracting solvent and to maximize solvent–analyte interactions.

Sour crude oil: crude oil containing an abnormally large amount of sulfur compounds.

SO_x: oxides of sulfur.

Soxhlet extraction: an extraction technique for solids in which the sample is contacted with solvent repeatedly over several hours, increasing the extraction efficiency.

Specific gravity: the mass (or weight) of a unit volume of any substance at a specified temperature compared to the mass of an equal volume of pure water at a standard temperature; see also **Density**.

Spent catalyst: catalyst that has lost much of its activity due to the deposition of coke and metals.

Stabilization: the removal (i.e., stripping) of volatile constituents from a higher-boiling fraction or product; the production of a product which to all intents and purposes, does not undergo any further reaction when exposed to the air.

Stationary phase: in chromatography, the porous solid or liquid phase through which an introduced sample passes. The different affinities the stationary phase has for a sample allow the components in the sample to be separated, or resolved.

Sulfonic acids: acids obtained from petroleum or a petroleum product with strong sulfuric acid.

Sulfuric acid alkylation: an alkylation process in which olefins (C_3, C_4, and C_5) combine with isobutane in the presence of a catalyst (sulfuric acid) to form branched-chain hydrocarbons used especially in gasoline blending stock.

Supercritical fluid: an extraction method in which the extraction fluid is present at a pressure and temperature above its critical point.

SW-846: an EPA multivolume publication entitled *Test Methods for Evaluating Solid Waste, Physical/Chemical Methods*; the official compendium of analytical and sampling methods that have been evaluated and approved for use in complying with the RCRA regulations and that functions primarily as a guidance document setting forth acceptable, although not required methods for the regulated and regulatory communities to use in responding to RCRA-related sampling and analysis requirements. SW-846 changes over time as new information and data are developed.

Target analyte: target analytes are compounds that are required analytes in U.S. EPA analytical methods. BTEX and PAHs are examples of petroleum-related compounds that are target analytes in U.S. EPA methods.

Thermal cracking: a process that decomposes, rearranges, or combines hydrocarbon molecules by the application of heat, without the aid of catalysts.

Thin-layer chromatography (TLC): a chromatographic technique employing a porous medium of glass coated with a stationary phase. An extract is spotted near the bottom of the medium and placed in a chamber with solvent (mobile phase). The solvent moves up the medium and separates the components of the extract, based on affinities for the medium and solvent.

Total petroleum hydrocarbons (TPHs): a family of several hundred chemical compounds that originally come from petroleum.

TPH-D(DRO): gas chromatographic test for TPH diesel range organics.

TPH E: gas chromatographic test for TPH extractable organic compounds.

TPH-G(GRO): gas chromatographic test for TPH gasoline range organics.

TPH V: gas chromatographic test for TPH volatile organic compounds.

Trace element: those elements that occur at very low levels in a given system.

Ultimate analysis: elemental composition.

Unresolved complex: the thousands of compounds that a gas chromatograph *mixture* (UCM) is unable to fully separate.

Upgrading: the conversion of petroleum to value-added salable products.

Vacuum distillation: distillation (*q.v.*) under reduced pressure.

Vacuum residuum: a residuum (*q.v.*) obtained by distillation of a crude oil under vacuum (reduced pressure); that portion of petroleum that boils above a selected temperature such as 510°C (950°F) or 565°C (1050°F).

VGC (viscosity–gravity constant): an index of the chemical composition of crude oil defined by the general relation between specific gravity (sp gr) at 60°F and Saybolt Universal viscosity (SUV) at 100°F.

VI (Viscosity index): an arbitrary scale used to show the magnitude of viscosity changes in lubricating oils with changes in temperature.

Visbreaking: a process for reducing the viscosity of heavy feedstocks by controlled thermal decomposition.

Viscosity: a measure of the ability of a liquid to flow or a measure of its resistance to flow; the force required to move a plane surface of area 1 m^2 over another parallel plane surface 1 m away at a rate of 1 m/s when both surfaces are immersed in the fluid.

Viscosity–gravity constant: see **VGC**.

Viscosity index: see **VI**.

VOC (VOCs): volatile organic compound(s); volatile organic compounds are regulated because they are precursors to ozone; carbon-containing gases and vapors from incomplete gasoline combustion and from the evaporation of solvents.

Volatile compounds: a relative term that may mean (1) any compound that will purge, (2) any compound that will elute before the solvent peak (usually, those below C_6), or (3) any compound that will not evaporate during a solvent removal step.

Watson characterization factor: see **Characterization factor**.

Wax: wax of petroleum origin consists primarily of normal paraffins; wax of plant origin consists of esters of unsaturated fatty acids.

Weathered crude oil: crude oil which, due to natural causes during storage and handling, has lost an appreciable quantity of its more volatile components; also indicates uptake of oxygen.

Wobbe index (or Wobbe number): the calorific value of a gas divided by the specific gravity.

Zeolite: a crystalline aluminosilicate used as a catalyst and having a particular chemical and physical structure.

INDEX

Abiotic factors, 4, 5
Absorption spectrophotometry, 46
Accuracy, 171, 180
Acid number of asphalt, 286
Acid rain, 7, 90, 244
Acids, 81
Actions of emissions cleanup, 237
Acute hazardous waste, 115
Adsorption, 113
Adsorption chromatography, 43
Air emissions, 306
Air pollution, 131
Aliphatic naphtha, 68
Alkalis, 80
Alkanes, 33
  potential isomers, 35, 259
Alkenes, 33
Alkylation, 64, 82, 101
Alkylation emissions and waste, 102
Alkylation unit, 67, 102
Analysis, 185
  asphalt, 283
  gaseous effluents, 237
  liquid effluents, 257
  solid effluents, 283
Analytical methods, 185
  gaseous effluents, 247
  organic compounds in water, 218
Anthropogenic stress, 5
API gravity, 41
  asphalt, 291
Aromatic naphtha, 68
Aromatics, 13, 38
  multiring, 34
  single-ring, 33
Ash, 27
Ash in coke, 296
Asphalt, 35, 77, 283
  acid number, 286
  analysis, 283
  API gravity, 291

  asphaltene content, 287
  carbon disulfide insolubles, 288
  composition, 289
  density, 290
  elemental analysis, 292
  float test, 293
  manufacture, 285
  softening point, 293
  viscosity, 294
  weathering, 294
Asphalt base crude oil, 13, 209
Asphalt emulsion, 77, 285
Asphaltene content of asphalt, 287
Asphaltene content of fuel oil, 269
Asphaltenes, 38
Assessment of methods, 230
Atmospheric distillation, 35, 53, 60
Atmospheric equivalent boiling point, 53
Atmospheric residuum, 35
Atmospheric tower, 36
Automotive gasoline, 69
Aviation fuel, 70
Aviation gasoline, 70
Aviation turbine fuel, 70

Beavon process, 246, 308
Benzene, 34
Bias, 233
Biota, 4
Biotic factors, 4, 5
Bitumen, 11
Blending process, environmental impact, 131
Boiling fractions, 53
Brine, 27
BTEX, 210
Bunker C fuel oil, 74
Butane isomerization unit, 66, 106

Calcined coke, 295
Calorific value of gases, 248

Environmental Analysis and Technology for the Refining Industry, by James G. Speight
Copyright © 2005 John Wiley & Sons, Inc.

Carbenes, 38
Carboids, 38
Carbon distribution, 15
Carbon disulfide insolubles, 288
Carbon-13 magnetic resonance spectroscopy, 46
Carcinogen, 6
  classification, 7
Carcinogenic, 4
Catalysts coke, 295
Catalyst dust, 27
Catalysts, 83
Catalytic cracking, 61, 98
Catalytic cracking waste, 98, 99
Catalytic reforming, 105
Catalytic reforming unit, 105
Catalytic reforming waste, 105
CERCLA, 130, 142
  definition of hazardous waste, 26
Chain of custody, 153
Characteristics of chemical waste, 21
Characterization gravity, 14
Calcining, 83
Catalytic dewaxing, 77
Catalytic reactor shutdown, 23
Chemical composition of petroleum, 12, 33, 34
Chemical composition of petroleum products, 34
Chemical release, timing of, 6
Chemical waste, 8
  characteristics, 21
  F-category waste, 26
  K-category waste, 26
  origin, 20
  types, 9
Classification of petroleum, 11
  carbon distribution, 15
  characterization gravity, 14
  correlation index, 13
  density, 14
  UOP characterization factor, 16
  viscosity–gravity constant, 15
Classification of petroleum products, 19
  boiling range, 19, 258
  environmental behavior, 19
Claus process, 307
Clay catalysts, 83
Clay-gel adsorption, 42
Clean Air Act, 126, 133
Clean Water Act, 128, 140
Cloud point, 44
Coefficient of variance, 180

Coke, 77, 294, 295
  ash, 296
  composition, 297
  density, 299
  dust control material, 300
  hardness, 300
  manufacture, 295
  metals, 300
  sulfur, 301
Coke forms, 296
Coking, 61, 96
Coking waste, 96, 98
Combustibility, 22
Combustible chemical, 22
Component streams for gasoline, 69
Composition by fractionation, 37
Composition by volatility, 35
Composition of asphalt, 289
Composition of coke, 297
Composition of fuel oil, 270
Composition of gaseous effluents, 249
Composition of petroleum, 31, 32
Composition of naphtha, 261
Condensate releases, 160
Congealing point, 44
Contaminant, 8, 125
  in water, 223
Conversion processes, 60
  environmental impact, 131
Correlation index, 13
Corrosive substances, 24
Corrosivity, 21, 23, 114, 138
Crankcase oil, 74
Crude oil, see Petroleum
Cutback asphalt, 77
Cut point, 53
Cycloalkanes, 33
Cycloparaffins, 13, 33

Darwinism, 5
Deasphalting, 106
Deasphalting emissions and waste, 107
Deasphalting unit, 107
Definitions, 3
Delayed coking unit, 62, 97, 295
Demet process, 85
Density and specific gravity, 14, 40
  asphalt, 290
  coke, 299
  fuel oil, 270
  gaseous effluents, 252
  naphtha, 266
Desalting, 64, 92
Desalting unit, 93

INDEX 345

Desalting wastes, 93
Dewaxing, 106
Dewaxing unit, 76, 108
Dewaxing waste, 107
Diesel fuel, 71, 72
Diesel-like products, 19
Diesel range organics, 213
Dispersion, 111
Dissolution, 111
Distillate fuel oil, 71, 72
Distillation, 35, 284
   atmospheric pressure, 53, 94
   reduced pressure, 53, 95
Distillation emissions and waste, 94, 95
Domestic fuel oil, 72
Drilling mud, 27
Dropping point, 44
Dust control material, 300
Dust explosions, 23

Ecology, 4, 5
Ecosystem, 5, 7
EIA, 6
Elemental analysis of asphalt, 292
Elemental analysis of fuel oil, 272
Elemental analysis of petroleum, 41
Elemental composition of petroleum, 32, 41
Emissions, 88
   actions for cleanup, 237
   fugitive, 238
   nonpoint source, 238
   point source, 238
Emulsification, 112
Environment, 4
Environmental behavior, 25
Environmental impact assessment, 6
Environmental impact, mitigation, 6
Environmental leakage, sources, 20
Environmental pollution, 6
Environmental regulations, 125, 132
   chronology of, 134
Environmental release, 110
Environmental technology, 4
EPA test methods for total petroleum hydrocarbons, 192
Equivalent carbon number index, 212
Error of the mean, 180
Evaporation, 122
Evaporation rate, 266
Extract cleaning, 169, 186
Extract concentration, 168, 186
Extractable hydrocarbon, 189
Extraction, 185
Extraction methods, 157, 161, 163

F-category waste, 26
   origin, 27
Field screening, 214
Finishing processes, 65
Fire point, 52
Flame photometry, 46
Flammability, 22, 23, 114
Flammability limit, 23
Flammability of organic liquids, 22
Flammability range, 23
Flammable chemical, 22
Flammable compressed gas, 23
Flammable liquid, 22
Flammable solid, 22
Flash point, 22, 52
   fuel oil, 276
   naphtha, 266
Float test for asphalt, 293
Fluid catalytic cracking unit, 63, 98
Fluid coking unit, 63, 97
Fluorescent indicator adsorption, 42
Fourier transform infrared (FTIR) spectroscopy, 45
Fractionation by chromatography, 42
Fractionation by volatility, 53
Fractions, 205
Freezing point, 44
Freon extractable material, 212
F-type chemicals, 25
Fuel gas, 68
Fuel oil, 71, 268
   asphaltene content, 269
   composition, 270
   density, 270
   elemental analysis, 272
   flash point, 276
   metals content, 276
   pour point and viscosity, 277
   stability, 278
Fugitive emissions, 238
Furnace fuel oil, 72

Gas chromatographic methods for total petroleum group analysis, 201, 204
Gas chromatographic methods for total petroleum hydrocarbons, 191, 202
Gaseous effluents, 237
   analysis, 247
   calorific value, 248
   composition, 249
   density, 252
   environmental effects, 245
   sampling, 247
   sulfur content, 253

346 INDEX

Gaseous effluents (*continued*)
    volatility and vapor pressure, 253
Gases
    inorganic, 237
    organic, 237
Gasoline, 68
    automotive, 69
    aviation, 70
    component streams, 69
Gasoline products, 19
Gasoline range organics, 213
Gel permeation chromatography, 42
Gravimetric methods for total petroleum
        hydrocarbons, 196
Grease, 75
Green coke, 295
Greenhouse effect, 7
Greenhouse gases, 237
Gross consignment, 167
Gross lot, 167

Handling petroleum products, 108
Hardness of coke, 300
Harmonious symbiotic relationship, 5
Hazardous chemicals, 115
    petroleum exclusion, 125
Hazardous Materials Transportation Act, 146
Hazardous waste, 8, 115
    acute, 115
    CERCLA definition, 26
    definition, 21
Hazardous waste number, 25
Headspace analysis, 163, 215
Heavy fuel oil, 72, 73
Heavy gas oil, 36
Heavy oil, 10
Hexane extractable material, 212
Heteroatoms, 44
Higher-boiling products, 57, 114
High performance liquid chromatography
        (HPLC), 43, 199
High performance liquid chromatography for
        total petroleum group analysis, 203
Hydrocarbon, 12
    classification, 13
Hydrocarbon oil, 189
Hydrocracking, 61
Hydrocracking unit, 64, 100
Hydrocracking waste, 99
Hydrodesulfurization unit, 66
Hydrogen fluoride, 82
Hydrotreating, 100
Hydrotreating unit, 101
Hydrotreating waste, 101

Ignitability, 21, 114, 138
Ignitable chemicals, 22
Immunoassay methods for total petroleum
        group analysis, 201
Immunoassay methods for total petroleum
        hydrocarbons, 198
Infrared spectroscopy, 45
Infrared spectroscopy methods for total
        petroleum hydrocarbons, 195
Innage gauge, 171
Inorganic chemicals, 151
Insecticides, 75
Insulating oil, 75
Interfacial tension, 48
Interlaboratory precision, 174
Intermediate precision, 175
Intermediate products, 19
Intralaboratory precision, 174
Ion exchange chromatography, 43
Isomerization, 105
Isomerization unit, 66, 106
Isomerization waste, 106
Isomers, 35, 259

Jet fuel, 70

K-category waste, 26
    origin, 27
Kerosene, 71
Kerosine, *see* Kerosene
Kinematic viscosity, 50
K-type chemicals, 25

Lamarckism, 5
Leachability, 186
Leaching, 112
Leaking underground storage tanks, 209
Liquefaction and solidification, 44
Liquefied petroleum gas (LPG), 64, 67, 239
Liquid chromatography, *see* Adsorption
        chromatography
Liquid effluents
    analysis, 257
    naphtha, 258
Low-API fuel oil, 19
Lower-boiling products, 57, 114
Lubricating oil, 74

Manufacture of asphalt, 285
Manufacture of coke, 295
Mass spectrometry, 46
Melting point, 44
Metals, 35, 44
Metals content of coke, 300

Metals content of fuel oil, 276
Method assessment, 230
Method dependence for total petroleum hydrocarbons, 186
Method detection limit, 182
Method validation, 174
Methods for total petroleum hydrocarbons, 187, 188
Met-X process, 85
Middle distillates, 18, 57
Mineral oil, 189
Mitigation, 6
Montmorillonite, 83
Motor oil, 74
Multiring aromatics, 34
Mutagenic, 4

Naphtha, 68, 258
  composition, 261
  density, 266
  evaporation rate, 266
  flash point, 266
  odor and color, 267
  volatility, 267
Naphthenes, 13, 33
Natural gas, 64, 240
Negative bias, 233
No. 1 fuel oil, 73
No. 2 fuel oil, 73
No. 3 fuel oil, 73
No. 4 fuel oil, 73
No. 5 fuel oil, 73
No. 6 fuel oil, 73, 74
Nonpoint source, 238
Nonspecific sources, 27
Nonvolatile compounds, 161, 228
Nuclear magnetic resonance spectroscopy, 45

Occupational safety and Health Act (OSHA), 144
Odor and color of naphtha, 267
Oil and grease, 189
Oil Pollution Act, 128, 143
Oily sludge, 101
Olefins, 33
Organic chemicals, 151
Organic compounds in water, analysis for, 218
Outage gauge, 171
Oxidizing agents, 25

Paraffin base crude oil, 13, 209
Paraffin wax, 76
Paraffins, 13, 33
Particulate matter, 244

Petrochemicals, 59, 78
Petrolatum, 76
Petroleum, 9, 10
  asphalt base, 13
  chemical composition, 12
  composition, 31, 32
  elemental composition, 32
  paraffin base, 13
  properties, 31
  specific gravity, 12
Petroleum chemicals, 151
Petroleum constituents, 209
Petroleum ether, 258
Petroleum fractions, 205
Petroleum group analysis, 198
  gas chromatographic methods, 201, 204
  high performance liquid chromatography, 203
  immunoassay, 201
  thin-layer chromatography, 200
Petroleum jelly, 76
Petroleum products, 16, 59
  classification, 19
  property measurement, 170
Petroleum refinery, 17, 58, 91
Petroleum solvents, *see* Solvents
Physical factors, 4
PINA analysis, 39
PIONA analysis, 39, 199
Pitch, 285
PNA analysis, 39
Point source, 20, 238
Pollutant, 125
  primary, 8
  secondary, 8
Pollution, 7
Pollution prevention, 305, 311
Polymerization, 64
Polymerization emissions and waste, 102
Polymerization unit, 67, 103
Polynuclear aromatic hydrocarbons, 4, 34, 118
  formation in nature, 4
PONA analysis, 39
Porphyrins, 35
Positive bias, 233
Potential isomers of alkanes, 35, 259
Pour point, 44
Pour point of fuel oil, 277
Precision, 173, 180
Primary refining process, 35
Primary standards, 152
Principal components analysis, 173
Process emissions, 88
Process gas, 241

Process wastes, 90
Propane deasphalting, 106, 285
Propane deasphalting unit, 107
Properties of petroleum, 31, 40
    measurement of, 170
Proton magnetic resonance spectroscopy, 45
P-type chemicals, 25
Purge and trap methods, 159, 168, 213, 215
Purgeable hydrocarbons, 120
Pyrophoric substances, 23

Quality assurance, 175, 179, 181
Quality assurance plan, 182
Quality control, 175, 176, 179

Reactive chemicals, 24
Reactivity, 21, 23, 114, 139
Receptor, 8
Recovery, 181
Reduced crude, 285
Refinery, 17, 58
Refinery chemicals, 80
Refinery gas, 68, 241
Refinery processes, 60
Refinery products, 60
Refinery waste, 20, 87
    management, 148
    treatment, 306
Regulated compounds, 151
Regulations, 125
Reid vapor pressure, 171
Relative percent difference, 180
Relative standard deviation, 180
Representative samples, 157
Reserves, 6
Resource, 6
Resource Conservation and Recovery Act (RCRA), 128, 137
Resid, 11
    analysis, 283
Residual fuel oil, 71, 74
Residual products, 20
Resins, 38
Road oil, 77
Ruggedness of test method, 175

Safe Drinking Water Act, 130, 141
Sample cleanup, 170
Sample collection, 153, 154, 185
    gaseous effluents, 247
Sample preparation, 166
Sample separation, 162
Sampling handling protocols, 153
Sampling log, 153

Sampling procedures, 166
Sampling protocols, 155
SARA analysis, 38, 199, 270
Saturates, 38
Saybolt furol viscosity (SFS), 49
Saybolt universal viscosity (SUS), 49
SCOT process, 246, 308
Screening measures for total petroleum hydrocarbons, 208
Scrubber sludge, 27
Secondary standards, 152
Sediment, 165
Sedimentation, 113
Semivolatile compounds, 150, 151, 216, 228
Separation processes, 60
    environmental impact, 131
Silica gel adsorption, 42
Simulated distillation (simdis), 53
Single-ring aromatics, 34
Skeletal alteration processes, 66
Slack wax, 76
Sludge, 21, 101
Snyder column, 168
Softening point, 293
Soil pollution, 132
Solid phase extraction, 163, 216
Solids, 165
Solid waste, 80
Solvent dewaxing unit, 76, 108
Solvent extraction procedures, 159
    sensitivity of, 167
Solvents, 68, 71
    effects on organisms, 18
Sonication, 216
Soxhlet extraction, 164, 216
Specific sources, 27
Spectroscopic properties, 45
Spike recovery, 181
Spreading, 113
Stability of fuel oil, 278
Standard deviation, 180
Standard deviation of sampling, 167
Standard error of sampling, 167
Standard error of the mean, 180
Static tension, 48
Still gas, 64
Stoddard solvent, 71
Storage of petroleum products, 108
Stove oil, 72
Straight-run fractions, 36
Sulfonic acids, 82
Sulfur content of coke, 301
Sulfur content of gaseous effluents, 253
Sulfuric acid sludge, 82

Supercritical fluid chromatography, 43
Supercritical fluid extraction, 164
Superfund (CERCLA), 130
Surface tension, 47

Target compounds, 199
Teratogen, 7
Teratogenic, 4
Terminology, 3
Thin-layer chromatography, 200
Total petroleum hydrocarbons, 119, 189, 207, 284
  EPA test methods, 192
  gas chromatographic methods, 191, 202
  gravimetric methods, 196
  immunoassay methods, 198
  infrared spectroscopy methods, 195
  method dependence, 186
  methods, 187, 188
  screening measures, 208
Total recoverable petroleum hydrocarbons, 212, 232
Toxicity, 21, 23, 24, 113, 139, 186
  polynuclear aromatic hydrocarbons, 4, 118
Toxicity characteristic leaching procedure, 186
Toxic Substances Control Act, 130, 144
Treating processes, environmental impact, 131

Ultimate analysis of petroleum, *see* Elemental analysis of petroleum
Unregulated compounds, 151
UOP characterization factor, 16
Urea dewaxing, 77
U-type chemicals, 25

Vacuum distillation, 36, 53, 60
Vaporizing tendency, 51

Vapor pressure, 52
  gaseous effluents, 253
Visbreaking, 61
Visbreaking unit, 72, 96
Visbreaking waste, 96
Visbroken feedstocks, 38
Viscosity-gravity constant, 15
Viscosity index, 51
Viscosity of asphalt, 294
Viscosity of fuel oil, 277
Volatile compounds, 158, 216
Volatile organic compounds, 151, 238
Volatility, 51
  gaseous effluents, 253
  naphtha, 267

Waste, *see* Refinery waste
Wastewater, 80, 90, 95, 120, 279
Wastewater treatment, 309
Water analysis, 165
  contaminants in, 223
  organic compounds, 218
Water pollution, 132
Wax, 76
Wax recrystallization, 76
Weathering
  asphalt, 294
  effect on composition, 210
Wellman–Lord process, 246, 309
White oil, 75
White spirit, 258
Wind, 113
Wobbe number, 248

X-ray fluorescence spectrometry, 45

Zeolite catalysts, 85
Zero-headspace procedures, 159

CHEMICAL ANALYSIS

A SERIES OF MONOGRAPHS ON ANALYTICAL CHEMISTRY AND ITS APPLICATIONS

Series Editor
J. D. WINEFORDNER

Vol. 1 **The Analytical Chemistry of Industrial Poisons, Hazards, and Solvents.** *Second Edition.* By the late Morris B. Jacobs
Vol. 2 **Chromatographic Adsorption Analysis.** By Harold H. Strain (*out of print*)
Vol. 3 **Photometric Determination of Traces of Metals.** *Fourth Edition*
Part I: General Aspects. By E. B. Sandell and Hiroshi Onishi
Part IIA: Individual Metals, Aluminum to Lithium. By Hiroshi Onishi
Part IIB: Individual Metals, Magnesium to Zirconium. By Hiroshi Onishi
Vol. 4 **Organic Reagents Used in Gravimetric and Volumetric Analysis.** By John F. Flagg (*out of print*)
Vol. 5 **Aquametry: A Treatise on Methods for the Determination of Water.** *Second Edition* (*in three parts*). By John Mitchell, Jr. and Donald Milton Smith
Vol. 6 **Analysis of Insecticides and Acaricides.** By Francis A. Gunther and Roger C. Blinn (*out of print*)
Vol. 7 **Chemical Analysis of Industrial Solvents.** By the late Morris B. Jacobs and Leopold Schetlan
Vol. 8 **Colorimetric Determination of Nonmetals.** *Second Edition.* Edited by the late David F. Boltz and James A. Howell
Vol. 9 **Analytical Chemistry of Titanium Metals and Compounds.** By Maurice Codell
Vol. 10 **The Chemical Analysis of Air Pollutants.** By the late Morris B. Jacobs
Vol. 11 **X-Ray Spectrochemical Analysis.** *Second Edition.* By L. S. Birks
Vol. 12 **Systematic Analysis of Surface-Active Agents.** *Second Edition.* By Milton J. Rosen and Henry A. Goldsmith
Vol. 13 **Alternating Current Polarography and Tensammetry.** By B. Breyer and H.H.Bauer
Vol. 14 **Flame Photometry.** By R. Herrmann and J. Alkemade
Vol. 15 **The Titration of Organic Compounds** (*in two parts*). By M. R. F. Ashworth
Vol. 16 **Complexation in Analytical Chemistry: A Guide for the Critical Selection of Analytical Methods Based on Complexation Reactions.** By the late Anders Ringbom
Vol. 17 **Electron Probe Microanalysis.** *Second Edition.* By L. S. Birks
Vol. 18 **Organic Complexing Reagents: Structure, Behavior, and Application to Inorganic Analysis.** By D. D. Perrin
Vol. 19 **Thermal Analysis.** *Third Edition.* By Wesley Wm.Wendlandt
Vol. 20 **Amperometric Titrations.** By John T. Stock
Vol. 21 **Reflctance Spectroscopy.** By Wesley Wm.Wendlandt and Harry G. Hecht
Vol. 22 **The Analytical Toxicology of Industrial Inorganic Poisons.** By the late Morris B. Jacobs
Vol. 23 **The Formation and Properties of Precipitates.** By Alan G.Walton
Vol. 24 **Kinetics in Analytical Chemistry.** By Harry B. Mark, Jr. and Garry A. Rechnitz
Vol. 25 **Atomic Absorption Spectroscopy.** *Second Edition.* By Morris Slavin
Vol. 26 **Characterization of Organometallic Compounds** (*in two parts*). Edited by Minoru Tsutsui
Vol. 27 **Rock and Mineral Analysis.** *Second Edition.* By Wesley M. Johnson and John A. Maxwell
Vol. 28 **The Analytical Chemistry of Nitrogen and Its Compounds** (*in two parts*). Edited by C. A. Streuli and Philip R.Averell
Vol. 29 **The Analytical Chemistry of Sulfur and Its Compounds** (*in three parts*). By J. H. Karchmer
Vol. 30 **Ultramicro Elemental Analysis.** By Güther Toölg
Vol. 31 **Photometric Organic Analysis** (*in two parts*). By Eugene Sawicki
Vol. 32 **Determination of Organic Compounds: Methods and Procedures.** By Frederick T. Weiss
Vol. 33 **Masking and Demasking of Chemical Reactions.** By D. D. Perrin

Vol. 34. **Neutron Activation Analysis**. By D. De Soete, R. Gijbels, and J. Hoste
Vol. 35. **Laser Raman Spectroscopy**. By Marvin C. Tobin
Vol. 36. **Emission Spectrochemical Analysis**. By Morris Slavin
Vol. 37. **Analytical Chemistry of Phosphorus Compounds**. Edited by M. Halmann
Vol. 38. **Luminescence Spectrometry in Analytical Chemistry**. By J. D.Winefordner, S. G. Schulman, and T. C. O'Haver
Vol. 39. **Activation Analysis with Neutron Generators**. By Sam S. Nargolwalla and Edwin P. Przybylowicz
Vol. 40. **Determination of Gaseous Elements in Metals**. Edited by Lynn L. Lewis, Laben M. Melnick, and Ben D. Holt
Vol. 41. **Analysis of Silicones**. Edited by A. Lee Smith
Vol. 42. **Foundations of Ultracentrifugal Analysis**. By H. Fujita
Vol. 43. **Chemical Infrared Fourier Transform Spectroscopy**. By Peter R. Griffiths
Vol. 44. **Microscale Manipulations in Chemistry**. By T. S. Ma and V. Horak
Vol. 45. **Thermometric Titrations**. By J. Barthel
Vol. 46. **Trace Analysis: Spectroscopic Methods for Elements**. Edited by J. D.Winefordner
Vol. 47. **Contamination Control in Trace Element Analysis**. By Morris Zief and James W. Mitchell
Vol. 48. **Analytical Applications of NMR**. By D. E. Leyden and R. H. Cox
Vol. 49. **Measurement of Dissolved Oxygen**. By Michael L. Hitchman
Vol. 50. **Analytical Laser Spectroscopy**. Edited by Nicolo Omenetto
Vol. 51. **Trace Element Analysis of Geological Materials**. By Roger D. Reeves and Robert R. Brooks
Vol. 52. **Chemical Analysis by Microwave Rotational Spectroscopy**. By Ravi Varma and Lawrence W. Hrubesh
Vol. 53. **Information Theory as Applied to Chemical Analysis**. By Karl Eckschlager and Vladimir Stepanek
Vol. 54. **Applied Infrared Spectroscopy: Fundamentals, Techniques, and Analytical Problemsolving**. By A. Lee Smith
Vol. 55. **Archaeological Chemistry**. By Zvi Goffer
Vol. 56. **Immobilized Enzymes in Analytical and Clinical Chemistry**. By P. W. Carr and L. D. Bowers
Vol. 57. **Photoacoustics and Photoacoustic Spectroscopy**. By Allan Rosencwaig
Vol. 58. **Analysis of Pesticide Residues**. Edited by H. Anson Moye
Vol. 59. **Affity Chromatography**. By William H. Scouten
Vol. 60. **Quality Control in Analytical Chemistry**. *Second Edition*. By G. Kateman and L. Buydens
Vol. 61. **Direct Characterization of Fineparticles**. By Brian H. Kaye
Vol. 62. **Flow Injection Analysis**. By J. Ruzicka and E. H. Hansen
Vol. 63. **Applied Electron Spectroscopy for Chemical Analysis**. Edited by Hassan Windawi and Floyd Ho
Vol. 64. **Analytical Aspects of Environmental Chemistry**. Edited by David F. S. Natusch and Philip K. Hopke
Vol. 65. **The Interpretation of Analytical Chemical Data by the Use of Cluster Analysis**. By D. Luc Massart and Leonard Kaufman
Vol. 66. **Solid Phase Biochemistry: Analytical and Synthetic Aspects**. Edited by William H. Scouten
Vol. 67. **An Introduction to Photoelectron Spectroscopy**. By Pradip K. Ghosh
Vol. 68. **Room Temperature Phosphorimetry for Chemical Analysis**. By Tuan Vo-Dinh
Vol. 69. **Potentiometry and Potentiometric Titrations**. By E. P. Serjeant
Vol. 70. **Design and Application of Process Analyzer Systems**. By Paul E. Mix
Vol. 71. **Analysis of Organic and Biological Surfaces**. Edited by Patrick Echlin
Vol. 72. **Small Bore Liquid Chromatography Columns: Their Properties and Uses**. Edited by Raymond P.W. Scott
Vol. 73. **Modern Methods of Particle Size Analysis**. Edited by Howard G. Barth
Vol. 74. **Auger Electron Spectroscopy**. By Michael Thompson, M. D. Baker, Alec Christie, and J. F. Tyson
Vol. 75. **Spot Test Analysis: Clinical, Environmental, Forensic and Geochemical Applications**. By Ervin Jungreis
Vol. 76. **Receptor Modeling in Environmental Chemistry**. By Philip K. Hopke

Vol. 77	**Molecular Luminescence Spectroscopy: Methods and Applications** (*in three parts*). Edited by Stephen G. Schulman	
Vol. 78	**Inorganic Chromatographic Analysis.** Edited by John C. MacDonald	
Vol. 79	**Analytical Solution Calorimetry.** Edited by J. K. Grime	
Vol. 80	**Selected Methods of Trace Metal Analysis: Biological and Environmental Samples.** By Jon C.VanLoon	
Vol. 81	**The Analysis of Extraterrestrial Materials.** By Isidore Adler	
Vol. 82	**Chemometrics.** By Muhammad A. Sharaf, Deborah L. Illman, and Bruce R. Kowalski	
Vol. 83	**Fourier Transform Infrared Spectrometry.** By Peter R. Griffiths and James A. de Haseth	
Vol. 84	**Trace Analysis: Spectroscopic Methods for Molecules.** Edited by Gary Christian and James B. Callis	
Vol. 85	**Ultratrace Analysis of Pharmaceuticals and Other Compounds of Interest.** Edited by S. Ahuja	
Vol. 86	**Secondary Ion Mass Spectrometry: Basic Concepts, Instrumental Aspects, Applications and Trends.** By A. Benninghoven, F. G. Rüenauer, and H.W.Werner	
Vol. 87	**Analytical Applications of Lasers.** Edited by Edward H. Piepmeier	
Vol. 88	**Applied Geochemical Analysis.** By C. O. Ingamells and F. F. Pitard	
Vol. 89	**Detectors for Liquid Chromatography.** Edited by Edward S.Yeung	
Vol. 90	**Inductively Coupled Plasma Emission Spectroscopy: Part 1: Methodology, Instrumentation, and Performance; Part II: Applications and Fundamentals.** Edited by J. M. Boumans	
Vol. 91	**Applications of New Mass Spectrometry Techniques in Pesticide Chemistry.** Edited by Joseph Rosen	
Vol. 92	**X-Ray Absorption: Principles,Applications,Techniques of EXAFS, SEXAFS, and XANES.** Edited by D. C. Konnigsberger	
Vol. 93	**Quantitative Structure-Chromatographic Retention Relationships.** By Roman Kaliszan	
Vol. 94	**Laser Remote Chemical Analysis.** Edited by Raymond M. Measures	
Vol. 95	**Inorganic Mass Spectrometry.** Edited by F.Adams,R.Gijbels, and R.Van Grieken	
Vol. 96	**Kinetic Aspects of Analytical Chemistry.** By Horacio A. Mottola	
Vol. 97	**Two-Dimensional NMR Spectroscopy.** By Jan Schraml and Jon M. Bellama	
Vol. 98	**High Performance Liquid Chromatography.** Edited by Phyllis R. Brown and Richard A. Hartwick	
Vol. 99	**X-Ray Fluorescence Spectrometry.** By Ron Jenkins	
Vol. 100	**Analytical Aspects of Drug Testing.** Edited by Dale G. Deustch	
Vol. 101	**Chemical Analysis of Polycyclic Aromatic Compounds.** Edited by Tuan Vo-Dinh	
Vol. 102	**Quadrupole Storage Mass Spectrometry.** By Raymond E. March and Richard J. Hughes (*out of print: see Vol. 165*)	
Vol. 103	**Determination of Molecular Weight.** Edited by Anthony R. Cooper	
Vol. 104	**Selectivity and Detectability Optimization in HPLC.** By Satinder Ahuja	
Vol. 105	**Laser Microanalysis.** By Lieselotte Moenke-Blankenburg	
Vol. 106	**Clinical Chemistry.** Edited by E. Howard Taylor	
Vol. 107	**Multielement Detection Systems for Spectrochemical Analysis.** By Kenneth W. Busch and Marianna A. Busch	
Vol. 108	**Planar Chromatography in the Life Sciences.** Edited by Joseph C. Touchstone	
Vol. 109	**Fluorometric Analysis in Biomedical Chemistry: Trends and Techniques Including HPLC Applications.** By Norio Ichinose, George Schwedt, Frank Michael Schnepel, and Kyoko Adochi	
Vol. 110	**An Introduction to Laboratory Automation.** By Victor Cerdá and Guillermo Ramis	
Vol. 111	**Gas Chromatography: Biochemical, Biomedical, and Clinical Applications.** Edited by Ray E. Clement	
Vol. 112	**The Analytical Chemistry of Silicones.** Edited by A. Lee Smith	
Vol. 113	**Modern Methods of Polymer Characterization.** Edited by Howard G. Barth and Jimmy W. Mays	
Vol. 114	**Analytical Raman Spectroscopy.** Edited by Jeanette Graselli and Bernard J. Bulkin	
Vol. 115	**Trace and Ultratrace Analysis by HPLC.** By Satinder Ahuja	
Vol. 116	**Radiochemistry and Nuclear Methods of Analysis.** By William D. Ehmann and Diane E.Vance	

Vol. 117 **Applications of Fluorescence in Immunoassays**. By Ilkka Hemmila
Vol. 118 **Principles and Practice of Spectroscopic Calibration**. By Howard Mark
Vol. 119 **Activation Spectrometry in Chemical Analysis**. By S. J. Parry
Vol. 120 **Remote Sensing by Fourier Transform Spectrometry**. By Reinhard Beer
Vol. 121 **Detectors for Capillary Chromatography**. Edited by Herbert H. Hill and Dennis McMinn
Vol. 122 **Photochemical Vapor Deposition**. By J. G. Eden
Vol. 123 **Statistical Methods in Analytical Chemistry**. By Peter C. Meier and Richard Züd
Vol. 124 **Laser Ionization Mass Analysis**. Edited by Akos Vertes, Renaat Gijbels, and Fred Adams
Vol. 125 **Physics and Chemistry of Solid State Sensor Devices**. By Andreas Mandelis and Constantinos Christofides
Vol. 126 **Electroanalytical Stripping Methods**. By Khjena Z. Brainina and E. Neyman
Vol. 127 **Air Monitoring by Spectroscopic Techniques**. Edited by Markus W. Sigrist
Vol. 128 **Information Theory in Analytical Chemistry**. By Karel Eckschlager and Klaus Danzer
Vol. 129 **Flame Chemiluminescence Analysis by Molecular Emission Cavity Detection**. Edited by David Stiles, Anthony Calokerinos, and Alan Townshend
Vol. 130 **Hydride Generation Atomic Absorption Spectrometry**. Edited by Jiri Dedina and Dimiter L. Tsalev
Vol. 131 **Selective Detectors: Environmental, Industrial, and Biomedical Applications**. Edited by Robert E. Sievers
Vol. 132 **High-Speed Countercurrent Chromatography**. Edited by Yoichiro Ito and Walter D. Conway
Vol. 133 **Particle-Induced X-Ray Emission Spectrometry**. By Sven A. E. Johansson, John L. Campbell, and Klas G. Malmqvist
Vol. 134 **Photothermal Spectroscopy Methods for Chemical Analysis**. By Stephen E. Bialkowski
Vol. 135 **Element Speciation in Bioinorganic Chemistry**. Edited by Sergio Caroli
Vol. 136 **Laser-Enhanced Ionization Spectrometry**. Edited by John C. Travis and Gregory C. Turk
Vol. 137 **Fluorescence Imaging Spectroscopy and Microscopy**. Edited by Xue Feng Wang and Brian Herman
Vol. 138 **Introduction to X-Ray Powder Diffractometry**. By Ron Jenkins and Robert L. Snyder
Vol. 139 **Modern Techniques in Electroanalysis**. Edited by Petr Vanýek
Vol. 140 **Total-Reflction X-Ray Fluorescence Analysis**. By Reinhold Klockenkamper
Vol. 141 **Spot Test Analysis: Clinical, Environmental, Forensic, and Geochemical Applications**. *Second Edition*. By Ervin Jungreis
Vol. 142 **The Impact of Stereochemistry on Drug Development and Use**. Edited by Hassan Y. Aboul-Enein and Irving W.Wainer
Vol. 143 **Macrocyclic Compounds in Analytical Chemistry**. Edited by Yury A. Zolotov
Vol. 144 **Surface-Launched Acoustic Wave Sensors: Chemical Sensing and Thin-Film Characterization**. By Michael Thompson and David Stone
Vol. 145 **Modern Isotope Ratio Mass Spectrometry**. Edited by T. J. Platzner
Vol. 146 **High Performance Capillary Electrophoresis: Theory, Techniques, and Applications**. Edited by Morteza G. Khaledi
Vol. 147 **Solid Phase Extraction: Principles and Practice**. By E. M. Thurman
Vol. 148 **Commercial Biosensors: Applications to Clinical, Bioprocess and Environmental Samples**. Edited by Graham Ramsay
Vol. 149 **A Practical Guide to Graphite Furnace Atomic Absorption Spectrometry**. By David J. Butcher and Joseph Sneddon
Vol. 150 **Principles of Chemical and Biological Sensors**. Edited by Dermot Diamond
Vol. 151 **Pesticide Residue in Foods: Methods, Technologies, and Regulations**. By W. George Fong, H. Anson Moye, James N. Seiber, and John P. Toth
Vol. 152 **X-Ray Fluorescence Spectrometry**. *Second Edition*. By Ron Jenkins
Vol. 153 **Statistical Methods in Analytical Chemistry**. *Second Edition*. By Peter C. Meier and Richard E. Züd
Vol. 154 **Modern Analytical Methodologies in Fat- and Water-Soluble Vitamins**. Edited by Won O. Song, Gary R. Beecher, and Ronald R. Eitenmiller
Vol. 155 **Modern Analytical Methods in Art and Archaeology**. Edited by Enrico Ciliberto and Guiseppe Spoto

Vol. 156 **Shpol'skii Spectroscopy and Other Site Selection Methods: Applications in Environmental Analysis, Bioanalytical Chemistry and Chemical Physics**. Edited by C. Gooijer, F. Ariese and J.W. Hofstraat
Vol. 157 **Raman Spectroscopy for Chemical Analysis**. By Richard L. McCreery
Vol. 158 **Large (C> = 24) Polycyclic Aromatic Hydrocarbons: Chemistry and Analysis**. By John C. Fetzer
Vol. 159 **Handbook of Petroleum Analysis**. By James G. Speight
Vol. 160 **Handbook of Petroleum Product Analysis**. By James G. Speight
Vol. 161 **Photoacoustic Infrared Spectroscopy**. By Kirk H. Michaelian
Vol. 162 **Sample Preparation Techniques in Analytical Chemistry**. Edited by Somenath Mitra
Vol. 163 **Analysis and Purification Methods in Combination Chemistry**. Edited by Bing Yan
Vol. 164 **Chemometrics: From Basics to Wavelet Transform**. By Foo-tim Chau, Yi-Zeng Liang, Junbin Gao, and Xue-guang Shao
Vol. 165 **Quadrupole Ion Trap Mass Spectrometry**. *Second Edition.* By Raymond E. March and John F. J. Todd
Vol. 166 **Handbook of Coal Analysis**. By James G. Speight
Vol. 167 **Introduction to Soil Chemistry: Analysis and Instrumentation**. By Alfred R. Conklin
Vol. 168 **Environmental Analysis and Technology for the Refining Industry**. By James G. Speight